CAMBRIDGE TRACTS IN MATHEMATICS

General Editors

---

193 Distribution Modulo One and Diophantine Approximation

CAMBRIDGE TRACTS IN MATHEMATICS

A complete list of books in the series can be found at www.cambridge.org/mathematics. Recent titles include the following:

163. Linear and Projective Representations of Symmetric Groups. By A. KLESHCHEV
164. The Covering Property Axiom, CPA. By K. CIESIELSKI and J. PAWLIKOWSKI
165. Projective Differential Geometry Old and New. By V. OVSIENKO and S. TABACHNIKOV
166. The Lévy Laplacian. By M. N. FELLER
167. Poincaré Duality Algebras, Macaulay's Dual Systems, and Steenrod Operations. By D. MEYER and L. SMITH
168. The Cube-A Window to Convex and Discrete Geometry. By C. ZONG
169. Quantum Stochastic Processes and Noncommutative Geometry. By K. B. SINHA and D. GOSWAṂI
170. Polynomials and Vanishing Cycles. By M. TIBĂR
171. Orbifolds and Stringy Topology. By A. ADEM, J. LEIDA, and Y. RUAN
172. Rigid Cohomology. By B. LE STUM
173. Enumeration of Finite Groups. By S. R. BLACKBURN, P. M. NEUMANN, and G. VENKATARAMAN
174. Forcing Idealized. By J. ZAPLETAL
175. The Large Sieve and its Applications. By E. KOWALSKI
176. The Monster Group and Majorana Involutions. By A. A. IVANOV
177. A Higher-Dimensional Sieve Method. By H. G. DIAMOND, H. HALBERSTAM, and W. F. GALWAY
178. Analysis in Positive Characteristic. By A. N. KOCHUBEI
179. Dynamics of Linear Operators. By F. BAYART and É. MATHERON
180. Synthetic Geometry of Manifolds. By A. KOCK
181. Totally Positive Matrices. By A. PINKUS
182. Nonlinear Markov Processes and Kinetic Equations. By V. N. KOLOKOLTSOV
183. Period Domains over Finite and p-adic Fields. By J.-F. DAT, S. ORLIK, and M. RAPOPORT
184. Algebraic Theories. By J. ADÁMEK, J. ROSICKÝ, and E. M. VITALE
185. Rigidity in Higher Rank Abelian Group Actions I: Introduction and Cocycle Problem. By A. KATOK and V. NIŢICĂ
186. Dimensions, Embeddings, and Attractors. By J. C. ROBINSON
187. Convexity: An Analytic Viewpoint. By B. SIMON
188. Modern Approaches to the Invariant Subspace Problem. By I. CHALENDAR and J. R. PARTINGTON
189. Nonlinear Perron–Frobenius Theory. By B. LEMMENS and R. NUSSBAUM
190. Jordan Structures in Geometry and Analysis. By C.-H. CHU
191. Malliavin Calculus for Lévy Processes and Infinite-Dimensional Brownian Motion. By H. OSSWALD
192. Normal Approximations with Malliavin Calculus. By I. NOURDIN and G. PECCATI
193. Distribution Modulo One and Diophantine Approximation. By Y. BUGEAUD
194. Mathematics of Two-Dimensional Turbulence. By S. KUKSIN and A. SHIRIKYAN
195. A Universal Construction for Groups Acting Freely on Real Trees. By I. CHISWELL and T. MÜLLER
196. The Theory of Hardy's Z-Function. By A. IVIĆ
197. Induced Representations of Locally Compact Groups. By E. KANIUTH and K. F. TAYLOR
198. Topics in Critical Point Theory. By K. PERERA and M. SCHECHTER

Distribution Modulo One and Diophantine Approximation

YANN BUGEAUD
Université de Strasbourg

CAMBRIDGE
UNIVERSITY PRESS

CAMBRIDGE
UNIVERSITY PRESS

University Printing House, Cambridge CB2 8BS, United Kingdom

One Liberty Plaza, 20th Floor, New York, NY 10006, USA

477 Williamstown Road, Port Melbourne, VIC 3207, Australia

4843/24, 2nd Floor, Ansari Road, Daryaganj, Delhi - 110002, India

79 Anson Road, #06-04/06, Singapore 079906

Cambridge University Press is part of the University of Cambridge.

It furthers the University's mission by disseminating knowledge in the pursuit of education, learning and research at the highest international levels of excellence.

www.cambridge.org
Information on this title: www.cambridge.org/9780521111690

First published 2012

A catalogue record for this publication is available from the British Library

Library of Congress Cataloging in Publication data
Bugeaud, Yann, 1971–
Distribution modulo one and diophantine approximation / Yann Bugeaud.
p. cm. – (Cambridge tracts in mathematics ; 193)
Includes bibliographical references and index.
ISBN 978-0-521-11169-0 (hardback)
1. Diophantine analysis. 2. Distribution modulo one. I. Title.
QA242.B84 2012
512.7′4–dc23
2012013417

ISBN 978-0-521-11169-0 Hardback

Contents

Preface

> primitive Mathematik
> hohe Kunst
>
> THOMAS BERNHARD

> Un chercheur universitaire
> est un individu qui en sait toujours plus
> sur un sujet toujours moindre,
> en sorte qu'il finit par savoir tout de rien.
>
> SIMON LEYS

Three of the main questions that motivate the present book are the following:

- ▷ *Is there a transcendental real number α such that $\|\alpha^n\|$ tends to* 0 *as n tends to infinity?*
- ▷ *Is the sequence of fractional parts $\{(3/2)^n\}$, $n \geq 1$, dense in the unit interval?*
- ▷ *What can be said on the digital expansion of an irrational algebraic number?*

The latter question amounts to the study of the sequence $(\xi 10^n)_{n\geq 1}$ modulo one, where ξ is an irrational algebraic number. More generally, for given real numbers $\xi \neq 0$ and $\alpha > 1$, we are interested in the distribution of the sequences $(\{\xi\alpha^n\})_{n\geq 1}$ and $(\|\xi\alpha^n\|)_{n\geq 1}$, where $\{\cdot\}$ (resp., $\|\cdot\|$) denotes the fractional part (resp., the distance to the nearest integer). The situation is very well understood from a metrical point of view. However, for a given pair (ξ, α), our knowledge on $(\{\xi\alpha^n\})_{n\geq 1}$ is extremely limited, except in very few cases. For instance when $\xi = 1$

and α is a Pisot number, that is, an algebraic integer (an algebraic integer is an algebraic number whose minimal defining polynomial over $\mathbb{Z}$ is monic) all of whose Galois conjugates (except itself) are lying in the open unit disc, it is not difficult to show that $\|\alpha^n\|$ tends to 0 as n tends to infinity. A classical example is given by $\alpha = (1 + \sqrt{5})/2$.

The first chapter is devoted to basic results from the theory of uniform distribution modulo one. We state Weyl's criterion and use it to establish several classical metrical statements. We show that, if $\alpha > 1$ is fixed, then $(\xi\alpha^n)_{n\geq 1}$ is uniformly distributed modulo one for almost all (unless otherwise specified, almost all always refers to the Lebesgue measure) positive real numbers ξ. Likewise, if $\xi \neq 0$ is fixed, then $(\xi\alpha^n)_{n\geq 1}$ is uniformly distributed modulo one for almost all real numbers $\alpha > 1$. We conclude this chapter with a few words on uniform distribution of multidimensional sequences.

Chapter 2 starts with a sufficient condition, proved by Pisot in 1938, on the sequence $(\|\xi\alpha^n\|)_{n\geq 1}$ which implies that the real number α is a Pisot number. We show that this condition can be weakened if α is assumed to be an algebraic number. The chapter continues with various constructions of pairs (ξ, α) such that $(\xi\alpha^n)_{n\geq 1}$ is not dense modulo one. Among other results, we follow a method introduced by Peres and Schlag in 2010 to establish, for every given real number $\alpha > 1$, the existence of real numbers ξ for which $\inf_{n\geq 1} \|\xi\alpha^n\|$ is positive. Furthermore, we prove that, for every positive real number ε, there exist uncountably many real numbers $\alpha > 1$ such that $\|\alpha^n\| < \varepsilon$ for every $n \geq 1$. When α is an integer, say b, there are plenty of irrational real numbers ξ such that $(\xi b^n)_{n\geq 1}$ is not dense modulo one, take for example any irrational real number ξ whose b-ary expansion has no two consecutive zeros. This is no longer true when the sequence $(b^n)_{n\geq 1}$ is replaced by the sequence $(r^m s^n)_{m,n\geq 0}$, where r and s are multiplicatively independent positive integers. Then, Furstenberg established in 1967 that, for every irrational number ξ, the set of real numbers $\{r^m s^n \xi\}$, $m, n \geq 0$, is dense in $[0, 1]$. We end this section with a short survey on a still open conjecture of de Mathan and Teulié, who asked whether, for every real number ξ and every prime number p, we have

$$\inf_{q\geq 1} q \cdot \|q\xi\| \cdot |q|_p = 0,$$

where $|.|_p$ denotes the usual p-adic absolute value normalized in such a way that $|p|_p = p^{-1}$.

The special case where α is an algebraic number is investigated in Chapter 3. In 1968, Mahler asked for the existence of positive real numbers ξ for which $\{\xi(3/2)^n\} < 1/2$ for every $n \geq 0$. He proved that there are at most countably many real numbers with the latter property and we still do not know whether there is at least one such number. Chapter 3 is partly devoted to the study of the sequences $(\{\xi(p/q)^n\})_{n\geq 1}$ and $(||\xi(p/q)^n||)_{n\geq 1}$, for a non-zero real number ξ and coprime integers p, q with $p > q \geq 1$. Among other results, it is established that, assuming that $q \geq 2$ or that ξ is irrational, we always have

$$\limsup_{n\to+\infty} \left\{\xi\left(\frac{p}{q}\right)^n\right\} - \liminf_{n\to+\infty} \left\{\xi\left(\frac{p}{q}\right)^n\right\} \geq \frac{1}{p}.$$

This result was proved in 1995 by Flatto, Lagarias and Pollington, and reproved in 2006 by Dubickas, by means of a different, simpler approach. We further establish that, for every non-zero real number ξ, the sequence $(||\xi(3/2)^n||)_{n\geq 1}$ has a limit point greater than 0.238117 and a limit point smaller than 0.285648, as was shown by Dubickas in 2006. The proof involves combinatorics on words and properties of the Thue–Morse infinite word. We complement these results with various constructions of Pollington and Dubickas of real numbers ξ for which $||\xi(3/2)^n|| < 1/3$ for every $n \geq 1$ and of real numbers ξ for which $\inf_{n\geq 1} ||\xi(3/2)^n||$ is quite large.

In Chapter 4, we introduce the notion of normality to an integer base in accordance with Émile Borel's original definition given in 1909 and establish his fundamental theorem that almost all real numbers are normal to all integer bases. We show that Borel's definition is redundant in part and state several equivalent definitions. For an integer $b \geq 2$ and positive integers r and s, we show that normality to base b^r is equivalent to normality to base b^s. Furthermore, we prove that a real number ξ is normal to base b if, and only if, the sequence $(\xi b^n)_{n\geq 1}$ is uniformly distributed modulo one. Combined with one of the metrical results established in Chapter 1, this gives an alternative proof of Borel's theorem. Replacing b by a real number $\alpha > 1$, the above criterion allows us to define the notion of normality to a non-integer base α. The first explicit example of a normal number was given in 1933 by Champernowne, who proved that the real number (now usually called the Champernowne number)

$$0.12345678910111213 14\ldots,$$

whose sequence of decimals is the increasing sequence of all positive integers, is normal to base 10. This statement has been extended in

1946 by Copeland and Erdős. Their result implies in particular that the real number

$$0.235711131719232931\ldots,$$

whose sequence of decimals is the increasing sequence of all prime numbers, is also normal to base 10. We further introduce the notions of block complexity, richness and entropy, which are useful to measure the complexity of the b-ary expansion of a real number. The chapter ends with the study of the rational approximation to a family of real numbers including the Champernowne number.

Further explicit examples of numbers normal to a given base are constructed in Chapter 5, following a method, developed in 2002 by Bailey and Crandall, which rests on estimates for exponential sums. We discuss the problem of the construction of real numbers which are absolutely normal, that is, normal to every integer base. Furthermore, we present an explicit example of a real number which is normal to no integer base. This chapter ends with a few words on a theory of Bailey and Crandall to explain random behaviour for the digits in the integer expansions of fundamental mathematical constants.

The general question investigated in Chapter 6 is: What can be said on the expansions of a given real number to several bases? The existence of real numbers being normal to some integer base and non-normal to other integer bases was confirmed by Cassels and, independently, by W. M. Schmidt in the years 1959–1960. We reproduce Cassels' proof establishing that almost all elements of the middle third Cantor set (in the sense of the Cantor measure) are normal to every integer base which is not a power of 3. For non-integer bases, we follow works of Brown, Moran and Pollington to establish various results on the existence of real numbers normal to some base $\alpha > 1$, but not normal to another base $\beta > 1$. Their method uses suitable Riesz product measures. We then show that, given two coprime integers $r \geq 2$ and $s \geq 2$, any irrational real number cannot have too many zeros both in its r-ary and in its s-ary expansion. The chapter ends with a short discussion on the representation of integers in two different bases.

In Chapter 7, for an integer $b \geq 2$, we introduce exponents of Diophantine approximation to measure the accuracy with which a given real number ξ is approximated by rational numbers whose denominators are integer powers of b or are of the form $b^r(b^s - 1)$ for integers $r \geq 0$, $s \geq 1$. Such rational numbers occur naturally when one searches for good rational approximations to ξ by simply looking at its b-ary expansion. We use the (α, β)-games introduced by W. M. Schmidt in 1966 to prove the existence of real numbers, all of whose integer expansions have blocks of zeros of bounded length. We further give several results on Diophantine approximation on the middle third Cantor set, including a construction of real numbers lying in this set and having a prescribed irrationality exponent. We conclude this chapter with the computation of the Hausdorff dimension of sets of real numbers with specific digital properties.

Chapter 8 is mainly concerned with digital expansions of algebraic, irrational real numbers ξ. We first show, following Adamczewski and Bugeaud, that the number of distinct subblocks of n digits occurring in the b-ary expansion of ξ, viewed as an infinite word on $\{0, 1, \ldots, b-1\}$, cannot be bounded by a constant times n. The proof combines elementary combinatorics on words with deep tools from Diophantine approximation that are gathered in Appendix E. Next, we follow a skilful approach of Bailey, Borwein, Crandall and Pomerance to get a lower bound for the number of non-zero digits in the b-ary expansion of ξ. The chapter ends with a discussion on a problem of Mahler on the digits of the integer multiples of a given irrational real number.

In Chapter 9, we discuss analogous questions for continued fraction expansions and for β-expansions. We present the construction of a normal continued fraction and mention several transcendence criteria for continued fractions. We survey without proof various results on β-expansions.

Chapter 10 offers a list of open questions. We hope that these will motivate further research.

The ten chapters are completed by six appendices, which, mostly without proofs, gather classical results from combinatorics on words, measure theory, continued fractions, Diophantine approximation, among others.

The chapters are largely independent of each other.

The purpose of the exercises is primarily to give complementary results, thus many of them are an adaptation of an original research work to which the reader is directed.

We have tried, in the end-of-chapter notes, to be as exhaustive as possible and to quote less-known papers. Of course, exhaustivity is an impossible task, and it is clear that the choice of the references concerning works at the border of the main topic of this book reflects the personal taste and the limits of the knowledge of the author.

There exist already many textbooks dealing, in part, with the subject of the present one, e.g., by Koksma [389], Niven [543], Salem [619], Kuipers and Niederreiter [411], Rauzy [605], Schmidt [635], Bertin *et al.* [80], Drmota and Tichy [232], Harman [335], Strauch and Porubský [678].

However, the intersection never exceeds one or two chapters. Most of the results presented here were proved after the year 2000 and have not yet appeared in a book, as is also the case for many of the older results.

Many colleagues sent me comments, remarks and suggestions. I am very grateful to all of them. Special thanks are due to Toufik Zaïmi, who very carefully read several parts of this book.

The present book will be regularly updated on my institutional web page:

`http://www-irma.u-strasbg.fr/~bugeaud/Book2.html`

Frequently used notation

$\lfloor x \rfloor$: greatest integer $\leq x$.
$\lceil \cdot \rceil$: smallest integer $\geq x$.
$\lfloor x \rceil$: greatest integer $< x$.
$\{\cdot\}$: fractional part.
$\| \cdot \|$: distance to the nearest integer.
positive : strictly positive.
$\log_b$: logarithm with respect to the base b; in particular, $\log_{\mathrm{e}} = \log$.
An empty sum is equal to 0 and an empty product is equal to 1.
$\mathbb{T}$: the torus $[0, 1)$ with 0 and 1 identified.
$\underline{x}, \underline{y}$: d-dimensional vectors with real or integral entries.
Card : the cardinality (of a finite set).
r, s: (often) two multiplicatively independent integers, which means that $r, s \geq 2$ and $(\log r)/(\log s)$ is irrational.
D_N : discrepancy, Ch. 1.
deg : degree of a polynomial or of an algebraic number.
$\mathrm{Tr}(\alpha)$: trace of the algebraic number α, Ch. 2.
$H(\alpha)$: naïve height of the algebraic number α, App. E.
$L(\alpha)$, $\ell(\alpha)$: length, reduced length of the algebraic number α, Ch. 3.
$| \cdot |_p$: p-adic absolute value, Ch. 2 and App. E.
$(t_n)_{n \geq 1}, (m_n)_{n \geq 1}$: increasing sequence of positive real numbers, of positive integers, Ch. 2.
Z-number, $Z_\alpha(s, s + t)$: Ch. 3.
b : an integer ≥ 2 (the base).
$\mathcal{A}$: a finite or infinite alphabet, often equal to $\{0, 1, \ldots, b - 1\}$.
$(c)_b, \ell_b(c)$: the word $d_\ell d_{\ell-1} \ldots d_1 d_0$ on $\{0, 1, \ldots, b-1\}$ representing the positive integer c in base b, that is, such that $c = d_\ell b^\ell + \cdots + d_1 b + d_0$ and $d_\ell \neq 0$; then, $\ell_b(c) = \ell$.
$U, V, W, \ldots, \mathbf{a}, \mathbf{d}$: finite words.

$\boldsymbol{a}, \boldsymbol{w}, \boldsymbol{x}, \ldots$: infinite words.

$\boldsymbol{a} = a_1a_2a_3\ldots$: the b-ary expansion of a real number ξ, thus $\xi = \lfloor\xi\rfloor + \sum_{k\geq 1} a_k b^{-k} = \lfloor\xi\rfloor + 0 \cdot a_1a_2\ldots$, Ch. 3, 4, 6, 7 & 8.

$\mathrm{Dio}(\boldsymbol{a})$: the Diophantine exponent of the word $\boldsymbol{a}$, App. A.

$\boldsymbol{t} = abbabaabbaababba\ldots$: the Thue–Morse infinite word on $\{a, b\}$, App. A.

$A_b(d, N, \xi), A_b(D_k, N, \xi)$: Ch. 4.

$p_b(n, k)$: Ch. 4.

$p(\cdot, \boldsymbol{a}, b), p(\cdot, \xi, b), p(\cdot, \xi), p_\infty(\cdot, \xi, b)$: complexity function, Ch. 4, 9 & 10, App. A.

$E(\boldsymbol{a}, b)$, $E(\xi, b)$, $E(\xi)$: entropy, Ch. 4 & 9.

$\xi_{\mathbf{c}} = 0.123456789101112\ldots$: the Champernowne number, Ch. 4.

$\mathcal{N}(b), \mathcal{N}(\alpha)$: set of real numbers normal to base b, to base α, Ch. 4 & 6.

$V(\xi, b)$, $V_b(\xi)$: Ch. 4 & 6.

$\mathcal{DC}(\cdot, \xi, b)$, $\mathcal{DC}(\cdot, b)$: Ch. 6 & 8.

$\mathcal{NZ}(\cdot, \xi, b)$, $\mathcal{NZ}(\cdot, b)$: Ch. 6 & 8.

λ : the Lebesgue measure on the real line.

$\lambda(I) = |I|$: the Lebesgue measure of an interval I.

$B(x, \rho)$: the open interval $(x - \rho, x + \rho)$, Ch. 7 and App. C.

K : the middle third Cantor set, Ch. 7 and App. C.

μ_K : the standard measure on K, App. C.

μ : a measure (not the Lebesgue one).

$\hat{\mu}$: the Fourier transform of the measure μ.

v_1, v_b, v'_b, v'_T : Ch. 7 & 9.

$\Lambda_b(\xi)$: Ch. 8 & 10.

$T_{\mathcal{G}}, \mu_{\mathcal{G}}$: Gauss map, Gauss measure, Ch. 9.

T_b, T_β : Ch. 9.

$A_\beta(D, N, x)$: Ch. 9.

$\mathcal{D}(\beta)$: Ch. 9.

$\mathrm{ord}_p(a)$, $\mathrm{ord}(a, p^h)$: App. B.

dim : Hausdorff dimension, App. C.

$\mathcal{H}^s$: s-dimensional Hausdorff measure, App. C.

μ : irrationality exponent, App. E.

$w_n(\xi)$, $w_n^*(\xi)$: App. E.

A-, S-, T-, U-, A^*-, S^*-, T^*-, U^*-number : App. E.

1
Distribution modulo one

In this chapter, we give a brief account of the theory of uniform distribution modulo one. For complements, and for additional bibliographic references, the reader is directed to the monographs [232, 411].

1.1 Weyl's criterion

In the years 1909–1910 there appeared three papers by Bohl [109], Sierpiński [653] and Weyl [731] devoted partly to the distribution of real sequences. A few years later, Weyl [732, 733] developed the results of these papers, giving birth to the study of uniform distribution.

DEFINITION 1.1. The sequence $(x_n)_{n\geq 1}$ of real numbers is *dense modulo one* if every interval of positive length included in $[0, 1]$ contains at least one element of $(\{x_n\})_{n\geq 1}$. The sequence $(x_n)_{n\geq 1}$ is *uniformly distributed modulo one* if, for every real numbers u, v with $0 \leq u < v \leq 1$, we have

$$\lim_{N\to+\infty} \frac{\mathrm{Card}\{n : 1 \leq n \leq N, u \leq \{x_n\} < v\}}{N} = v - u.$$

Weyl developed his theory and stated various criteria ensuring that a given sequence is uniformly distributed modulo one. The last statement of Theorem 1.2 is often referred to as 'Weyl's criterion'.

THEOREM 1.2. *The sequence $(x_n)_{n\geq 1}$ of real numbers is uniformly distributed modulo one if, and only if, for every complex-valued, 1-periodic continuous function f we have*

$$\lim_{N\to+\infty} \frac{1}{N} \sum_{n=1}^{N} f(x_n) = \int_0^1 f(x)\,\mathrm{d}x,$$

that is, if, and only if,

$$\lim_{N\to+\infty} \frac{1}{N}\sum_{n=1}^{N} \mathrm{e}^{2i\pi h x_n} = 0 \quad \textit{holds for every non-zero integer } h.$$

PROOF. The first statement follows from the definition of the Riemann integral. The second one is an immediate application of the approximation theorem of Stone–Weierstrass stating that, for the supremum norm, finite linear combinations of functions $x \mapsto \mathrm{e}^{2i\pi hx}$, $h \in \mathbb{Z}$, with complex coefficients, are dense in the space of step functions. □

We display an immediate application of Theorem 1.2.

THEOREM 1.3. *For any irrational real number α, the sequence $(n\alpha)_{n\ge1}$ is uniformly distributed modulo one.*

PROOF. This follows from Theorem 1.2 and the inequalities

$$\left|\sum_{n=1}^{N} \mathrm{e}^{2i\pi hn\alpha}\right| \le \left|\frac{\mathrm{e}^{2i\pi hN\alpha}-1}{\mathrm{e}^{2i\pi h\alpha}-1}\right| \le \left|\frac{2}{\mathrm{e}^{2i\pi h\alpha}-1}\right|, \tag{1.1}$$

valid for every non-zero integer N and h. □

Proofs of Theorem 1.3 which do not involve the exponential function were previously given in [109, 653, 731]; see also [605].

The next result is an extension of Theorem 1.3. It was first proved by Weyl [732, 733].

THEOREM 1.4. *Let $P(X) = a_d X^d + \cdots + a_1 X + a_0$ be a real polynomial of degree $d \ge 1$. If at least one coefficient among $a_1, \ldots, a_d$ is irrational, then the sequence $(P(n))_{n\ge1}$ is uniformly distributed modulo one.*

We present a proof of Theorem 1.4 based on van der Corput's lemma [205]. We begin with stating and proving the van der Corput inequality.

LEMMA 1.5. *Let a and N be positive integers. Let $u_1, \ldots, u_N$ be complex numbers and L be an integer with $1 \le aL \le N$. Then we have*

$$L^2\left|\sum_{n=1}^{N} u_n\right|^2 \le L(N + a(L-1))\sum_{n=1}^{N} |u_n|^2 + 2(N + a(L-1))\sum_{\ell=1}^{L-1}(L-\ell)\,\mathrm{Re}\sum_{n=1}^{N-a\ell} u_n \overline{u}_{n+a\ell},$$

where Re *denotes the real part.*

PROOF. Setting $u_n = 0$ for $n \le 0$ and for $n > N$, we have

$$L \sum_{n=1}^{N} u_n = \sum_{p=1}^{N+a(L-1)} \sum_{\ell=0}^{L-1} u_{p-a\ell}.$$

The Cauchy–Schwarz inequality then gives

$$\begin{aligned}
L^2 \Big| \sum_{n=1}^{N} u_n \Big|^2 &\le (N + a(L-1)) \sum_{p=1}^{N+a(L-1)} \Big| \sum_{\ell=0}^{L-1} u_{p-a\ell} \Big|^2 \\
&= (N + a(L-1)) \sum_{p=1}^{N+a(L-1)} \Big(\sum_{\ell=0}^{L-1} u_{p-a\ell} \Big) \Big(\sum_{j=0}^{L-1} \overline{u}_{p-aj} \Big) \\
&= (N + a(L-1)) \sum_{p=1}^{N+a(L-1)} \sum_{\ell=0}^{L-1} |u_{p-a\ell}|^2 \\
&\quad + 2\big(N + a(L-1)\big) \operatorname{Re} \sum_{p=1}^{N+a(L-1)} \sum_{\ell=1}^{L-1} \sum_{j=0}^{\ell-1} u_{p-a\ell} \overline{u}_{p-aj} \\
&= L\big(N + a(L-1)\big) \sum_{n=1}^{N} |u_n|^2 + 2\big(N + a(L-1)\big) \Sigma_1,
\end{aligned}$$

where we have set

$$\Sigma_1 := \operatorname{Re} \sum_{p=1}^{N+a(L-1)} \sum_{\ell=1}^{L-1} u_{p-a\ell} \big(\overline{u}_p + \cdots + \overline{u}_{p-a(\ell-1)} \big).$$

We check that, for $\ell = 1, \ldots, L-1$ and $p = a\ell + 1, \ldots, N$, the product $u_{p-a\ell}\overline{u}_p$ occurs exactly $L - \ell$ times in the latter double sum. This proves the lemma. □

Lemma 1.5 is the key tool for the proof of the next theorem of Korobov and Postnikov [403], which generalizes a fundamental result of van der Corput [205] treating the case $a = b = 1$.

THEOREM 1.6. *Let $(x_n)_{n \ge 1}$ be a given sequence of real numbers. Let a and b be positive integers. If for every positive integer ℓ the sequence $(x_{n+a\ell} - x_n)_{n \ge 1}$ is uniformly distributed modulo one, then $(x_{bn})_{n \ge 1}$ is uniformly distributed modulo one.*

PROOF. Let h be a non-zero integer. Let N be a positive integer and observe that

$$\sum_{n=1}^{N} \mathrm{e}^{2i\pi h x_{bn}} = \frac{1}{b} \sum_{j=1}^{b} \sum_{n=1}^{bN} \mathrm{e}^{2i\pi h x_n}\, \mathrm{e}^{2i\pi (jn/b)}, \tag{1.2}$$

thus,

$$\left|\sum_{n=1}^{N} \mathrm{e}^{2i\pi h x_{bn}}\right| \leq \max_{1\leq j\leq b}\left|\sum_{n=1}^{bN} \mathrm{e}^{2i\pi h x_n}\,\mathrm{e}^{2i\pi(jn/b)}\right|.$$

Let $j = 1, \ldots, b$. Fix a positive integer L and assume that N exceeds aL. We apply Lemma 1.5 with $u_n = \mathrm{e}^{2i\pi(hx_n + jn/b)}$ to obtain, after division by $b^2L^2N^2$, that

$$\begin{aligned}\left|\frac{1}{bN}\sum_{n=1}^{bN} \mathrm{e}^{2i\pi(hx_n+jn/b)}\right|^2 &\leq \frac{bN + a(L-1)}{bLN} \\ &+ 2\sum_{\ell=1}^{L-1} \frac{(bN+a(L-1))(L-\ell)(bN-a\ell)}{b^2L^2N^2} \\ &\qquad \times \left|\frac{1}{bN-a\ell}\sum_{n=1}^{bN-\ell} \mathrm{e}^{2i\pi(h(x_{n+a\ell}-x_n)-ja\ell/b)}\right|.\end{aligned}$$

Let ℓ be a positive integer. Since the sequence $(x_{n+a\ell} - x_n)_{n\geq 1}$ is uniformly distributed modulo one, we get by Theorem 1.2 that

$$\lim_{N\to+\infty} \frac{1}{bN - aL}\sum_{n=1}^{bN-a\ell} \mathrm{e}^{2i\pi h(x_{n+a\ell}-x_n)} = 0,$$

which implies that

$$\limsup_{N\to+\infty}\left|\frac{1}{bN}\sum_{n=1}^{bN} \mathrm{e}^{2i\pi(hx_n+jn/b)}\right|^2 \leq \frac{1}{L}.$$

Since the latter inequality is true for any arbitrary L and $j = 1, \ldots, b$, we get from (1.2) that

$$\lim_{N\to+\infty}\left|\frac{1}{N}\sum_{n=1}^{N} \mathrm{e}^{2i\pi h x_{bn}}\right| = 0,$$

and it follows from Weyl's criterion (Theorem 1.2) that $(x_{bn})_{n\geq 1}$ is uniformly distributed modulo one. □

PROOF OF THEOREM 1.4. The case $d = 1$ reduces to Theorem 1.3. Assume that $d \geq 2$ and that $a_2, \ldots, a_d$ are all rational numbers. Set $R(X) = P(X) - a_1X - a_0$. Let D be a positive integer such that $Da_2, \ldots, Da_d$ are integers. Observe that $\{R(Dk+t)\} = \{R(t)\}$ for $k \geq 0$ and $t \geq 1$. Consequently, for every non-zero integer h, we have

$$
\begin{aligned}
\frac{1}{N}\sum_{n=1}^{N} e^{2i\pi hP(n)} &= \frac{1}{N}\sum_{n=\lfloor N/D\rfloor D+1}^{N} e^{2i\pi hP(n)} \\
&\quad + \frac{1}{N}\sum_{t=1}^{D}\sum_{k=0}^{\lfloor N/D\rfloor-1} e^{2i\pi h(R(Dk+t)+a_1(Dk+t)+a_0)} \\
&= \frac{1}{N}\sum_{n=\lfloor N/D\rfloor D+1}^{N} e^{2i\pi hP(n)} \\
&\quad + \Big(\sum_{t=1}^{D} e^{2i\pi h(R(t)+a_1t+a_0)}\Big)\cdot\Big(\frac{1}{N}\sum_{k=0}^{\lfloor N/D\rfloor-1} e^{2i\pi ha_1Dk}\Big).
\end{aligned}
\tag{1.3}
$$

Since a_1 is irrational, arguing as in the proof of Theorem 1.3, we get that the last sum is bounded. Consequently, the left-hand side of (1.3) tends to 0 as N tends to infinity. By Theorem 1.2, this proves that $(P(n))_{n\geq 1}$ is uniformly distributed modulo one.

To complete the proof for an arbitrary polynomial

$$P(X) = a_dX^d + \cdots + a_1X + a_0,$$

we proceed by induction on the largest index ℓ such that a_ℓ is irrational. We have already established the case $\ell = 1$. Let $P(X)$ be a real polynomial of degree $d \geq 2$ and assume that the largest index ℓ such that a_ℓ is irrational satisfies $\ell \geq 2$. Let h be a positive integer and set

$$
\begin{aligned}
Q_h(X) &= P(X+h) - P(X) \\
&= a_d\big((X+h)^d - X^d\big) + \cdots + a_\ell\big((X+h)^\ell - X^\ell\big) + \cdots + a_1h.
\end{aligned}
$$

The coefficients of $X^{d-1}, \ldots, X^\ell$ in $Q_h(X)$ are rational numbers, but the coefficient of $X^{\ell-1}$ is irrational. Applying the inductive assumption shows that $(Q_h(n))_{n\geq 1}$ is uniformly distributed modulo one. It then follows from Theorem 1.6 applied with $a = b = 1$ that $(P(n))_{n\geq 1}$ is uniformly distributed modulo one. □

1.2 Metrical results

In this section, we present several metrical statements on the distribution of sequences of real numbers.

THEOREM 1.7. *Let $(x_n)_{n\geq 1}$ be a sequence of real numbers satisfying*

$$\liminf_{n\to+\infty}(x_{n+1} - x_n) > 0.$$

Then, for almost all real numbers ξ, the sequence $(\xi x_n)_{n\geq 1}$ is uniformly distributed modulo one.

We establish Theorem 1.7, proved by Weyl [733], by means of an auxiliary lemma of Davenport, Erdős and LeVeque [216].

LEMMA 1.8. *Let S be a set and μ a measure on S. Let $(X_n)_{n\geq 1}$ be a bounded sequence of measurable functions defined on S. If the series*

$$\sum_{N\geq 1} \frac{1}{N} \int_S \left| \frac{1}{N} \sum_{1\leq n\leq N} X_n \right|^2 \mathrm{d}\mu$$

converges, then μ-almost all elements s of S satisfy

$$\lim_{N\to+\infty} \frac{1}{N} \sum_{1\leq n\leq N} X_n(s) = 0.$$

PROOF. For a positive real number $\varepsilon < \frac{1}{2}$ and a positive integer N, set

$$A_N(\varepsilon) := \left\{ s \in S : \left| \frac{1}{N} \sum_{1\leq n\leq N} X_n(s) \right| \geq \varepsilon \right\}.$$

Since

$$\varepsilon^2 \mu(A_N(\varepsilon)) \leq \int_S \left| \frac{1}{N} \sum_{1\leq n\leq N} X_n \right|^2 \mathrm{d}\mu,$$

the assumption implies that the series

$$\sum_{N\geq 1} \frac{\mu(A_N(\varepsilon))}{N}$$

converges. Set $N_1 = 1$ and

$$N_{k+1} = \lceil N_k/(1-\varepsilon) \rceil + 1, \quad \text{for } k \geq 1.$$

For any positive integer k, let M_k be an integer satisfying

$$N_k \leq M_k < N_{k+1}, \quad \frac{\mu(A_{M_k}(\varepsilon))}{M_k} = \min_{N_k \leq N < N_{k+1}} \frac{\mu(A_N(\varepsilon))}{N}.$$

We deduce from

$$\sum_{N_k \leq N < N_{k+1}} \frac{\mu(A_N(\varepsilon))}{N} \geq (N_{k+1} - N_k) \frac{\mu(A_{M_k}(\varepsilon))}{M_k} \geq \varepsilon \mu(A_{M_k}(\varepsilon))$$

that the series

$$\sum_{k\geq 1} \mu(A_{M_k}(\varepsilon))$$

converges. Lemma C.1 then implies that μ-almost all elements of S belong to only finitely many sets $A_{M_k}(\varepsilon)$. This means that, for μ-almost all elements s of S, we have

$$\left|\frac{1}{M_k}\sum_{1\le n\le M_k} X_n(s)\right| < \varepsilon,$$

as soon as k is sufficiently large.

Let s be in S and N be a positive integer. Let k be the unique integer defined by the inequalities $N_k \le N < N_{k+1}$. Let c be a common upper bound for all the functions $|X_n|$. Since

$$\begin{aligned}\frac{1}{N}\sum_{1\le n\le N} X_n(s) &- \frac{1}{M_k}\sum_{1\le n\le M_k} X_n(s)\\ &= \frac{1}{N}\Big(\sum_{1\le n\le N} X_n(s) - \sum_{1\le n\le M_k} X_n(s)\Big) + \Big(\frac{1}{N}-\frac{1}{M_k}\Big)\sum_{1\le n\le M_k} X_n(s),\end{aligned}$$

we get that

$$\begin{aligned}\left|\frac{1}{N}\sum_{1\le n\le N} X_n(s) - \frac{1}{M_k}\sum_{1\le n\le M_k} X_n(s)\right| &\le 2c\frac{N_{k+1}-N_k}{N_k}\\ &\le \frac{2c\varepsilon}{1-\varepsilon} + \frac{4c}{N_k} \le 5c\varepsilon,\end{aligned}$$

if k is large enough. Consequently, for μ-almost all elements s of S, we have

$$\left|\frac{1}{N}\sum_{1\le n\le N} X_n(s)\right| < (1+5c)\varepsilon,$$

as soon as N is sufficiently large. This proves the lemma. □

PROOF OF THEOREM 1.7. Let a, b be real numbers with $a < b$. Let h be a non-zero integer. Without any loss of generality, we assume that there exists a positive real number c such that $x_{n+1} - x_n \ge c$ for $n \ge 1$. Since, for any integer $N \ge 3$, we have

$$\begin{aligned}\int_a^b \left|\frac{1}{N}\sum_{n=1}^{N} \mathrm{e}^{2i\pi h\xi x_n}\right|^2 \mathrm{d}\xi &\le \frac{b-a}{N} + \frac{1}{N^2}\int_a^b \sum_{n=1}^{N}\sum_{m=1}^{n-1} \mathrm{e}^{2i\pi h\xi(x_n-x_m)}\,\mathrm{d}\xi \\ &\le \frac{b-a}{N} + \frac{1}{N^2}\cdot\sum_{n=1}^{N}\sum_{m=1}^{n-1}\frac{1}{\pi|h|(x_n-x_m)} \\ &\le \frac{b-a}{N} + \frac{1}{N^2}\cdot\sum_{n=1}^{N}\sum_{m=1}^{n-1}\frac{1}{\pi|h|(n-m)c} \\ &\le \frac{b-a}{N} + \frac{1}{\pi|h|cN^2}\cdot\left(N+\frac{N}{2}+\cdots+\frac{N}{N-1}\right) \\ &\le \frac{b-a}{N} + \frac{2\log N}{\pi|h|cN},\end{aligned}$$

it follows that the series

$$\sum_{N\ge1}\frac{1}{N}\int_a^b \left|\frac{1}{N}\sum_{n=1}^{N} \mathrm{e}^{2i\pi h\xi x_n}\right|^2 \mathrm{d}\xi$$

converges. We then deduce from Lemma 1.8 and the Weyl criterion (Theorem 1.2) that $(\xi x_n)_{n\ge1}$ is uniformly distributed modulo one for almost all ξ in $[a,b]$. This completes the proof of the theorem. □

We display an immediate consequence of Theorem 1.7.

COROLLARY 1.9. *Let α be a real number greater than 1. Then, for almost all real numbers ξ, the sequence $(\xi\alpha^n)_{n\ge1}$ is uniformly distributed modulo one.*

We complement this corollary by a metrical result of Koksma [388].

THEOREM 1.10. *Let ξ be a non-zero real number. Then, for almost all real numbers α greater than 1, the sequence $(\xi\alpha^n)_{n\ge1}$ is uniformly distributed modulo one.*

PROOF. Let a, b, m, n be integers with $1 \le a < b$ and $1 \le m < n$. Let h be a non-zero integer and set

$$I_{h,m,n} = \int_a^b \mathrm{e}^{2i\pi h\xi(\alpha^n-\alpha^m)}\,\mathrm{d}\alpha.$$

The function $\Phi : x \mapsto x^n - x^m$ is strictly increasing on $[a,b]$, and let Ψ be its reciprocal function. Observe that

$$I_{h,m,n} = \int_{a^n-a^m}^{b^n-b^m} \mathrm{e}^{2i\pi h\xi u}\Psi'(u)\,\mathrm{d}u,$$

where, for $u \in [a^n - a^m, b^n - b^m]$,

$$\Psi'(u) = \frac{1}{n\Psi(u)^{n-1} - m\Psi(u)^{m-1}} \leq \frac{1}{n-m}. \tag{1.4}$$

Since Ψ' is positive, decreasing on $[a^n - a^m, b^n - b^m]$, we get from (1.4) and the extended mean value theorem that there exists c in the interval $[a^n - a^m, b^n - b^m]$ such that

$$|I_{h,m,n}| = \Psi'(a^n - a^m) \left| \int_a^c \mathrm{e}^{2i\pi h\xi u} \, \mathrm{d}u \right| \leq \frac{1}{\pi h\xi(n-m)}.$$

For a positive integer N, we then have

$$\begin{aligned} I_N := \int_a^b \left| \frac{1}{N} \sum_{n=1}^{N} \mathrm{e}^{2i\pi h\xi\alpha^n} \right|^2 \mathrm{d}\alpha &\leq \frac{b-a}{N} + \frac{2}{N^2} \left| \sum_{1 \leq m < n \leq N} I_{h,m,n} \right| \\ &\leq \frac{b-a}{N} + \frac{3}{\pi h\xi} \cdot \frac{\log N}{N}. \end{aligned}$$

Thus, the sum of I_N/N over the positive integers N converges, and we deduce from Lemma 1.8 that

$$\lim_{N \to +\infty} \frac{1}{N} \sum_{1 \leq n \leq N} \mathrm{e}^{2i\pi h\xi\alpha^n} = 0, \quad \text{for almost all } \alpha \in [a,b].$$

The theorem follows from Weyl's criterion (Theorem 1.2). □

Consequently, the sequence $(\xi\alpha^n)_{n\geq 1}$ is uniformly distributed modulo one for almost all real pairs (ξ, α) with $\alpha > 1$ and ξ in $\mathbb{R}$.

1.3 Discrepancy

In the first sections of this chapter, we have considered uniform distribution from a qualitative point of view and were merely interested in deciding whether a given sequence is or is not uniformly distributed modulo one. However, a quick look at several sequences shows that the rate of convergence in Definition 1.1 can vary greatly. To measure the deviation of a uniformly distributed sequence from an 'ideal' distribution, the notion of *discrepancy* was introduced. According to [411], the first paper in which this concept is studied in its own right was published in 1936 by Bergström [76]. However, investigations for several uniformly distributed sequences had been carried out earlier. The first extensive study of discrepancy was undertaken in 1939 by van der Corput and Pisot [206].

Here, we content ourselves to state the definition and some basic results. The interested reader is directed to the monographs [232, 411].

DEFINITION 1.11. Let N be a positive integer. Let $x_1, \ldots, x_N$ be real numbers. The number

$$D_N(x_1, \ldots, x_N) := \sup_{0 \le u < v \le 1} \left| \frac{\mathrm{Card}\{n : 1 \le n \le N, u \le \{x_n\} < v\}}{N} - (v - u) \right|$$

is called the *discrepancy* of $x_1, \ldots, x_N$. For an infinite sequence $\boldsymbol{x}$ of real numbers, the discrepancy $D_N(\boldsymbol{x})$ is the discrepancy of the first N terms of $\boldsymbol{x}$.

The concept of discrepancy gives a natural criterion to decide whether or not a given sequence is uniformly distributed modulo one, whose proof is left as an exercise.

THEOREM 1.12. *The sequence $\boldsymbol{x}$ is uniformly distributed modulo one if, and only if,* $\lim_{N \to +\infty} D_N(\boldsymbol{x}) = 0$.

The discrepancy of a real sequence $\boldsymbol{x}$ cannot be too small and it satisfies

$$\frac{1}{N} \le D_N(\boldsymbol{x}) \le 1 \quad (N \ge 1). \tag{1.5}$$

The left-hand side inequality of (1.5) can be considerably improved for arbitrarily large values of N, as was proved by W. M. Schmidt [634].

THEOREM 1.13. *For any infinite sequence $\boldsymbol{x}$ of real numbers, there are arbitrarily large integers N such that*

$$D_N(\boldsymbol{x}) \ge \frac{\log N}{25N}.$$

Theorem 1.13 was proved in [634] with 25 replaced by a larger numerical constant; see [232, 411] for a proof. The van der Corput sequence $\boldsymbol{v}$, defined below, shows that Theorem 1.13 is best possible up to the numerical constant 25. Indeed, for $n \ge 1$, let $n - 1 = \sum_{j=0}^{m} a_j 2^j$ be the representation of $n - 1$ in base 2 and set

$$v_n := \sum_{j=0}^{m} a_j 2^{-j-1}.$$

The sequence $\boldsymbol{v} = (v_n)_{n\geq 1}$ is then contained in the unit interval and its discrepancy satisfies

$$D_N(\boldsymbol{v}) \leq \frac{\log(N+1)}{(\log 2)N} \quad (N \geq 1). \tag{1.6}$$

The special case of the sequence $(n\alpha)_{n\geq 1}$ has been thoroughly studied; see [232, Section 1.4] and the references given therein. The discrepancy of many sequences $\boldsymbol{x}$ of real numbers satisfies

$$D_N(\boldsymbol{x}) = O\big(N^{-1/2}(\log\log N)^{1/2}\big);$$

see [232, 411].

1.4 Distribution functions

For the sake of completeness, we briefly give the definition of a distribution function, following [411].

DEFINITION 1.14. A *distribution function* f is any non-decreasing function which satisfies $f(0) = 0$ and $f(1) = 1$ and maps the unit interval into itself. Let $(x_n)_{n\geq 1}$ be a sequence of points in the unit interval and f a distribution function. We say that $(x_n)_{n\geq 1}$ has f as its asymptotic distribution function if, for every a in $[0,1]$, we have

$$\lim_{N\to+\infty} \frac{1}{N} \sum_{n=1}^{N} \chi_{[0,a]}(x_n) = f(a),$$

where $\chi_{[0,a]}$ is the characteristic function of the interval $[0,a]$. If the indicated limit fails to exist for some a, then the sequence $(x_n)_{n\geq 1}$ does not have an asymptotic distribution function.

If the sequence $(x_n)_{n\geq 1}$ has the function $f(x) = x$ as its asymptotic distribution function, then it is uniformly distributed modulo one. The converse is obviously true.

DEFINITION 1.15. If $(x_n)_{n\geq 1}$ is a sequence of points in the unit interval and f is a distribution function, we say that f is a *distribution function of* $(x_n)_{n\geq 1}$ if there exists an increasing sequence $N_1, N_2, \ldots$ of positive integers such that, for every a in $[0,1]$, we have

$$\lim_{j\to+\infty} \frac{1}{N_j} \sum_{n=1}^{N_j} \chi_{[0,a]}(x_n) = f(a).$$

Any sequence $(x_n)_{n\geq 1}$ of points in the unit interval has at least one distribution function.

1.5 The multidimensional case

The definition of uniform distribution modulo one for real sequences extends in a natural way to sequences in $\mathbb{R}^d$, by replacing intervals $[u, v)$ by d-dimensional parallelepipeds.

DEFINITION 1.16. The sequence $(\underline{x}_n)_{n\geq 1} = ((x_{n,1}, \ldots, x_{n,d}))_{n\geq 1}$ of elements of $\mathbb{R}^d$ is said to be *uniformly distributed modulo one* if, for every real number $u_1, \ldots, u_d, v_1, \ldots, v_d$ with $0 \leq u_j < v_j \leq 1$ for $j = 1, \ldots, d$, we have

$$\lim_{N\to+\infty} \frac{\operatorname{Card}\{n : 1 \leq n \leq N, u_j \leq \{x_{n,j}\} < v_j \text{ for } j = 1, \ldots, d\}}{N} = \prod_{j=1}^{d} (v_j - u_j).$$

The scalar product of the d-dimensional real vectors $\underline{x} = (x_1, \ldots, x_d)$ and $\underline{y} = (y_1, \ldots, y_d)$ is

$$\langle \underline{x}, \underline{y} \rangle = x_1 y_1 + \cdots + x_d y_d.$$

Theorem 1.17 extends Theorem 1.2 and its proof is similar to that of Theorem 1.2; see e.g. [232].

THEOREM 1.17. *The sequence $(\underline{x}_n)_{n\geq 1}$ of elements of $\mathbb{R}^d$ is uniformly distributed modulo one if, and only if, for every complex-valued, 1-periodic continuous function f defined on $\mathbb{R}^d$ we have*

$$\lim_{N\to+\infty} \frac{1}{N} \sum_{n=1}^{N} f(\underline{x}_n) = \int_{[0,1]^d} f(\underline{x})\,\mathrm{d}\underline{x},$$

that is, if, and only if, for every non-zero integer vector $\underline{h}$, one has

$$\lim_{N\to+\infty} \frac{1}{N} \sum_{n=1}^{N} \mathrm{e}^{2i\pi\langle \underline{h}, \underline{x}_n \rangle} = 0.$$

As an easy application of Theorem 1.17, Weyl [733] obtained a new proof (and a strengthening) of Kronecker's approximation theorem.

THEOREM 1.18. *Let $\alpha_1, \ldots, \alpha_d$ be real numbers such that $1, \alpha_1, \ldots, \alpha_d$ are linearly independent over the field of rational numbers. Then, the* *sequence* $(n\alpha_1, \ldots, n\alpha_d)_{n\geq 1}$ *is uniformly distributed (thus, in particular, dense) modulo one.*

PROOF. Our assumption implies that, for every non-zero integer vector $(h_1, \ldots, h_d)$, the real number $\alpha := h_1\alpha_1 + \cdots + h_d\alpha_d$ is irrational. We conclude by applying Theorem 1.17 and using (1.1) with $h = 1$. □

For later use, we display the following easy result.

THEOREM 1.19. *Let* $(\underline{x}_n)_{n\geq 1} = ((x_{n,1}, \ldots, x_{n,d}))_{n\geq 1}$ *be a sequence of elements of* $\mathbb{R}^d$ *which is uniformly distributed modulo one. Let* ϕ *be a 1-periodic, continuous, real function. Then, the sequence*

$$\big(\phi(x_{n,1}) + \cdots + \phi(x_{n,d})\big)_{n\geq 1}$$

is uniformly distributed modulo one if, and only if,

$$\int_0^1 \mathrm{e}^{2i\pi h\phi(x)}\,\mathrm{d}x = 0$$

holds for every non-zero integer h.

PROOF. Since $(\underline{x}_n)_{n\geq 1}$ is uniformly distributed modulo one and the function ϕ is continuous, we get from Theorem 1.17 that

$$\lim_{N\to+\infty} \frac{1}{N} \sum_{n=1}^{N} \mathrm{e}^{2i\pi h(\phi(x_{n,1})+\cdots+\phi(x_{n,d}))} = \Big(\int_0^1 \mathrm{e}^{2i\pi h\phi(x)}\,\mathrm{d}x\Big)^d$$

holds for every non-zero integer h. We conclude by using Theorem 1.2. □

We end this section by the definition of complete uniform distribution, a notion introduced by Korobov [396].

DEFINITION 1.20. The sequence $(x_n)_{n\geq 1}$ of real numbers is *completely uniformly distributed modulo one* if, for every $d \geq 1$, the d-dimensional sequence $((x_{n+1}, \ldots, x_{n+d}))_{n\geq 1}$ is uniformly distributed modulo one.

Constructions of completely uniformly distributed sequences are given in [396, 399, 426, 661, 662]; see also the survey [583] and the monograph [402]. Franklin [312] established that, for almost all real numbers $\alpha > 1$, the sequence $(\alpha^n)_{n\geq 1}$ is completely uniformly distributed.

1.6 Exercises

Exercise 1.1. Let $(x_n)_{n\geq 1}$ be a sequence uniformly distributed modulo one. Let $(y_n)_{n\geq 1}$ be a converging sequence of real numbers. Prove that the sequence $(x_n + y_n)_{n\geq 1}$ is uniformly distributed modulo one.

Exercise 1.2. Let α be a real number smaller than -1. Prove that, for almost all real numbers ξ, the sequence $(\xi\alpha^n)_{n\geq 1}$ is uniformly distributed modulo one.

Exercise 1.3. Prove Theorem 1.12.

Exercise 1.4. Prove (1.5) and (1.6).

1.7 Notes

▷ Theorem 1.18 was originally proved by Kronecker [410]; see [389, p. 83] for further references and [334] for alternative proofs.

▷ It follows from the pointwise ergodic theorem and the ergodicity of the multiplication by an integer $b \geq 2$ in the torus that for almost all positive real numbers ξ the sequence $(\xi b^n)_{n\geq 1}$ is uniformly distributed modulo one; see e.g. [726] and Section C.4.

2

On the fractional parts of powers of real numbers

We established in the previous chapter several metrical statements on the distribution modulo one of sequences $(\xi\alpha^n)_{n\geq 1}$. However, very little is known for *given* real numbers ξ and α. This chapter and the next one are mainly concerned with the following general questions.

(Hardy, 1919) *Do there exist a transcendental real number $\alpha > 1$ and a non-zero real number ξ such that $||\xi\alpha^n||$ tends to 0 as n tends to infinity?*

(Mahler, 1968) *Given a real number $\alpha > 1$ and an interval $[s, s+t)$ included in $[0, 1)$, is there a non-zero real number ξ such that $s \leq \{\xi\alpha^n\} < s+t$ for all integers $n \geq 0$? What is the smallest possible t for which such a ξ does exist?*

The second of these questions was asked by Mahler in the particular case where $\alpha = 3/2$ and $[s, s+t) = [0, 1/2)$.

Section 2.1 is devoted to classical results of Pisot and of Vijayaraghavan, which are also presented in Salem's monograph [619] and in [80]. In the next two sections, we investigate the set of pairs (ξ, α) for which the sequence $(\{\xi\alpha^n\})_{n\geq 1}$ avoids an interval of positive length included in $[0, 1]$. Among other results, we show in Section 2.3 that, however close to 1 the real number $\alpha > 1$ can be, there always exist non-zero real numbers ξ such that the sequence $(\{\xi\alpha^n\})_{n\geq 1}$ enjoys the latter property. The situation is opposite for the sequence $(\{\xi 2^m 3^n\})_{m,n\geq 0}$, which is dense in $[0, 1]$ for every irrational number ξ, by a celebrated result of Furstenberg presented in Section 2.5. We conclude this chapter with a few words on a conjecture of de Mathan and Teulié, also called the mixed Littlewood conjecture.

2.1 Thue, Hardy, Pisot and Vijayaraghavan

Let $\alpha > 1$ and $\xi \neq 0$ be real numbers. In 1912, Thue [686] proved that if there exist real numbers C and ρ with $0 < \rho < 1$ such that

$||\xi\alpha^n|| < C\rho^n$ for every $n \geq 1$, then α must be an algebraic number. His proof rests on a clever application of the *Schubfachprinzip*; see Exercise 2.1. The same result was stated without proof by Hardy [333, Theorem C], who wrote that this is a special case of a theorem of Borel [113]. In addition, Hardy established that α must be an algebraic integer (that is, an algebraic number whose minimal defining polynomial over $\mathbb{Z}$ is monic), all of whose Galois conjugates (except α itself) are lying in the open unit disc. Hardy's result was independently rediscovered by Pisot [563], who was not aware of Thue's and Hardy's works. It was subsequently considerably improved by Pisot [564, 565] in 1937.

THEOREM 2.1. *Let $\alpha > 1$ and $\xi \neq 0$ be real numbers such that*

$$\sum_{n\geq 0} ||\xi\alpha^n||^2 \quad \textit{converges.} \tag{2.1}$$

Then, α is an algebraic integer and all its Galois conjugates (except α itself) are lying in the open unit disc. Furthermore, ξ lies in the number field $\mathbb{Q}(\alpha)$.

Before giving the proof of Theorem 2.1, we state a very useful auxiliary lemma. Throughout this section, we need several results on recurrence sequences gathered in Appendix F.

LEMMA 2.2. *Let $\alpha > 1$ and $\xi \neq 0$ be real numbers. For $n \geq 0$, write $\xi\alpha^n = a_n + \varepsilon_n$, where a_n is an integer and $|\varepsilon_n| \leq 1/2$. Assume that there is a non-negative integer n_0 such that the sequence $(a_n)_{n\geq n_0}$ is a recurrence sequence. Then, α is an algebraic integer all of whose Galois conjugates (except α itself) are lying in the closed unit disc and ξ is an element of $\mathbb{Q}(\alpha)$. If, furthermore, the minimal defining polynomial $Q(X) = X^d + q_{d-1}X^{d-1} + \cdots + q_1X + q_0$ of α over $\mathbb{Z}$ satisfies $q_0a_n + \cdots + q_{d-1}a_{n+d-1} + a_{n+d} = 0$ for $n \geq n_0$ and if*

$$\limsup_{n\to+\infty} |\varepsilon_n| < \frac{\min_\sigma |\sigma(\xi Q'(\alpha))|}{2^d\alpha}, \tag{2.2}$$

where the minimum is taken over all the Galois embeddings $\sigma \neq \mathrm{Id}$ of $\mathbb{Q}(\alpha)$ into $\mathbb{C}$, then α has no Galois conjugates on the unit circle.

In view of Theorem 3.9, we cannot remove the dependence on ξ in (2.2).

PROOF. Since $(a_n)_{n\geq n_0}$ is a recurrence sequence of integers, it follows from a lemma of Fatou [300] (see Theorem F.2) that there are coprime polynomials $U(X)$ and $V(X)$ with integer coefficients such that $V(0) = 1$ and

$$\sum_{n \geq n_0} a_n z^n = \frac{z^{n_0} U(z)}{V(z)}.$$

Writing $V(X) = 1 + v_{k-1}X + \cdots + v_0 X^k$, we have

$$\begin{aligned} f(z) := \sum_{n \geq n_0} \varepsilon_n z^n = \sum_{n \geq n_0} \xi \alpha^n z^n - \sum_{n \geq n_0} a_n z^n \\ = \frac{\xi \alpha^{n_0} z^{n_0}}{1 - \alpha z} - \frac{z^{n_0} U(z)}{1 + v_{k-1} z + \cdots + v_0 z^k}. \end{aligned}$$

Since the radius of convergence of f is at least equal to 1, the polynomial $V(X)$ has only one root in the open unit disc, namely $1/\alpha$. Consequently, the reciprocal polynomial $X^k V(1/X) = X^k + v_{k-1}X^{k-1} + \cdots + v_0$ has a single root, namely α, with modulus greater than 1, all its other roots being in the closed unit disc. The computation of the residue of f at $1/\alpha$ shows that $\xi \alpha^{n_0} = -\alpha U(1/\alpha)/V'(1/\alpha)$, in particular we get that ξ is in $\mathbb{Q}(\alpha)$.

We establish the last assertion of the lemma. Let $\alpha = \alpha_1, \alpha_2, \ldots, \alpha_d$ denote the Galois conjugates of α and define the algebraic integers $\delta_0, \ldots, \delta_{d-2}$ by

$$S(X) := (X - \alpha)(X - \alpha_2) \cdots (X - \alpha_{d-1}) = X^{d-1} + \delta_{d-2} X^{d-2} + \cdots - \delta_0.$$

Our assumption implies that $k = d$ and $Q(X) = X^d V(1/X)$. It then follows from the partial fraction decomposition of $f(z)$ that

$$a_n = \xi \alpha^n + \xi_2 \alpha_2^n + \cdots + \xi_d \alpha_d^n, \quad \text{for } n \geq n_0,$$

where $\xi_2, \ldots, \xi_d$ are the respective Galois conjugates of ξ. Let ε be a positive real number and n_1 be an integer exceeding n_0 and such that $|\varepsilon_n| \leq \varepsilon$ for every $n \geq n_1$. We then combine $S(\alpha) = S(\alpha_2) = \cdots = S(\alpha_{d-1}) = 0$ with the fact that every Galois conjugate of α has modulus at most 1 to deduce from the equality

$$\begin{aligned} \xi \alpha^n S(\alpha) - (\varepsilon_{n+d-1} + \delta_{d-2}\varepsilon_{n+d-2} + \cdots + \delta_0 \varepsilon_n) \\ = \xi \alpha^n S(\alpha) + \xi_2 \alpha_2^n S(\alpha_2) + \cdots + \xi_d \alpha_d^n S(\alpha_d), \end{aligned}$$

that

$$\begin{aligned} |\xi_d \alpha_d^n S(\alpha_d)| = |\varepsilon_{n+d-1} + \delta_{d-2}\varepsilon_{n+d-2} + \cdots + \delta_0 \varepsilon_n| \\ \leq \varepsilon(1 + |\delta_0| + \cdots + |\delta_{d-2}|) \leq \varepsilon 2^d \alpha, \end{aligned} \tag{2.3}$$

for every $n \ge n_1$. With ε smaller than the right-hand side of (2.2), we get from (2.3) that $|\alpha_d^n| < 1$ for every large n, since $Q'(\alpha_d) = S(\alpha_d)$. The same holds for all the other Galois conjugates of α, except α itself. Consequently, none of the Galois conjugates of α lies on the unit circle. □

PROOF OF THEOREM 2.1. For $n \ge 0$, write $\xi\alpha^n = a_n + \varepsilon_n$, where a_n is a rational integer (sometimes, we use the terminology *rational integer*, rather than *integer*, to denote an element of $\mathbb{Z}$) and $|\varepsilon_n| \le 1/2$, and consider the Hankel determinant

$$\Delta_n = \begin{vmatrix} a_0 & a_1 & \dots & a_n \\ a_1 & a_2 & \dots & a_{n+1} \\ \dots & \dots & \dots & \dots \\ a_n & a_{n+1} & \dots & a_{2n} \end{vmatrix}.$$

For $m \ge 1$, set

$$\eta_m = a_m - \alpha a_{m-1} = \alpha\varepsilon_{m-1} - \varepsilon_m,$$

and observe that

$$\Delta_n = \begin{vmatrix} a_0 & \eta_1 & \dots & \eta_n \\ a_1 & \eta_2 & \dots & \eta_{n+1} \\ \dots & \dots & \dots & \dots \\ a_n & \eta_{n+1} & \dots & \eta_{2n} \end{vmatrix}. \tag{2.4}$$

We get from the definition of η_m that

$$\eta_m^2 \le (\alpha + 1)^2(\varepsilon_{m-1}^2 + \varepsilon_m^2),$$

and the assumption (2.1) implies that $\sum_{n \ge 1} \eta_n^2$ converges. For $h \ge 1$, set $R_h = \sum_{n \ge h} \eta_n^2$. Bounding $|\Delta_n|$ by means of the Hadamard inequality, we derive from (2.4) that

$$\begin{aligned} \Delta_n^2 &\le \Big(\sum_{m=0}^{n} a_m^2\Big)\Big(\sum_{m=1}^{n+1} \eta_m^2\Big) \cdots \Big(\sum_{m=n}^{2n} \eta_m^2\Big) \\ &\le C\alpha^{2n} R_1 \dots R_n \le C \prod_{h=1}^{n} (\alpha^2 R_h), \end{aligned}$$

for a positive real number C depending only on ξ and α. Since Δ_n is a rational integer and $\alpha^2 R_h$ tends to zero as h tends to infinity, the Hankel determinant Δ_n is equal to 0 for every large integer n, say for $n \ge n_0$. Theorem F.3 then implies that the sequence $(a_n)_{n \ge n_0}$ is a recurrence sequence of integers. By Lemma 2.2 we first deduce that α is an algebraic integer. Let $X^d + q_{d-1}X^{d-1} + \cdots + q_0$ denote its minimal polynomial over $\mathbb{Z}$. Then, for $n \geq 0$, we have

$$\begin{aligned}
0 &= \xi(\alpha^{n+d} + q_{d-1}\alpha^{n+d-1} + \cdots + q_0\alpha^n) \\
&= a_{n+d} + q_{d-1}a_{n+d-1} + \cdots + q_0 a_n \\
&\qquad\qquad + (\varepsilon_{n+d} + q_{d-1}\varepsilon_{n+d-1} + \cdots + q_0\varepsilon_n).
\end{aligned} \tag{2.5}$$

Since $\sum_{n\geq 0} \varepsilon_n^2$ converges, we have $\lim_{n\to+\infty} \varepsilon_n = 0$ and

$$|\varepsilon_{n+d} + q_{d-1}\varepsilon_{n+d-1} + \cdots + q_0\varepsilon_n| < 1$$

for every sufficiently large integer n. We thus get from (2.5) that $a_{n+d} + q_{d-1}a_{n+d-1} + \cdots + q_0 a_n = 0$ for every sufficiently large integer n and, since (2.2) is satisfied, the theorem is then a direct consequence of the last assertion of Lemma 2.2. □

Gelfond [319] proved that Pisot's assumption (2.1) can be replaced by an inequality of the type

$$\limsup_{n\to+\infty} \sqrt{n}\, ||\xi\alpha^n|| \leq c, \tag{2.6}$$

for a suitable positive constant c; see Exercise 2.2. Decomps-Guilloux and Grandet-Hugot [224] showed that one can take $c = 1/(2\sqrt{2}(1+\alpha)^2)$. This was improved by Hata [337] to $c = 0.9026/\alpha^2$.

Let α be a real algebraic integer of degree $d \geq 1$, with Galois conjugates $\alpha = \alpha_1, \alpha_2, \ldots, \alpha_d$ corresponding respectively to the embeddings $\sigma_1, \sigma_2, \ldots, \sigma_d$ of the field $\mathbb{Q}(\alpha)$ into $\mathbb{C}$. Let $X^d + q_{d-1}X^{d-1} + \cdots + q_1 X + q_0$ be its minimal defining polynomial over the integers. Let ξ be an algebraic number in the field $\mathbb{Q}(\alpha)$ and N a non-negative integer such that the traces (recall that the trace $\mathrm{Tr}(\eta)$ of an algebraic number η in $\mathbb{Q}(\alpha)$ is the sum $\sigma_1(\eta) + \cdots + \sigma_d(\eta)$)

$$\mathrm{Tr}(\xi\alpha^N), \mathrm{Tr}(\xi\alpha^{N+1}), \ldots, \mathrm{Tr}(\xi\alpha^{N+d-1})$$

are all rational integers. Using the linearity of the trace function, we see that, for every integer $n \geq N + d$, we have

$$\begin{aligned}
0 &= \mathrm{Tr}(\xi\alpha^{n-d}(\alpha^d + q_{d-1}\alpha^{d-1} + \cdots + q_1\alpha + q_0)) \\
&= \mathrm{Tr}(\xi\alpha^n) + q_{d-1}\mathrm{Tr}(\xi\alpha^{n-1}) + \cdots + q_0\mathrm{Tr}(\xi\alpha^{n-d}).
\end{aligned}$$

By induction, we derive that, for every $n \geq N$, the trace of $\xi\alpha^n$ is a rational integer and

$$||\xi\alpha^n|| \leq |\xi\alpha^n - \mathrm{Tr}(\xi\alpha^n)| \leq |\sigma_2(\xi)\alpha_2^n| + \cdots + |\sigma_d(\xi)\alpha_d^n|.$$

Consequently, if we assume that $\alpha_2, \dots, \alpha_d$ are all lying in the open unit disc, we deduce that

$$\lim_{n \to +\infty} ||\xi \alpha^n|| = 0. \tag{2.7}$$

This observation and Lemma 2.2 motivate the following definitions.

DEFINITION 2.3. A *Pisot number* is a real algebraic integer α greater than 1 all of whose Galois conjugates, except α itself, have their modulus strictly smaller than 1.

DEFINITION 2.4. A *Salem number* is a real algebraic integer α greater than 1 all of whose Galois conjugates, except α itself, have their modulus at most equal to 1, one at least having a modulus equal to 1.

We claim that if α is a Salem number, then $1/\alpha$ is a Galois conjugate of α and all its other Galois conjugates are lying on the unit circle. Indeed, by definition, α has a Galois conjugate, say β, on the unit circle. Hence, β^{-1}, the complex conjugate of β, is also a root of the minimal defining polynomial $Q(X)$ of α over $\mathbb{Z}$. This shows that $Q(X) = X^d Q(1/X)$, where d is the degree of $Q(X)$, and proves the claim.

Pisot [564] used the terminology 'nombre ρ' (literally, ρ-number) to denote algebraic integers whose Galois conjugates are all in the open unit disc. Vijayaraghavan [708] denoted by S the set of these numbers and proved that S contains algebraic integers of every degree, a result which was independently established in [686]. These numbers were previously considered by Thue [686] and by Hardy [333] and are usually termed 'Pisot–Vijayaraghavan numbers', or 'PV-numbers', or 'S numbers' (do not confuse these with one of the four classes in Mahler's classification recalled in Definition E.13), or 'Pisot numbers'. Perhaps, it would be more accurate to speak of 'Thue–Hardy–Pisot–Vijayaraghavan numbers'...

The strategy for proving Theorem 2.1 was firstly to establish that α is algebraic and, secondly, to get precise information on its Galois conjugates. It is reasonable to think that the assumption in Theorem 2.1 could be relaxed if α is assumed to be algebraic. This is indeed the case. Theorem 2.5 below shows that (2.7) does not hold if α is algebraic but not Pisot. This was proved by Hardy [333] and reproved by Pisot [565], independently; see also [607, 708]. The weaker assumption in the last assertion of Theorem 2.5 comes from Vijayaraghavan's paper [708]; see also [566].

THEOREM 2.5. *Let $\alpha > 1$ be an algebraic number and ξ be a non-zero real number such that*

$$\lim_{n\to+\infty} ||\xi\alpha^n|| = 0.$$

Then, α is a Pisot number and ξ lies in the field $\mathbb{Q}(\alpha)$. The same conclusion holds if the sequence $(\{\xi\alpha^n\})_{n\geq 1}$ has only finitely many limit points.

PROOF. Let $q_dX^d + \cdots + q_1X + q_0$ be the minimal defining polynomial of α over the integers. Let ε be a positive real number with $0 < \varepsilon < 1/(|q_0| + \cdots + |q_d|)$. Assume that the sequence $(\{\xi\alpha^n\})_{n\geq 1}$ has only finitely many limit points, denoted by $\zeta_1, \ldots, \zeta_r$. By Theorem 1.18, there exists a positive integer q such that $||q\zeta_i|| < \varepsilon/2$ for $i = 1, \ldots, r$. Consequently, there exists n_0 such that, for every $n \geq n_0$, we have $||q\xi\alpha^n|| \leq \varepsilon$.

Let $n \geq n_0$ be an integer. Write $q\xi\alpha^n = a_n + \varepsilon_n$, where a_n is a rational integer and $|\varepsilon_n| \leq \varepsilon$. Since

$$q\xi\alpha^n(q_0 + q_1\alpha + \cdots + q_d\alpha^d) = 0,$$

we get

$$\begin{aligned}|q_0a_n + q_1a_{n+1} + \cdots + q_da_{n+d}| &= |q_0\varepsilon_n + q_1\varepsilon_{n+1} + \cdots + q_d\varepsilon_{n+d}| \\ &\leq \varepsilon(|q_0| + \cdots + |q_d|) < 1,\end{aligned}$$

and, as the left-hand side of the latter inequality is a rational integer, we deduce that

$$q_0a_n + q_1a_{n+1} + \cdots + q_da_{n+d} = 0.$$

Consequently, the sequence $(a_n)_{n\geq n_0}$ is a linear recurrence sequence and we apply the first assertion of Lemma 2.2 to conclude that α is a Pisot or a Salem number, and ξ lies in the field $\mathbb{Q}(\alpha)$. Furthermore, the last assertion of Lemma 2.2 implies that, by choosing ε sufficiently small, we get the stronger conclusion that α is a Pisot number. □

We display an immediate consequence of Theorem 2.5.

COROLLARY 2.6. *Let ξ be a non-zero real number. Let p and q be coprime integers satisfying $p > q \geq 2$. Then the sequence $(\{\xi(p/q)^n\})_{n\geq 1}$ has infinitely many limit points.*

Corollary 2.6 for $\xi = 1$ has been established independently by Pisot [565], Vijayaraghavan [707] and Rédei [606]; see Exercise 2.5. An alternative proof of Corollary 2.6 can be found in [237]; see also [708].

In the course of the proof of Theorem 2.5, we have established the following statement, which was noticed by Pisot [565].

THEOREM 2.7. *Let $\alpha > 1$ be an algebraic number and ξ be a non-zero real number. Let $q_d X^d + \cdots + q_1 X + q_0$ be the minimal defining polynomial of α over the integers. If*

$$\limsup_{n \to +\infty} ||\xi \alpha^n|| < \frac{1}{|q_0| + \cdots + |q_d|}, \tag{2.8}$$

then α is either a Pisot or a Salem number, and ξ lies in the field $\mathbb{Q}(\alpha)$.

Refinements of Theorem 2.7 are given in Section 3.5.

2.2 On some exceptional pairs (ξ, α)

Following Pisot [564], we show that there are few pairs (ξ, α) such that $||\xi \alpha^n||$ tends to 0 as n tends to infinity.

THEOREM 2.8. *There are only countably many pairs (ξ, α) of real numbers such that $\xi \neq 0$, $\alpha > 1$, and*

$$\limsup_{n \to +\infty} ||\xi \alpha^n|| < \frac{1}{2(1+\alpha)^2}. \tag{2.9}$$

PROOF. Let $\xi \neq 0$ and $\alpha > 1$ be real numbers. For $n \geq 1$, write

$$\xi \alpha^n = a_n + \varepsilon_n,$$

where $|\varepsilon_n| \leq 1/2$ and a_n is an integer. If (2.9) holds, then there exist an integer n_0 and a positive real number ε with $2(1+\alpha)^2 \varepsilon < 1$ and $|\varepsilon_n| < \varepsilon$ for every $n \geq n_0$. Since, for $n \geq n_0$, we have

$$a_{n+2} - \frac{a_{n+1}^2}{a_n} = \frac{\varepsilon_n \varepsilon_{n+2} - \varepsilon_{n+1}^2}{a_n} - \frac{\xi \alpha^n}{a_n} (\varepsilon_{n+2} - 2\varepsilon_{n+1}\alpha + \varepsilon_n \alpha^2),$$

it follows that

$$\limsup_{n \to +\infty} \left| a_{n+2} - \frac{a_{n+1}^2}{a_n} \right| \leq \varepsilon (1 + 2\alpha + \alpha^2) < \frac{1}{2}.$$

Consequently, there exists an integer n_1 such that $|a_{n+2} - a_{n+1}^2/a_n| < 1/2$ for $n \geq n_1$. This implies that a_{n+2} is the nearest integer to a_{n+1}^2/a_n for $n \geq n_1$.

Let then Φ be the map which associates to every pair (ξ, α), with $\xi \neq 0$, $\alpha > 1$ and such that (2.9) holds, the triple (m, v, w) of integers, where m is the smallest integer for which $|a_{n+2} - a_{n+1}^2/a_n| < 1/2$ for $n \geq m$, and where $(v, w) = (a_m, a_{m+1})$. It remains for us to prove that Φ is injective. To do this, let (m, v, w) be a given triple of integers and assume that $(m, v, w) = \Phi(\xi, \alpha)$ for some $\xi \neq 0$ and $\alpha > 1$. There exists exactly one sequence $(a'_n)_{n \geq m}$ of integers satisfying $(v, w) = (a'_m, a'_{m+1})$, and $|a'_{n+2} - a'^2_{n+1}/a'_n| < 1/2$ for $n \geq m$. This sequence coincides with the sequence $(a_n)_{n \geq m}$ defined as above. It is thus uniquely determined by (m, v, w), and we have

$$\alpha = \lim_{n \to +\infty} \frac{a'_{n+1}}{a'_n} \quad \text{and} \quad \xi = \lim_{n \to +\infty} \frac{a'_n}{\alpha^n}.$$

Consequently, the pair (ξ, α) is also uniquely determined by (m, v, w). This concludes the proof. □

Independently, Vijayaraghavan [709] proved that there are only countably many real numbers α such that the sequence $(\{\alpha^n\})_{n \geq 1}$ has only finitely many limit points. A stronger result follows from Theorem 2.8.

COROLLARY 2.9. *There are only countably many pairs* (ξ, α) *of real numbers such that* $\xi \neq 0$, $\alpha > 1$, *and the sequence* $(\{\xi\alpha^n\})_{n \geq 1}$ *has only finitely many limit points.*

PROOF. Let $\xi \neq 0$ and $\alpha > 1$ be real numbers such that $(\{\xi\alpha^n\})_{n \geq 1}$ has only finitely many limit points, denoted by $\zeta_1, \ldots, \zeta_r$. By Theorem 1.18, there exists a positive integer q such that $||q\zeta_i|| < 1/(3(1+\alpha)^2)$ for $i = 1, \ldots, r$. Consequently, we get $\limsup_{n \to +\infty} ||q\xi\alpha^n|| < 1/(2(1+\alpha)^2)$ and the pair $(q\xi, \alpha)$ satisfies the assumption of Theorem 2.8. This proves the corollary. □

The first assertion of the next result was proved by Pisot [566], using Thue's method [686]; see Exercise 2.1.

THEOREM 2.10. *Let* ξ *and* α *be real numbers with* $\xi \geq 1$ *and* $\alpha > 1$. *If*

$$\sup_{n \geq 0} ||\xi\alpha^n|| \leq \frac{1}{2e\alpha(1+\alpha)(1+\log\xi)}, \tag{2.10}$$

then α *is either a Pisot or a Salem number, and* ξ *lies in the field* $\mathbb{Q}(\alpha)$. *Conversely, if* α *is either a Pisot or a Salem number, then there exists* ξ *in* $\mathbb{Q}(\alpha)$ *such that (2.10) holds.*

Before proceeding with the proof of Theorem 2.10, we state a useful auxiliary lemma. Note that Thue [686] proved that there exist Pisot numbers of arbitrary degree all of whose Galois conjugates are real.

LEMMA 2.11. *Let* $\mathbb{K}$ *be a real number field of degree* d. *Then there exist Pisot numbers of degree* d *in* $\mathbb{K}$.

PROOF. Let α be an algebraic integer such that $\mathbb{K} = \mathbb{Q}(\alpha)$. Let $\alpha = \alpha_1, \alpha_2, \ldots, \alpha_r, \alpha_{r+1}, \ldots, \alpha_{r+s}, \overline{\alpha_{r+1}}, \ldots, \overline{\alpha_{r+s}}$ be the Galois conjugates of α, where $\alpha_2, \ldots, \alpha_r$ are real numbers. Let $M > 1$ and $\delta < 1$ be positive real numbers. By Minkowski's first theorem (see e.g. [181, p. 151] or [146, Theorem B.2]), if M is sufficiently large, then there are rational integers $m_0, \ldots, m_{d-1}$, not all zero, such that

$$\left|\sum_{i=0}^{d-1} m_i \alpha^i\right| \leq M \quad \text{and} \quad \left|\sum_{i=0}^{d-1} m_i \alpha_j^i\right| \leq \delta, \quad 2 \leq j \leq r+s.$$

Setting $\theta := m_0 + m_1\alpha + \cdots + m_{d-1}\alpha^{d-1}$, we see that θ is a real algebraic integer, all of whose Galois conjugates (except θ itself) are of modulus at most δ. Since the norm of θ is at least equal to 1, we get immediately that θ or $-\theta$ is greater than 1. Consequently, θ or $-\theta$ is a Pisot number of degree d in $\mathbb{K}$. □

Proof of Theorem 2.10. Let $\xi \geq 1$ and $\alpha > 1$ be such that (2.10) holds. Set $s = \lfloor \log \xi \rfloor + 1$, $v = \lceil 2\alpha\xi^{1/s} \rceil - 1$, and check that

$$(s+1)\xi^{1/s} < \mathrm{e}(1 + \log \xi). \tag{2.11}$$

For $n \geq 0$, write $\xi\alpha^n = a_n + \varepsilon_n$, where $|\varepsilon_n| \leq 1/2$ and a_n is an integer. Set $\varepsilon = (2\mathrm{e}\alpha(1+\alpha)(1+\log\xi))^{-1}$ and observe that, by (2.10),

$$1/|\varepsilon_n| \geq 1/\varepsilon = 2\mathrm{e}\alpha(1+\alpha)(1+\log\xi).$$

For a tuple $\underline{v} = (v_0, \ldots, v_s)$ of rational integers from $[0, v]$, setting

$$u_{\underline{v},n} = v_0 a_n + v_1 a_{n+1} + \cdots + v_s a_{n+s},$$

we deduce from (2.11) that

$$\begin{aligned} 0 \leq u_{\underline{v},0} \leq (s+1)v(\xi\alpha^s + \varepsilon) &\leq (s+1)v\xi\alpha^s + \frac{1}{\alpha+1} \\ &< (s+1)(v+1)\xi\alpha^s - 1. \end{aligned}$$

Our choices of s and v imply also that

$$(v+1)^{s+1} \geq 2^s(v+1)\xi\alpha^s,$$

thus it follows from the *Schubfachprinzip* that there are integers $v_0', \ldots, v_s'$ in $[-v, v]$, not all zero, such that

$$v_0' a_0 + v_1' a_1 + \cdots + v_s' a_s = 0. \tag{2.12}$$

For $n \geq 1$, we have

$$\begin{aligned} |v_0' a_n + v_1' a_{n+1} + \cdots + v_s' a_{n+s} \\ - \alpha(v_0' a_{n-1} + v_1' a_n + \cdots + v_s' a_{n+s-1})| \\ = |v_0'(\alpha\varepsilon_{n-1} - \varepsilon_n) + v_1'(\alpha\varepsilon_n - \varepsilon_{n+1}) + \cdots + \\ + v_s'(\alpha\varepsilon_{n+s-1} - \varepsilon_{n+s})| \\ \leq (s+1)(\alpha+1)v\varepsilon < 1, \end{aligned} \tag{2.13}$$

by (2.11). Since $(a_n)_{n\geq 0}$ is a sequence of integers, an immediate induction based on (2.12) and (2.13) shows that $v_0' a_n + v_1' a_{n+1} + \cdots + v_s' a_{n+s} = 0$ for $n \geq 0$. Applying Lemma 2.2, we get the first assertion of the theorem.

We deal now with the second assertion. Let α be a Pisot or a Salem number of degree $d \geq 2$. By Lemma 2.11, there exists a Pisot number μ such that $\mathbb{Q}(\alpha) = \mathbb{Q}(\mu)$. Let δ be such that $0 < \delta < 1$ and the remaining Galois conjugates of α and μ satisfy $|\alpha_j| \leq 1$ and $|\mu_j| \leq \delta$, for $j = 2, \ldots, d$. Let m be a positive integer and set $\xi = \mu^m$. For $n \geq 1$, the number $\xi\alpha^n + \xi_2\alpha_2^n + \cdots + \xi_d\alpha_d^n$, where $\xi_j = \mu_j^m$ for $j = 2, \ldots, d$, is a rational integer and $|\xi_2\alpha_2^n + \ldots + \xi_d\alpha_d^n| \leq (d-1)\delta^m$. Selecting m large enough such that

$$(d-1)\delta^m \leq \frac{1}{2e\alpha(1+\alpha)(1+m\log\mu)},$$

we get (2.10). This finishes the proof of the theorem. □

Pisot [566] applied Theorem 2.10 to establish that, if the sequence $(\{\xi\alpha^n\})_{n\geq 1}$, where $\xi \neq 0$ and $\alpha > 1$, has only finitely many limit points, and if the speed of convergence to these limit points is in $o(n^{-k-1})$, where k denotes the number of irrational limit points of $(\{\xi\alpha^n\})_{n\geq 0}$, then α is a Pisot or a Salem number.

The assumption (2.10) was weakened by Cantor [177] to

$$\sup_{n\geq 0} ||\xi\alpha^n|| \leq \frac{1}{e(1+\alpha)^2(2+\sqrt{\log\xi})},$$

by means of an improved criterion for an integer sequence to satisfy a linear recurrence; see also [223, 224]. This is a special case of a more general result, which includes Theorems 2.1 and 2.10. Cantor's main statement answers a question of Pisot and Salem posed at the end of [567].

In the opposite direction to Theorem 2.10, Boyd [127] established that there are uncountably many pairs (ξ, α) such that all the numbers $||\xi\alpha^n||$, $n \geq 0$, are quite small.

THEOREM 2.12. *Let u, v be real numbers such that $3 < u < v$, and let a_0 be an integer satisfying $a_0 > (u+1)(u-1)^{-1}(v-u)^{-1}$. Then, there exist uncountably many α in $[u, v]$ for which there is a positive real number ξ_α in $(a_0 - 1/2, a_0 + 1/2)$ with*

$$\sup_{n \geq 0} ||\xi_\alpha \alpha^n|| \leq \frac{1}{(u-1)(\alpha-1)}.$$

We stress the following corollary of Theorem 2.12, which shows that, in some sense, Theorem 2.10 is not far from being best possible.

COROLLARY 2.13. *For any $c > 1$, there are $\alpha > 1$ and $\xi \geq 1$ with α transcendental and*

$$\sup_{n \geq 0} ||\xi \alpha^n|| \leq \frac{c + \log 2}{\alpha(1+\alpha)(1+\log \xi)}. \tag{2.14}$$

PROOF. Let $u \geq 4$ be an integer. Set $v = u + 2$ and $a_0 = 2$. By Theorem 2.12, there exist a transcendental real number α_u in $[u, v]$ and a real number ξ_u such that $||\xi_u|| = |\xi_u - 2| \leq (\alpha_u - 1)^{-1}(u-1)^{-1}$. In particular, we have $\xi_u > 1$ and the product $(u-1)^{-1}(\alpha_u - 1)^{-1}(\alpha_u(1 + \alpha_u)(1 + \log \xi_u))$ tends to $1 + \log 2$ as u tends to infinity. Consequently, $\alpha = \alpha_u$ and $\xi = \xi_u$ satisfy (2.14) when u is large enough in terms of c. □

PROOF OF THEOREM 2.12. The assumption on a_0 implies that

$$(a_0 v - (u-1)^{-1}) - (a_0 u + (u-1)^{-1}) > 1.$$

Therefore, the integer $a_1 = \lfloor a_0 u + (u-1)^{-1} \rfloor + 1$ satisfies

$$a_0 u + (u-1)^{-1} < a_1 < a_0 v - (u-1)^{-1}. \tag{2.15}$$

For $n \geq 1$, write

$$a_{n+1} = \lfloor a_n^2 / a_{n-1} \rfloor \quad \text{and} \quad \rho_n = a_n / a_{n-1}. \tag{2.16}$$

We get from (2.16) that

$$|\rho_{n+1} - \rho_n| = \frac{|a_{n+1} a_{n-1} - a_n^2|}{a_n a_{n-1}} \leq \frac{1}{a_n} \tag{2.17}$$

and

$$a_n^{-1} \leq \sum_{k=1}^{n} a_k^{-1} = a_0^{-1} \sum_{k=1}^{n} (\rho_k \rho_{k-1} \cdots \rho_1)^{-1}. \tag{2.18}$$

It follows from (2.15) that $u < \rho_1 < v$. Let n be a positive integer such that $u < \rho_j < v$ for $j = 1, \ldots, n$. Then, (2.17) and (2.18) give that

$$|\rho_{n+1} - \rho_1| < a_0^{-1} \sum_{k=1}^{n} u^{-k} < a_0^{-1}(u-1)^{-1},$$

hence,

$$\rho_1 - a_0^{-1}(u-1)^{-1} < \rho_{n+1} < \rho_1 + a_0^{-1}(u-1)^{-1}.$$

Combined with (2.15), this gives that $u < \rho_{n+1} < v$, since $\rho_1 = a_1/a_0$. Thus, we have $u < \rho_n < v$ for $n \geq 1$.

For positive integers n, m with $n < m$, we get from (2.18) that

$$\begin{aligned} |\rho_m - \rho_n| &\leq \sum_{j=n}^{m-1} |\rho_{j+1} - \rho_j| \\ &\leq \sum_{j=n}^{m-1} a_j^{-1} \leq \sum_{j=n}^{m-1} u^{-j+n-1} a_{n-1}^{-1} < a_{n-1}^{-1}(u-1)^{-1}. \end{aligned} \tag{2.19}$$

Since $a_n = \rho_n \ldots \rho_1 a_0 \geq u^n a_0$, the sequence $(a_n)_{n\geq 1}$ tends to infinity. Thus, the sequence $(\rho_n)_{n\geq 1}$ is a Cauchy sequence. Its limit, denoted by α, satisfies $u \leq \alpha \leq v$. Letting m tend to infinity in (2.19), we get

$$|\alpha - \rho_n| < a_{n-1}^{-1}(u-1)^{-1}, \tag{2.20}$$

hence, by multiplying both sides of (2.20) by $a_{n-1}\alpha^{-n}$,

$$|a_{n-1}\alpha^{-(n-1)} - a_n \alpha^{-n}| < \alpha^{-n}(u-1)^{-1}. \tag{2 21}$$

Since $\alpha > 1$, the sequence $(a_n \alpha^{-n})_{n\geq 1}$ is a Cauchy sequence. Let ξ denote its limit. We infer from (2.21) that

$$|\xi - a_n \alpha^{-n}| < (u-1)^{-1} \sum_{m \geq n} \alpha^{-m-1} = \alpha^{-n}(\alpha-1)^{-1}(u-1)^{-1}, \tag{2.22}$$

for $n \geq 0$. Since $3 < u \leq \alpha$, we have $(\alpha-1)^{-1}(u-1)^{-1} < 1/4$, thus

$$||\xi \alpha^n|| \leq (\alpha-1)^{-1}(u-1)^{-1},$$

as asserted. We further observe that (2.22) for $n = 0$ implies that $|\xi - a_0| < 1/4$ since $\alpha \geq u > 3$, proving that a_0 is the nearest integer to ξ.

To finish the proof, it remains to explain how to modify this construction to get uncountably many α with the same property. It suffices to replace the definition of a_{n+1} given in (2.16) for $n \geq 1$ by $a_{n+1}(f) = \lfloor a_n^2(f)/a_{n-1}(f) \rfloor + f(n)$, where $f(n) \in \{0, 1\}$. For a function $f : \{n : n \geq 1\} \to \{0, 1\}$, proceeding as above gives a real number

$\alpha(f)$ and it remains to show that $\alpha(f)$ and $\alpha(g)$ are different if f and g are different functions. This is left as Exercise 2.6. $\square$

2.3 On the powers of real numbers close to 1

In this section, we consider the following problem:

Let $\alpha > 1$ be a given real number. Do there exist a positive real number ξ and an interval I in $[0,1]$ such that $\{\xi\alpha^n\}$ avoids I for all $n \geq 0$? How large can I be?

This question is easier to answer when α is not too close to 1. We start with a result of Tijdeman [687] dealing with powers of real numbers greater than 2.

THEOREM 2.14. *Let α be a real number such that $\alpha > 2$ and let m be a positive integer. Then, there exists a real number ξ in $(m, m+1)$ such that*

$$\{\xi\alpha^n\} \in [0, 1/(\alpha-1)], \quad \textit{for all } n \geq 0.$$

PROOF. Set $x_0 = m$ and $x_{n+1} = \lfloor \alpha x_n \rfloor + 1$, for every integer $n \geq 0$. Observe that

$$0 < x_{n+1} - \alpha x_n \leq 1, \quad \text{for } n \geq 0. \tag{2.23}$$

For every non-negative integer n and k with $n > k$, we have

$$\alpha^{k-n} x_n - x_k = \sum_{j=k+1}^{n} \alpha^{k-j}(x_j - \alpha x_{j-1})$$

thus, by (2.23),

$$0 < \alpha^{k-n} x_n - x_k \leq \sum_{j=k+1}^{n} \alpha^{k-j} < \frac{1}{\alpha-1}.$$

Consequently, the sequence $(\alpha^{-n} x_n)_{n \geq 0}$ is increasing (take $k = n-1$) and bounded (take $k = 0$). Denote by ξ its limit and observe that, since $\alpha > 2$, we have

$$m = x_0 < \xi \leq x_0 + \frac{1}{\alpha-1} < m+1.$$

Let k be a non-negative integer. Since

$$\xi\alpha^k - x_k = \lim_{n \to +\infty} (\alpha^{k-n} x_n - x_k) \quad \text{is in } [0, 1/(\alpha-1)]$$

and x_k is an integer, the fractional part $\{\xi\alpha^k\}$ belongs to the interval $[0, 1/(\alpha-1)]$. This proves the theorem. $\square$

The idea of the proof of Theorem 2.14 can be applied to get various related statements, see e.g. [293, 308, 363, 687] and Exercise 2.7.

The problem posed at the beginning of this section is related to a question of Erdős, addressed near to the end of [279]:

Let $m_1 < m_2 < \ldots$ be an infinite sequence of integers satisfying $m_{n+1}/m_n > c > 1$ for $n \geq 1$. Is it true that there is always an irrational ξ for which the sequence $(\xi m_n)_{n \geq 1}$ is not dense modulo one?

A positive answer to Erdős' question was given independently by de Mathan [481, 482] and Pollington [570]; see also [679] for a weaker result. These authors were not aware of a remarkably rich paper of Khintchine [379] which had appeared in 1926 and in which Erdős' question was solved by means of a beautiful nested intervals construction; see Exercise 2.10. In all these papers, the existence of real numbers ξ such that the sequence $(\{\xi m_n\})_{n \geq 1}$ avoids an interval of $[0, 1]$ of positive length is proved.

We put forward a classical definition.

DEFINITION 2.15. An increasing sequence $(t_n)_{n \geq 1}$ of positive real numbers is called a *lacunary sequence* if

$$\liminf_{n \to +\infty} \frac{t_{n+1}}{t_n} > 1.$$

Clearly, for any real numbers $\alpha > 1$ and $\xi > 0$, the sequence $(\xi \alpha^n)_{n \geq 1}$ is lacunary. Several results on the latter sequence can be proved as well, and without additional difficulties, for lacunary sequences.

Peres and Schlag [556] developed a method based upon the Lovász local lemma in order to answer Erdős' question. As noted in [527], their method is very flexible and, unlike Khintchine's method, it applies to lacunary sequences of real numbers (not only of integers).

THEOREM 2.16. *Let ε be a real number with $0 < \varepsilon \leq 1/20$. Let $(\eta_n)_{n \geq 1}$ be a sequence of real numbers and $(t_n)_{n \geq 1}$ be a sequence of real numbers greater than 1 satisfying*

$$\frac{t_{n+1}}{t_n} \geq 1 + \varepsilon, \quad \text{for } n \geq 1.$$

Then, there exists a positive real number ξ such that

$$\| \xi t_n + \eta_n \| > 3 \cdot 10^{-3} \, \varepsilon \, |\log \varepsilon|^{-1}, \quad \text{for } n \geq 1. \tag{2.24}$$

Theorem 2.16 asserts in particular that, for every ε with $0 < \varepsilon \leq 1/20$ and for every lacunary sequence $(m_n)_{n \geq 1}$ of positive integers satisfying $m_{n+1}/m_n \geq 1 + \varepsilon$ for $n \geq 1$, there exists a real number ξ such that

$$\inf_{n \geq 1} ||\xi m_n|| > 3 \cdot 10^{-3} \, \varepsilon \, |\log \varepsilon|^{-1}. \tag{2.25}$$

Inequality (2.25) cannot be improved up to the factor $3 \cdot 10^{-3} \log(1/\varepsilon)$, as was shown by Peres and Schlag [556], using a connection, due to Katznelson [373] (see also [730, Chapter 5]), between this problem and a problem of Erdős on chromatic numbers. From a quantitative point of view, Khintchine's result [379] is stronger (regarding the dependence on ε, when ε is close to 1) than the results obtained in [482, 570]; see also [30, 243, 373] for slight improvements, none of them being as strong as (2.25).

PROOF. Without any loss of generality, we assume that $-1/2 \leq \eta_j < 1/2$ for $j \geq 1$. Set

$$\delta := 3 \cdot 10^{-3} \, \varepsilon \, |\log \varepsilon|^{-1} \quad \text{and} \quad h := \lceil 5\varepsilon^{-1} |\log \varepsilon| \rceil. \tag{2.26}$$

Observe that

$$\frac{t_{j+h}}{t_j} \geq (1 + \varepsilon)^h \geq \varepsilon^{-4} \geq \frac{1}{\delta}, \quad \text{for } j \geq 1. \tag{2.27}$$

Let $j \geq 1$ be an integer and set

$$\ell_j := \lfloor (\log(t_j/(2\delta))/\log 2) \rfloor,$$

in such a way that

$$2^{\ell_j+1}\delta \leq t_j < 2^{\ell_j+2}\delta. \tag{2.28}$$

For an integer a, set

$$E(j, a) := \left\{ \xi \in (1, 2) : \left| \xi - \frac{a}{t_j} + \frac{\eta_j}{t_j} \right| < \frac{\delta}{t_j} \right\}.$$

Since the length of $E(j, a)$ is at most equal to $2\delta/t_j$, it follows from (2.28) that there exists an integer b_a such that $E(j, a)$ is included in the open dyadic interval

$$\left(\frac{b_a}{2^{\ell_j}}, \frac{b_a + 2}{2^{\ell_j}} \right).$$

Denote by A_j the union of such intervals that covers

$$E(j, \lfloor t_j \rfloor) \cup E(j, \lfloor t_j \rfloor + 1) \cup \ldots \cup E(j, \lceil 2t_j \rceil).$$

Note that

$$\begin{aligned} \lambda(A_j) &\leq 2^{1-\ell_j} \left(\lceil 2t_j \rceil - \lfloor t_j \rfloor + 1 \right) \\ &\leq \left(\frac{8\delta}{t_j} \right) \left(\lceil 2t_j \rceil - \lfloor t_j \rfloor + 1 \right) \leq 24\delta, \end{aligned} \tag{2.29}$$

since $\lceil 2t_j \rceil - \lfloor t_j \rfloor + 1 \leq 3t_j$.

Let i be an integer with $i \geq 1$. Write

$$\bigcap_{1 \leq j \leq i} A_j^c = I_1 \cup \ldots \cup I_{U_i},$$

where $I_1, \ldots, I_{U_i}$ are distinct closed intervals of the form $[b/2^{\ell_i}, (b+1)/2^{\ell_i}]$ with b an integer. Then, we have

$$A_{i+h} \cap \bigcap_{1 \leq j \leq i} A_j^c = \bigcup_{\nu=1}^{U_i} (A_{i+h} \cap I_\nu). \tag{2.30}$$

Assume that the intersection $\cap_{j \leq i} A_j^c$ has positive Lebesgue measure. Then, $U_i \geq 1$. Let J be an open dyadic interval of length $2^{-\ell_{i+h}}$ and whose endpoints have denominator $2^{\ell_{i+h}}$. Let $I_\nu = [b/2^{\ell_i}, (b+1)/2^{\ell_i}]$ be an interval occurring in (2.30) such that $\lambda(J \cap I_\nu)$ is positive. This implies that J is contained in I_ν. Assume that the intersection of J and A_{i+h} has positive Lebesgue measure. Then, J is included in A_{i+h} and there exists an integer a with $\lfloor t_j \rfloor \leq a \leq \lceil 2t_j \rceil$ such that

$$\frac{a}{t_{i+h}} - \frac{\eta_{i+h}}{t_{i+h}} \in \Big(\frac{b}{2^{\ell_i}} - \frac{\delta}{t_{i+h}}, \frac{b+1}{2^{\ell_i}} + \frac{\delta}{t_{i+h}} \Big).$$

By (2.28) the same integer a can be associated to at most two different intervals J. Furthermore, the number W_ν of different possible values of a satisfies

$$W_\nu \leq t_{i+h} \Big(\frac{1}{2^{\ell_i}} + \frac{2\delta}{t_{i+h}} \Big) + 1 \leq 2^{-\ell_i} t_{i+h} + 2 \leq 2 \cdot 2^{-\ell_i} t_{i+h},$$

since

$$t_{i+h} \geq (1+\varepsilon)^{5|\log \varepsilon|/\varepsilon} t_i \geq \varepsilon^{-4} t_i \geq \delta \varepsilon^{-4} 2^{\ell_i + 1} \geq 2^{\ell_i+1},$$

by (2.27) and (2.28). This shows that

$$\begin{aligned} \lambda\Big(\bigcup_{\nu=1}^{U_i} (A_{i+h} \cap I_\nu) \Big) &\leq 2 U_i W_\nu 2^{-\ell_{i+h}} \\ &\leq U_i 2^{2-\ell_i-\ell_{i+h}} t_{i+h} \leq U_i 2^{4-\ell_i} \delta. \end{aligned}$$

However, by assumption,

$$\lambda\Big(\bigcap_{1 \leq j \leq i} A_j^c \Big) = U_i 2^{-\ell_i}.$$

Consequently, we get

$$\lambda\Big(A_{i+h} \cap \bigcap_{1 \leq j \leq i} A_j^c \Big) \leq 16\delta \lambda\Big(\bigcap_{1 \leq j \leq i} A_j^c \Big). \tag{2.31}$$

This is the key result for the last step of the proof, which consists in proving by induction that

$$\lambda\Big(\bigcap_{0\le j\le mh} A_j^c\Big) \ge \frac{1}{2}\lambda\Big(\bigcap_{0\le j\le (m-1)h} A_j^c\Big) > 0, \tag{2.32}$$

for $m \ge 1$, where we agree that $A_0^c = [1,2]$.

We first check that (2.32) holds for $m = 1$. To see this, note that

$$\begin{aligned}\lambda\Big(\bigcap_{0\le j\le h} A_j^c\Big) &\ge 1 - \sum_{j=1}^{h} \lambda(A_j)\\ &\ge 1 - 24\delta h \ge 1/2,\end{aligned}$$

by (2.29) and (2.26).

Let $m \ge 1$ be an integer such that (2.32) holds and write

$$\bigcap_{0\le j\le (m+1)h} A_j^c = \Big(\dots\Big(\Big(\bigcap_{0\le j\le mh} A_j^c\Big)\setminus A_{mh+1}\Big)\setminus\dots\Big)\setminus A_{(m+1)h},$$

thus, by (2.31),

$$\begin{aligned}\lambda\Big(\bigcap_{0\le j\le (m+1)h} A_j^c\Big) &\ge \lambda\Big(\bigcap_{0\le j\le mh} A_j^c\Big)\\ &\qquad - \sum_{u=1}^{h}\lambda\Big(A_{mh+u}\cap\Big(\bigcap_{0\le j\le mh} A_j^c\Big)\Big)\\ &\ge \lambda\Big(\bigcap_{0\le j\le mh} A_j^c\Big) - \sum_{u=1}^{h}\lambda\Big(A_{mh+u}\cap\Big(\bigcap_{0\le j\le (m-1)h+u} A_j^c\Big)\Big)\\ &\ge \lambda\Big(\bigcap_{0\le j\le mh} A_j^c\Big) - 16\delta\sum_{u=1}^{h}\lambda\Big(\bigcap_{0\le j\le (m-1)h+u} A_j^c\Big)\\ &\ge \lambda\Big(\bigcap_{0\le j\le mh} A_j^c\Big) - 16\delta h\,\lambda\Big(\bigcap_{0\le j\le (m-1)h} A_j^c\Big).\end{aligned}$$

The inductive assumption (2.32) shows that

$$\lambda\Big(\bigcap_{0\le j\le (m-1)h} A_j^c\Big) \le 2\lambda\Big(\bigcap_{0\le j\le mh} A_j^c\Big).$$

We then apply (2.26) to get that

$$\lambda\Big(\bigcap_{0\le j\le (m+1)h} A_j^c\Big) \ge (1-32h\delta)\lambda\Big(\bigcap_{0\le j\le mh} A_j^c\Big) \ge \frac{1}{2}\lambda\Big(\bigcap_{0\le j\le mh} A_j^c\Big),$$

which proves (2.32) with $m+1$ instead of m.

We have established that (2.32) holds for every positive integer m. Then, the sequence $(\bigcap_{0\le j\le mh} A_j^{\rm c})_{m\ge1}$ is a decreasing sequence of non-empty compact sets. The intersection $\bigcap_{j\ge0} A_j^{\rm c}$ is non-empty, and any real number ξ in this intersection avoids all the intervals $E(j,a)$, with $j\ge1$ and $a\ge1$, thus, by (2.26), it satisfies (2.24). □

The next corollary is a consequence of Theorem 2.16.

COROLLARY 2.17. *Let ε be a real number with $0<\varepsilon\le 1/20$ and η be a real number. Then, there exists a positive real number ξ such that*

$$\inf_{n\ge1} \|\xi(1+\varepsilon)^n+\eta\| > 3\cdot 10^{-3}\,\varepsilon\,|\log\varepsilon|^{-1}.$$

Furthermore, for every real number $\alpha\ge 21/20$ and η, there exists a positive real number ξ with

$$\inf_{n\ge1} \|\xi\alpha^n+\eta\| > 5\cdot 10^{-5}.$$

PROOF. The first statement is a particular case of Theorem 2.16. The second one follows immediately by taking $\varepsilon=1/20$. □

Notice that we have made no effort to get the best possible numerical constants in Corollary 2.17.

2.4 On the powers of some transcendental numbers

In this section, we survey several results related to Hardy's question [333] of the existence of a transcendental number $\alpha>1$ and a non-zero real number ξ such that $\|\xi\alpha^n\|$ tends to 0 as n tends to infinity. In an almost forgotten paper Vijayaraghavan [710] studied the set of real numbers $\alpha>1$ such that $(\alpha^n)_{n\ge1}$ is not dense modulo one. Among other results, he proved that, for every positive real number ε, the set of real numbers α such that $\|\alpha^n\|<\varepsilon$ for every $n\ge1$ is uncountable; see also [168, 571], where it is proved that this set has full Hausdorff dimension.

THEOREM 2.18. *For every sequence $(r_n)_{n\ge1}$ of real numbers and for every positive real number ε, there exist uncountably many real numbers $\alpha>1$ such that $\|\alpha^n-r_n\|\le\varepsilon$ for every positive integer n.*

We give two different proofs of Theorem 2.18. A first one is the Cantor-type construction as explained in [168, 710], while a second one, found by Dubickas [246], shows in a more explicit way the existence of real numbers $\alpha > 1$ whose powers are close (modulo one) to any arbitrary given sequence.

PROOF. Without loss of generality, we assume $\varepsilon < 1/2$ and $-1/2 \leq r_n < 1/2$ for $n \geq 1$. Put $H = \lceil 3/\varepsilon \rceil$, $a_n = r_n - \varepsilon$ and $b_n = r_n + \varepsilon$, for $n \geq 1$.

Set $E_1 = I_1 = [H + a_1, H + b_1]$. This is our Step 1. Since

$$(H + b_1)^2 - (H + a_1)^2 \geq 4\varepsilon H - 1 \geq 4,$$

there is an integer j_1 such that $j_1, \ldots, j_1 + 3$ are all lying in the interval $[(H + a_1)^2, (H + b_1)^2]$. For $h = j_1 + 1, j_1 + 2$, let $I_{2,h}$ denote the interval $[\sqrt{h + a_2}, \sqrt{h + b_2}]$. Since

$$(H + a_1)^2 \leq h + a_2 < h + b_2 \leq (H + b_1)^2,$$

the interval $I_{2,h}$ is included in I_1. By construction, every real number α in I_h is such that $\alpha - H$ and $\alpha^2 - h$ are in $[a_1, b_1]$ and $[a_2, b_2]$, respectively. Set $E_2 = I_{2,j_1+1} \cup I_{2,j_1+2}$. This completes Step 2.

We continue this process. Let $h_2 = j_1 + 1, j_1 + 2$. Since

$$\begin{aligned}(\sqrt{h_2 + b_2})^3 - (\sqrt{h_2 + a_2})^3 &\geq \big((\sqrt{h_2 + b_2})^2 - (\sqrt{h_2 + a_2})^2\big)\sqrt{h_2 + a_2} \\ &\geq 2\varepsilon H - 2 \geq 4,\end{aligned}$$

there is an integer j_2 such that $j_2, \ldots, j_2 + 3$ are in the interval $[(h_2 + a_2)^{3/2}, (h_2 + b_2)^{3/2}]$. For $h = j_2 + 1, j_2 + 2$, let $I_{3,h_2,h}$ be the interval $[(h + a_3)^{1/3}, (h + b_3)^{1/3}]$. By construction, $I_{3,h_2,h}$ is included in I_{2,h_2}. Proceeding in this way, we construct at Step 3 a union E_3 of four subintervals of I_1, whose elements α have the property that $\alpha - H, \alpha^2 - h_2$ and $\alpha^3 - h$ are in $[a_1, b_1]$, $[a_2, b_2]$ and $[a_3, b_3]$, respectively.

Continuing further in the same way, for $j \geq 4$, we construct at Step j a set E_j which is the union of 2^{j-1} closed intervals. This is a nested construction. The set $\bigcap_{j\geq 1} E_j$ is then a Cantor-type set, whose elements α have the property that, for $n \geq 1$, the fractional part of their nth power lies in $[h + a_n, h + b_n]$, for some integer h depending on α and on n, thus it satisfies $||\alpha^n - r_n|| \leq \varepsilon$.

We present an alternative proof, following [246] (see also [257]), of the existence of a real number $\alpha > 1$ with the requested property. Now, we assume $\varepsilon < 1/2$ and $0 \leq r_n < 1$ for $n \geq 1$. Let m be an integer with $m \geq 2 \log(1/\varepsilon) + 3$. Let $y_0 \geq 2$ be a real number and put

$$y_n := (\lceil y_{n-1}^{mn} \rceil + r_n)^{1/(mn)}, \quad \text{for } n \geq 1.$$

Since

$$y_n \geq (y_{n-1}^{mn} + r_n)^{1/(mn)} \geq y_{n-1},$$

for $n \geq 1$, the sequence $(y_n)_{n\geq 1}$ is non-decreasing. Let $n \geq 1$ be an integer. Observe that $y_n^{mn} - r_n$ is an integer, thus $\{y_n^{mn}\} = r_n$. Since $\lceil y_{n-1}^{mn} \rceil + r_n < y_{n-1}^{mn} + 2$, we get

$$\frac{y_n}{y_{n-1}} < (1 + 2y_{n-1}^{-mn})^{1/(mn)} < 1 + \frac{2}{mny_{n-1}^{mn}},$$

implying that

$$y_n - y_{n-1} < \frac{2}{mny_{n-1}^{mn-1}}.$$

Consequently,

$$\begin{aligned} y_n - y_0 &< \frac{2}{mny_{n-1}^{mn-1}} + \cdots + \frac{2}{2my_1^{2m-1}} + \frac{2}{my_0^{m-1}} \\ &\leq \frac{2}{m}\Big(\frac{1}{y_0} + \cdots + \frac{1}{y_0^n}\Big) \leq \frac{2y_0}{m}. \end{aligned}$$

This proves that

$$\beta := \lim_{n\to+\infty} y_n$$

exists.

Let k and n be integers with $n \leq k$. Since $y_k \geq 2$, we get

$$\begin{aligned} \Big(\frac{y_{k+1}}{y_k}\Big)^{mn} &< (1 + 2y_k^{-m(k+1)})^{n/(k+1)} < 1 + \frac{2n}{(k+1)y_k^{m(k+1)}} \\ &< 1 + \frac{2}{y_k^{m(k+1)}} \leq 1 + y_k^{-m(k+1)+1}. \end{aligned}$$

It follows that, for every fixed $n \geq 1$,

$$\Big(\frac{\beta}{y_n}\Big)^{mn} = \prod_{k=n}^{+\infty} \Big(\frac{y_{k+1}}{y_k}\Big)^{mn} < \prod_{k=n}^{+\infty} (1 + y_k^{-m(k+1)+1}) \leq 1 + y_n^{-m(n+1)+3}.$$

Consequently,

$$0 \leq \beta^{mn} - y_n^{mn} < y_n^{-m+3} \leq 2^{-m+3} < \varepsilon,$$

by our choice of m. Since $r_n = \{y_n^{mn}\}$, we conclude that $\|\beta^{mn} - r_n\| < \varepsilon$ for $n \geq 1$. This shows that $\alpha = \beta^m$ satisfies the conclusion of the theorem. □

The real numbers α constructed in the proofs of Theorem 2.18 are clearly greater than 2. At the end of [710], Vijayaraghavan refined his contruction to show that there are real numbers $\alpha > 1$ arbitrarily close to 1 and such that there exists an interval of length smaller than 1 containing all the fractional parts $\{\alpha^n\}$, $n \geq 1$. His result was improved by Bugeaud and Moshchevitin [168] as follows.

THEOREM 2.19. *Let ξ be a positive real number. For any sequence $(\eta_n)_{n\geq 1}$ of real numbers, there exist a positive real number γ, depending only on ξ and $(\eta_n)_{n\geq 1}$, and arbitrarily small positive real numbers ε such that*

$$\inf_{n\geq 1} \|\xi(1+\varepsilon)^n + \eta_n\| > \gamma\, \varepsilon\, |\log \varepsilon|^{-1}.$$

The proof of Theorem 2.19 is omitted, it uses a slight refinement of the method initiated by Peres and Schlag for the proof of Theorem 2.16.

Despite the fact that the sequence $(\alpha^n)_{n\geq 1}$ is uniformly distributed modulo one for almost all numbers $\alpha > 1$, we know only very few explicit examples of real numbers α with the weaker property that $(\alpha^n)_{n\geq 1}$ is dense modulo one. We conclude this section by a construction of Dubickas [246] of uncountably many real numbers $\alpha > 1$ enjoying the latter property.

THEOREM 2.20. *Let $(r_n)_{n\geq 1}$ be a sequence of real numbers in $[0,1)$ which is dense in $[0,1)$ and such that $r_n = 0$ for infinitely many n. Set $x_1 = 2$ and*

$$x_n := 2 + \lfloor (x_{n-1} + r_{n-1})^n - r_n \rfloor, \quad \text{for } n \geq 2.$$

Then, the limit

$$\alpha := \lim_{n\to+\infty} (x_n + r_n)^{1/n!}$$

exists and the sequence $(\alpha^{n!})_{n\geq 1}$ is dense modulo one.

PROOF. For $n \geq 2$, we have

$$x_n + r_n = 2 + \lfloor (x_{n-1} + r_{n-1})^n - r_n \rfloor + r_n > (x_{n-1} + r_{n-1})^n,$$

thus the sequence $\boldsymbol{x} = ((x_n + r_n)^{1/n!})_{n\geq 1}$ is increasing. Furthermore, for $n \geq 3$, we have

$$\begin{aligned}(x_{n-1} + r_{n-1} + (x_{n-1} + r_{n-1})^{-n+1})^n &\geq (x_{n-1} + r_{n-1})^n + n\\ &\geq (x_{n-1} + r_{n-1})^n + 3\\ &\geq x_n + r_n + 1\\ &> x_n + r_n + (x_n + r_n)^{-n}.\end{aligned}$$

It follows that the sequence $\boldsymbol{x}' = ((x_n + r_n + (x_n + r_n)^{-n})^{1/n!})_{n\geq 2}$ is decreasing. Consequently, both sequences $\boldsymbol{x}$ and $\boldsymbol{x}'$ converge and α is well defined. Let α' be the limit of $\boldsymbol{x}'$. For $n \geq 2$, we have

$$x_n + r_n < \alpha^{n!} \leq (\alpha')^{n!} < x_n + r_n + (x_n + r_n)^{-n},$$

from which we get that $\alpha = \alpha' > 1$.

Let y be in $(0, 1)$. Let ε be real with $0 < \varepsilon < 1 - y$. Let n be an integer such that r_n is lying in $(y, y + \varepsilon/2)$. Since x_n tends to infinity with n, we have

$$\begin{aligned} x_n + y < x_n + r_n &< \alpha^{n!} \\ &< x_n + r_n + x_n^{-n} < x_n + y + \varepsilon/2 + x_n^{-n} < x_n + y + \varepsilon, \end{aligned}$$

if n is large enough. This shows that infinitely many terms of the sequence $(\alpha^{n!})_{n\geq 1}$ belong to $[y, y + \varepsilon]$, which completes the proof of the theorem. □

2.5 A theorem of Furstenberg

It follows from Theorem 2.16 that, for every lacunary sequence $(m_n)_{n\geq 1}$ of positive integers, there exist ε with $0 < \varepsilon < 1/2$ and real numbers ξ such that all the fractional parts $\{\xi m_n\}$, $n \geq 1$, are contained in $[\varepsilon, 1 - \varepsilon]$. Does the same result hold if the sequence $(m_n)_{n\geq 1}$ is replaced by a sequence which grows slower than exponentially fast?

An important particular case was investigated by Furstenberg [315]. Recall that two non-zero real numbers r and s are *multiplicatively independent* if they are both different from 1 and the ratio $(\log r)/(\log s)$ of their logarithms is irrational; otherwise, r and s are *multiplicatively dependent*.

THEOREM 2.21. *Let r and s be multiplicatively independent integers. Then, for every irrational number ξ, the set of real numbers $\{\xi r^m s^n\}$, $m, n \geq 0$, is dense in the torus $\mathbb{T}$.*

We establish Theorem 2.21 by following the approach of Boshernitzan [122]. We begin by an easy auxiliary result.

DEFINITION 2.22. For a positive integer n, a set $X \subset \mathbb{T}$ is called n-invariant if nx (modulo 1) is in X whenever x belongs to X.

LEMMA 2.23. *Let $\mathcal{M} = (m_i)_{i\geq 1}$ be an infinite sequence of distinct positive integers, arranged in increasing order, such that m_{i+1}/m_i tends to 1 as i tends to infinity. Let X be a closed infinite subset of the torus*

$\mathbb{T}$ that is m_i-invariant for every $i \geq 1$. If 0 is a limit point of X, then $X = \mathbb{T}$.

PROOF. Let ε be a real number satisfying $0 < \varepsilon < 1/2$. Let h be an integer such that the inequality $m_{i+1}/m_i < 1 + \varepsilon$ holds for every $i \geq h$. Replacing X by $\{1 - x : x \in X\}$ if necessary, we can assume that there are non-zero elements of X arbitrarily close to 0. Let x be in X such that $0 < \{m_h x\} < \varepsilon$. Observe that for every $i \geq h$ such that $m_i \leq 1/x$ we have

$$0 < (m_{i+1} - m_i)x < m_i \varepsilon x \leq \varepsilon.$$

Consequently, the finite set

$$\{\{mx\} : m \in \mathcal{M}, m_h \leq m \leq 1/x\} \subset X$$

intersects every interval of length ε in $\mathbb{T}$. Since ε is arbitrary and X is closed, X is equal to $\mathbb{T}$. □

Observe that, if r and s are multiplicatively independent integers, then, for every positive integer u, the sequence $\mathcal{M}$ composed of all the integers of the form $r^{um}s^{un}$ with $m, n \geq 0$ arranged in increasing order satisfies the assumption of Lemma 2.23.

PROOF OF THEOREM 2.21. Let ε be a positive real number and t be an integer coprime with rs and satisfying $t > \max\{1/\varepsilon, 3\}$. Let u be a positive integer such that

$$r^u \equiv s^u \equiv 1 \mod t.$$

Denote by X the closure in the torus $\mathbb{T}$ of the set of real numbers $\{\xi r^m s^n\}$, $m, n \geq 0$.

First, assume that X does not contain rational numbers. Define

$$X = X_0 \supset X_1 \supset \ldots \supset X_{t-1}$$

by setting

$$X_{i+1} = \left\{x \in X_i : x + \frac{1}{t} \in X_i\right\}, \quad (0 \leq i \leq t-2).$$

Observe that, if x belongs to X_i for some $i = 0, \ldots, t-2$, and if X_i is r^u-invariant, then $r^u x$ is in X_{i+1}. Indeed, since X_i is r^u-invariant, $r^u(x + 1/t) = r^u x + (r^u - 1)/t + 1/t$ is in X_i and we use the fact that t divides $r^u - 1$ to get that $\{r^u x + 1/t\}$ belongs to X_i, hence, $r^u x$ is in X_{i+1}.

Consequently, we deduce from the fact that X_0 is simultaneously r^u- and s^u-invariant that the sets $X_1, \ldots, X_{t-1}$ are simultaneously r^u- and s^u-invariant, as well.

Furthermore, for $i = 1, \ldots, t-1$, the set X_i is closed if the set X_{i-1} is closed. Since X_0 is a closed set, all the sets $X_1, \ldots, X_{t-1}$ are closed.

The difficult point is to prove that X_i is never empty. To this end, for $i = 0, \ldots, t-1$, we introduce the set

$$D_i := X_i - X_i = \{x - x' : x, x' \in X_i\},$$

and we aim at establishing that it includes $1/t$. We argue by induction. Let $i = 0, \ldots, t-2$ be such that X_i is non-empty. Since X_i is compact, D_i is a closed set. The set D_i is simultaneously r^u- and s^u-invariant, hence, it is $(r^u)^m(s^u)^n$-invariant for every non-negative integer m, n. As X_i does not contain rational numbers, it is infinite and 0 is one of the limit points of D_i. Since r^u and s^u are multiplicatively independent, we deduce from Lemma 2.23 that $D_i = \mathbb{T}$, hence, $1/t$ is an element of $X_i - X_i$. This proves that X_{i+1} is non-empty.

Consequently, the set X_{t-1} is non-empty, and there is x_0 in X_{t-1} such that

$$x_0, \left\{x_0 + \frac{1}{t}\right\}, \left\{x_0 + \frac{2}{t}\right\}, \ldots, \left\{x_0 + \frac{t-1}{t}\right\} \quad \text{are all in } X.$$

Since $1/t < \varepsilon$ and ε is arbitrary, X is dense in $\mathbb{T}$. As X is a closed set, it must be equal to $\mathbb{T}$, a contradiction.

Thus, the set X must contain a rational point, say the point p/q, where p and q are coprime positive integers. Replacing if needed p/q by a suitable rational number of the form $(p/q)(rs)^\ell$, where $\ell \geq 0$, we can assume that q is coprime with rs. Let v be a positive integer such that

$$r^v \equiv s^v \equiv 1 \mod q.$$

Observe that the shifted set

$$X - p/q = \{\{x - p/q\} : x \in X\}$$

is simultaneously r^v- and s^v-invariant and has 0 for limit point. It then follows from Lemma 2.23 that $X = \mathbb{T}$. The theorem is established. □

2.6 A conjecture of de Mathan and Teulié

A famous open problem in simultaneous Diophantine approximation is the Littlewood conjecture [447], which claims that, for every given pair (ξ, η) of real numbers, we have

$$\inf_{q \geq 1} q \cdot \|q\xi\| \cdot \|q\eta\| = 0. \tag{2.33}$$

For references on this fascinating question, the reader is directed to [146, Section 10.1], to Venkatesh's article [705], or to Queffélec's survey [598].

In analogy with (2.33), de Mathan and Teulié [483] proposed in 2004 a 'mixed Littlewood conjecture'. For any prime number p, we normalize the usual p-adic value $|\cdot|_p$ in such a way that $|p|_p = p^{-1}$.

For every real number ξ and every prime number p, we have

$$\inf_{q \geq 1} q \cdot \|q\xi\| \cdot |q|_p = 0. \tag{2.34}$$

Obviously, the above conjecture holds if ξ is rational or has unbounded partial quotients, or if $\inf_{n \geq 1} \|\xi p^n\| = 0$. Thus, we only consider the case when ξ is an element of the set $\{\eta \in \mathbb{R} \setminus \mathbb{Q} : \inf_{q \geq 1} q \cdot \|q\eta\| > 0\}$ of badly approximable real numbers (see Definition D.8). Real numbers for which $\inf_{n \geq 1} \|\xi p^n\|$ is positive do exist (for example, real numbers without three consecutive digits 0 and without three consecutive digits $p-1$ in their p-ary expansion) and they form a set whose intersection with the set of badly approximable real numbers has Hausdorff dimension one; see Section 7.3. Furthermore, the set of badly approximable real numbers ξ for which (2.34) holds for any prime p has as well Hausdorff dimension one; see Exercise 7.7.

We briefly survey (without proofs) some results towards a proof of the de Mathan–Teulié conjecture. Einsiedler and Kleinbock [277] showed that a weaker form of it easily follows from Furstenberg's Theorem 2.21.

THEOREM 2.24. *Let p_1 and p_2 be distinct prime numbers. Then, the equality*

$$\inf_{q \geq 1} q \cdot \|q\xi\| \cdot |q|_{p_1} \cdot |q|_{p_2} = 0$$

holds for every real number ξ.

PROOF. The result is clear if ξ is rational. For an irrational number ξ, apply Theorem 2.21 with $r = p_1$ and $s = p_2$ to get that the sequence $(\xi p_1^m p_2^n)_{m,n \geq 0}$ is dense modulo one. This shows that, for every $\varepsilon > 0$, there are integers m and n such that $0 < \{\xi p_1^m p_2^n\} < \varepsilon$, that is, such that, setting $q = p_1^m p_2^n$, we have $q \cdot \|q\xi\| \cdot |q|_{p_1} \cdot |q|_{p_2} = \|q\xi\| < \varepsilon$. □

The next theorem, proved in [160], asserts that (2.34) holds for every pair (ξ, p) provided that the sequence of partial quotients of ξ is quasi-periodic in the following sense.

THEOREM 2.25. *Let $\xi = [a_0; a_1, a_2, \ldots]$ be a badly approximable real number. Let $t \geq 1$ be an integer and $b_1, \ldots, b_t$ be positive integers. If there exist two sequences $(m_k)_{k\geq 1}$ and $(h_k)_{k\geq 1}$ of positive integers with $(h_k)_{k\geq 1}$ being unbounded and, for $k \geq 1$,*

$$a_{m_k+j+nt} = b_j, \quad \text{for } j = 1, \ldots, t \text{ and } n = 0, \ldots, h_k - 1,$$

then

$$\inf_{q\geq 1} q \cdot \|q\xi\| \cdot |q|_p = 0$$

for every prime number p.

An immediate consequence of Theorem 2.25 is that (2.34) holds for every prime p and every quadratic number ξ, a result already proved in [483].

Einsiedler and Kleinbock [277] established that the set of possible exceptions to the de Mathan–Teulié conjecture is, from the metric point of view, very small.

THEOREM 2.26. *Let p be a prime number. The set of real numbers ξ which do not satisfy (2.34) has Hausdorff dimension zero.*

More precisely, it is proved in [277] that (2.34) holds for every prime number p and every real number ξ with positive entropy (see Definition 9.11).

Further recent results can be found in [52, 164]. Improving an earlier result of [167], Badziahin and Velani [52] established the following theorem.

THEOREM 2.27. *Let p be a prime number. The set of real numbers ξ such that*

$$\liminf_{q\to+\infty} q \cdot \log q \cdot \log\log q \cdot \|q\xi\| \cdot |q|_p > 0$$

has full Hausdorff dimension.

2.7 Exercises

EXERCISE 2.1 (*cf.* [686]). Let $\xi \neq 0$ and $\alpha > 1$ be real numbers. Assume that there are real numbers $c > 1$ and ρ such that $0 < \rho < 1$ and $\|\xi\alpha^n\| < c\rho^n$, for $n \geq 1$. Prove that α is an algebraic number. [Hint. For $n \geq 1$, set $\xi\alpha^n = a_n + \varepsilon_n$, where a_n is an integer and $|\varepsilon_n| \leq 1/2$. Choose $m \geq 2$ such that $\rho^{-m} > \alpha$. Let $c_1 < 1, c_2 > 1$ be positive real numbers to be suitably chosen in terms of ξ, c, α, m. Let ℓ be a large integer and N be an integer with $c_1\rho^{-\ell} > N > c_2\alpha^{\ell/m}$. Use the

Schubfachprinzip to show that, for a suitable choice of c_2, there are integers $u_0, \ldots, u_m$, not all zero, such that $|u_j| \le N$ for $j = 0, \ldots, m$ and $u_m a_\ell + \cdots + u_0 a_{\ell+m} = 0$. Prove by induction that, for a suitable choice of c_1, we have $u_m a_n + \cdots + u_0 a_{n+m} = 0$ for $n \ge \ell$. Deduce that $u_m + u_{m-1}\alpha + \cdots + u_0\alpha^n = 0$.]

EXERCISE 2.2. Prove that the conclusion of Theorem 2.1 remains true under the assumption (2.6) for a suitable positive real number c.

EXERCISE 2.3 (*cf.* [619, Exercise 2]). Prove that the conclusion of Theorem 2.1 remains true with (2.1) replaced by $\sum_{n=1}^{N} n||\xi\alpha^n||^2 = o(N)$.

EXERCISE 2.4 (*cf.* [181, p. 134]). Let α be a Pisot number of degree d. Prove that if the non-zero real number ξ satisfies $\lim_{n\to+\infty} ||\xi\alpha^n|| = 0$, then there are a non-negative integer N and an algebraic number ζ in $\mathbb{Q}(\alpha)$ such that $\mathrm{Tr}(\alpha^j\zeta)$ is a rational integer for $j = 0, \ldots, d-1$ and $\xi = \alpha^{-N}\zeta$.

EXERCISE 2.5 (Direct proof, by A. Weil [707], of Corollary 2.6 for $\xi = 1$). For $n \ge 1$, set $y_n = \{(p/q)^n\}$ and assume that $(y_n)_{n\ge1}$ has only finitely many limit points, denoted by $\zeta_1, \ldots, \zeta_k$. Assume that ζ_1 is irrational and consider a strictly increasing sequence $(n_r)_{r\ge1}$ such that y_{n_r} tends to ζ_1 as r tends to infinity. Let m be a positive integer and observe that

$$y_{n_r+m} \equiv (p/q)^m y_{n_r} \pmod{1/q^m}.$$

By letting r tend to infinity, obtain a contradiction.

Assume now that $\zeta_1, \ldots, \zeta_k$ are all rational numbers. Let then M be a positive integer such that $M\zeta_1, \ldots, M\zeta_k$ are rational integers, and observe that the sequence $(||M(p/q)^n||)_{n\ge1}$ tends to 0. Conclude.

EXERCISE 2.6. Complete the proof of Theorem 2.12.

EXERCISE 2.7 (*cf.* [687, Theorem 2]). Let $\alpha > 2$ be a real number such that 2α is an odd integer. Refine the proof of Theorem 2.14 to prove that, for every positive integer m, there exists ξ in $(m, m+1)$ such that $\{\xi\alpha^n\}$ lies in $[0, 1/(2\alpha-2)]$ for every integer $n \ge 0$.

EXERCISE 2.8 (*cf.* [363]). Let $\alpha > 1$ be a real number. Prove that the set of real numbers ξ satisfying

$$\limsup_{n\to+\infty} ||\xi\alpha^n|| < 1/(2\alpha+2)$$

is at most countable.

EXERCISE 2.9 (*cf.* [363]). Let ξ be a non-zero real number. Let δ and M be positive real numbers. Prove that the set of real numbers $\alpha > M$ satisfying

$$\limsup_{n\to+\infty} ||\xi\alpha^n|| < (1+\delta)/(2\alpha)$$

is at least countable and that the set of real numbers $\alpha > M$ satisfying

$$\limsup_{n\to+\infty} ||\xi\alpha^n|| < (1+\delta)/\alpha$$

is uncountable. [Hint. Take R a large positive integer, set $z_1 = R^2$, $z_{n+1} = \lfloor \xi^{-1/n} z_n^{(n+1)/n} \rfloor$ and $\beta_n = \xi^{-1/n} z_n^{(n+1)/n}$ for $n \ge 1$. Then, set $\alpha := \lim_{n\to+\infty} \beta_n$.]

EXERCISE 2.10 (*cf.* [379]). Let $\varepsilon < 1/4$ be a positive real number, t an integer, and $(m_n)_{n\ge1}$ an increasing sequence of positive integers satisfying $m_{n+t}/m_n > 1+\varepsilon$, for $n \ge 1$. Prove that there exists a real number ξ such that $||\xi m_n|| \ge (\varepsilon/3)^2(\log(t/\varepsilon))^{-2}$ for $n \ge 1$. [Hint. Let s be a positive integer such that $(1+\varepsilon)^s > 3ts+3$ and set $k = 3ts+3$. Observe that $m_{n+ts}/m_n > k$ for $n \ge 1$. Let I_1 be a closed real interval of length $1/k$ and divide it into k closed intervals of the same length $1/k^2$. Prove that there exist three consecutive such intervals which contain no rational of the form p/m_n with $1 \le m_n < k$. Let I_2 denote the middle interval. Divide it into k closed intervals of the same length $1/k^3$ and prove that there exist three consecutive such intervals which contain no rational of the form p/m_n with $k \le m_n < k^2$. Let I_3 denote the middle interval. Continue this process to construct an infinite nested sequence $I_1 \supset I_2 \supset \dots$ of closed intervals I_h, $h \ge 1$, with the property that, for $h \ge 1$, the length of I_h equals $1/k^h$ and the distance between I_{h+1} and any rational of the form p/m_n with $k^{h-1} \le m_n < k^h$ is at least equal to $1/k^{h+1}$. Let ξ be the point belonging to every interval I_h, $h \ge 1$. Prove that $||m_n\xi|| \ge 1/k^2$ for $n \ge 1$, and conclude.]

EXERCISE 2.11. Deduce from Theorem 2.16 that there is a positive constant κ such that, for every sufficiently large integer n and every real number $t_1, \dots, t_n$ satisfying $t_1 > 1$ and $t_{j+1}/t_j \ge 1 + (\kappa \log n)/n$ for $j = 1, \dots, n-1$, there exists a positive real number ξ such that $||\xi t_i|| \ge 1/(n+1)$ for $i = 1, \dots, n$.

EXERCISE 2.12 (*cf.* [246]). Let $(\delta_n)_{n\ge1}$ be a sequence of positive numbers with $\delta_n^{1/n}$ tending to 0 as n tends to infinity. Assume that there exist an infinite set $\mathcal{N}$ of positive integers n such that the real number $\alpha > 1$

satisfies $||\alpha^n|| < \delta_n$ for n in $\mathcal{N}$. Prove that α is either transcendental, or is a root of a rational integer. [Hint. Assume that α is algebraic with Galois conjugates $\alpha_1 = \alpha, \alpha_2, \ldots, \alpha_d$ and let $a_d \geq 1$ denote the leading coefficient of its minimal defining polynomial. For n in $\mathcal{N}$, show that $P_n := a_d^n \prod_{j=1}^d (\alpha_j^n - \lfloor \alpha^n + 1/2 \rfloor)$ is a rational integer, and thus satisfies either $P_n = 0$ or $|P_n| \geq 1$.] Deduce that the number α constructed in Theorem 2.20 is transcendental.

EXERCISE 2.13 (*cf.* [160]). Use the Folding Lemma (Theorem D.3) to give explicit examples of pairs (ξ, p) for which (2.34) and even the stronger inequality

$$\liminf_{q \to +\infty} q^2 \cdot ||q\xi|| \cdot |q|_p \leq 1$$

hold.

2.8 Notes

▷ Let $\alpha > 1$ and $\xi \neq 0$ be real numbers. By means of completely different arguments than those in the proof of Theorem 2.1, Feng [304] established that α is a Pisot number when (2.1) is replaced by the stronger assumption that $\sum_{n \geq 0} ||\xi\alpha^n||$ converges.

▷ Inspired by ideas from [566], Mignotte [514] proved, among other results, that a necessary and sufficient condition for a real number $\alpha > 1$ to be a rational integer is to satisfy $||\alpha^n|| \leq ((\alpha+1)(\alpha+2))^{-1}$ for every $n \geq 1$.

▷ Környei [394] extended some results of Pisot [566] to sequences of the form $(P_1(n)\alpha_1^n + \cdots + P_k(n)\alpha_k^n)_{n \geq 1}$, where the $P_i(X)$ are polynomials and the α_i are complex numbers.

▷ Kwapisz [414] gave a geometric proof of the first assertion of Theorem 2.5.

▷ Zaïmi [743] established, among other results, that a real number $\alpha > 1$ is a Pisot number if, and only if, there exists a non-zero real number ξ such that $||\xi\tilde{\alpha}|| < 1/3$ for every $\tilde{\alpha}$ of the form $a_0 + a_1\alpha + \cdots + a_\ell \alpha^\ell$, with $\ell \geq 1$ and $a_j \in \{0, 1\}$ for $0 \leq j \leq \ell$.

▷ For a Pisot or Salem number α, the set of real numbers ξ such that (2.8) holds has been described by Zaïmi [744].

▷ Cantor [176] slightly generalized Theorem 2.10.

▷ The frequency of an infinite set $\mathcal{N}$ of positive integers is the supremum of the real numbers A for which there are arbitrarily large integers n such that every integer in the interval $[n, An]$ belongs to $\mathcal{N}$. Let $\alpha > 1$ be a real algebraic number and ξ be a non-zero real number. Rauzy [601] proved that if $\mathcal{N}$ has infinite frequency and if $||\xi\alpha^n||$ tends to 0 as n tends to infinity along $\mathcal{N}$, then α is either a Pisot, or a Salem number. If, moreover, ξ is algebraic, then α is a Pisot number. Conversely, he established that, for any Pisot or Salem number α, there exist a set $\mathcal{N}$ of integers having infinite frequency and uncountably many real numbers ξ such that $||\xi\alpha^n||$ tends to 0 as n tends to infinity along $\mathcal{N}$.

▷ Erdős and Taylor [281] proved that, for any given lacunary sequence of integers $(m_n)_{n\geq 1}$, the set of real numbers ξ such that $(\xi m_n)_{n\geq 1}$ is not uniformly distributed modulo one has full Hausdorff dimension. The same result holds for the set of real numbers ξ such that $(\xi m_n)_{n\geq 1}$ has no distribution function, as established by Helson and Kahane [341].

▷ To complement Theorem 2.16, Boshernitzan [123] proved that, for any unbounded sequence of positive real numbers $(t_n)_{n\geq 1}$ such that $\limsup_{n\to+\infty} t_{n+1}/t_n$ is finite, the set of real numbers ξ such that the sequence $(\xi t_n)_{n\geq 1}$ is not dense modulo one is a countable union of sets of Hausdorff dimension less than 1. This set has zero Hausdorff measure if, moreover, $\lim_{n\to+\infty} t_{n+1}/t_n = 1$. Ajtai, Havas and Komlós [26] showed that, for any given sequence $(\varepsilon_n)_{n\geq 1}$ of positive real numbers tending to 0, there exists a sequence $(m_n)_{n\geq 1}$ of positive integers such that $m_{n+1}/m_n > 1 + \varepsilon_n$ for $n \geq 1$ and the sequence $(\xi m_n)_{n\geq 1}$ is uniformly distributed modulo one for every irrational number ξ; see also [121]. On the other hand, it is proved in [322] that, for any lacunary sequence $(m_n)_{n\geq 1}$ of positive integers, the set of real numbers ξ such that $(\xi m_n)_{n\geq 1}$ is uniformly distributed modulo one is a meagre set; see also [323].

▷ Various results on the distribution of the sequence $(\{\xi t_n\})_{n\geq 1}$ for fast-growing sequences $(t_n)_{n\geq 1}$ of positive real numbers are given in [253].

▷ Results on sublacunary sequences (that is, increasing sequences growing slower than any lacunary sequence) of real numbers have been obtained in [31, 526, 527].

▷ The method introduced by Peres and Schlag [556] and discussed in Section 2.3 has many other interesting applications to Diophantine problems. It has been applied by Rochev [612] to the distribution of fractional parts of values of linear forms.

$\triangleright$ Let ε be a positive real number. Pollington [571] proved that, for every non-zero real number ξ, the set of real numbers α such that $\{\xi\alpha^n\} < \varepsilon$ for every $n \geq 1$ has Hausdorff dimension one. Furthermore, the set of pairs of real numbers (ξ, α) with $\alpha > 1$ and such that $\{\xi\alpha^n\} < \varepsilon$ for every $n \geq 1$ has Hausdorff dimension two.

$\triangleright$ For a real number $\alpha > 1$, the quantity $\liminf_{n\to+\infty} ||\alpha^n||^{1/n}$ has been studied by Mahler and Szekeres [474]; see also [162]. Let $(\varepsilon_n)_{n\geq1}$ be a sequence of real numbers from $[0, 1/2]$. Koksma [391] showed that if the series $\sum_{n\geq1} \varepsilon_n$ converges then, for almost every real number $\alpha > 1$, there exists an integer $n_0(\alpha)$ such that $||\alpha^n|| \geq \varepsilon_n$ for $n \geq n_0(\alpha)$. Moreover, if $(\varepsilon_n)_{n\geq1}$ is non-increasing and the series $\sum_{n\geq1} \varepsilon_n$ diverges then, for almost every real number $\alpha > 1$, there exist arbitrarily large integers n such that $||\alpha^n|| \leq \varepsilon_n$.

$\triangleright$ Levin [430] (see also [429, 431] and Kulikova [412]) constructed real numbers α such that $(\alpha^n)_{n\geq1}$ is uniformly distributed modulo one and whose discrepancy is $O(N^{-1/2}(\log N)^4)$. This is almost as good as the corresponding metric result; see Section 1.3. In an unpublished manuscript, Lerma [424] gave a fairly complicated, alternative construction of real numbers greater than one with the same property.

$\triangleright$ Zame [745] established that if f is any distribution function and $(n_k)_{k\geq1}$ is a sequence of real numbers such that $n_{k+1} - n_k$ tends to infinity with k, then there exists a real number α such that the sequence $(\{\alpha^{n_k}\})_{k\geq1}$ has f as its distribution function.

$\triangleright$ Furstenberg's Theorem 2.21 was extended by Kra [405], Urban [693–699] and Gorodnik and Kadyrov [326]. See also Berend [71, 72], Rudolph [615], Johnson [355], Feldman [301], Johnson and Rudolph [356], Host [348, 349], Meiri [495] and Lindenstrauss [444].

$\triangleright$ A quantitative version of Theorem 2.21 has been established in [125]. It straightforwardly implies an improvement of Theorem 2.24 [164, 336].

$\triangleright$ The first lower bound for $||\mathrm{e}^n||$ was obtained in 1953 by Mahler [464] under the form

$$||\mathrm{e}^n|| \geq \mathrm{e}^{-cn\log n},$$

for every sufficiently large n, with $c = 40$. Subsequently, he [467] was able to slightly decrease the numerical constant to $c = 33$. In 1974,

Mignotte [512] gave a lower bound with $c = 17.7$, but Wielonsky [734] pointed out that Mignotte's proof contains errors and the corrected value is $c = 21.012$. In the same paper, he gave a new bound, namely with $c = 19.183$. This was improved by Hata [338], who obtained the value $c = 15.727$.

▷ For real numbers $\xi > 0$, $\alpha > 1$ and $t \in [0,1)$, let $M_{\xi,\alpha}(t)$ denote the number of times the value t occurs in the sequence $(\{\xi\alpha^n\})_{n\geq 1}$. It was proved independently in [274, 581, 680] that $M_{1,\alpha}(t) \leq 2$, unless α is a root of an integer, in which case $M_{1,\alpha}(0) = +\infty$. Subsequently, Dubickas [239] established that, for real numbers $\xi \neq 0$ and $\alpha > 1$, the sequence of fractional parts $(\{\xi\alpha^n\})_{n\geq 1}$ takes some value infinitely many times if, and only if, there exist integers $q \geq 2$, $d \geq 1$, $\ell \in \{0, 1, \ldots, d-1\}$, $u \neq 0$ and $v \geq 1$ such that $\alpha = q^{1/d}$ and $\xi = uq^{\ell/d}/v$. See Lemma 3.17 and [199, 263] for further related results.

▷ Let $(g_n)_{n\geq 1}$ be a sequence of positive real numbers satisfying $g_n \geq 1$ for $n \geq 1$ and $\lim_{n\to+\infty} g_n = +\infty$. Bugeaud [152] proved that, for any irrational real number ξ, there exists an increasing sequence of positive integers $(m_n)_{n\geq 1}$ satisfying $m_n \leq ng_n$ for $n \geq 1$ and such that the sequence of fractional parts $(\{\xi m_n\})_{n\geq 1}$ tends to 0 as n tends to infinity. This result is best possible in the sense that the condition $\lim_{n\to+\infty} g_n = +\infty$ cannot be weakened, as proved in [249].

▷ Let n be a positive integer and let $t_1 < t_2 < \ldots < t_n$ be n positive real numbers. The *lonely runner conjecture* asserts that there is a positive number ξ such that

$$\inf_{i=1,\ldots,n} ||\xi t_i|| \geq 1/(n+1).$$

It originally comes from a paper of Wills [735], where it is stated for integers t_i. It has been solved for $n \leq 6$, but remains open for every n greater than or equal to 7. We refer the reader to [261] for bibliographic references. In that paper, Dubickas observed that the lonely runner conjecture is closely related to Theorem 2.16 (see Exercise 2.11) and he applied his results from [247] to confirm the conjecture for every $n \geq 16342$ under the assumption that t_{i+1}/t_i exceeds $1 + (33\log n)/n$ for $i = 1, \ldots, n-1$.

3

On the fractional parts of powers of algebraic numbers

In this chapter we focus on the sequences $(\{\xi\alpha^n\})_{n\geq 1}$ and $(\|\xi\alpha^n\|)_{n\geq 1}$, where ξ is a non-zero real number and α is a real algebraic number greater than 1. We first observe that, if α is a Pisot number, then the sequence $(\{\alpha^n\})_{n\geq 1}$ has at most two limit points, namely 0 and 1. For example, 0 and 1 are the limit points of the sequence $(\{((1+\sqrt{5})/2)^n\})_{n\geq 1}$, while $\{(3+2\sqrt{2})^n\}$ tends to 1 as n tends to infinity. The situation is rather different if α is a Salem number, since $(\alpha^n)_{n\geq 1}$ is then dense modulo one, as was proved by Pisot and Salem; see Theorem 3.7. Very little is known about $(\{\alpha^n\})_{n\geq 1}$ when the algebraic number α is neither a Pisot number, nor a Salem number.

We begin in Section 3.1 with the case where α is a rational integer and continue in the next two sections with the cases where α is a rational number and an algebraic number, respectively, culminating in the statement and proof of Theorem 3.5. It gives an explicit, positive lower bound for the difference between the greatest and the smallest limit point of $(\{\xi\alpha^n\})_{n\geq 1}$, provided that ξ does not belong to $\mathbb{Q}(\alpha)$ when α is either a Pisot or a Salem number. The necessity of the latter assumption is discussed in Section 3.4. The next section is devoted to the study of the sequence $(\|\xi\alpha^n\|)_{n\geq 1}$ when α is a rational number. In Section 3.6 we provide various constructions of pairs (ξ, α), with $\alpha > 1$ being a rational number, such that all the points of the sequence $(\{\xi\alpha^n\})_{n\geq 1}$ are contained in a certain prescribed interval or finite union of intervals. The final sections are devoted to a brief exposition of the Waring problem and to the sequences of integer parts of powers of algebraic numbers.

3.1 The integer case

In this section we treat the easiest case, namely when α is a rational integer.

Let $b \geq 2$ be an integer. We first recall that any real number ξ has a unique b-ary expansion, that is, it can be written uniquely as

$$\xi = \lfloor \xi \rfloor + \sum_{k \geq 1} \frac{a_k}{b^k} = \lfloor \xi \rfloor + 0 \cdot a_1 a_2 \ldots,$$

where the digits $a_1, a_2, \ldots$ are integers from $\{0, 1, \ldots, b-1\}$ and an infinity of the a_k are not equal to $b-1$.

THEOREM 3.1. *Let $b \geq 2$ be an integer. For any irrational real number ξ, the numbers $\{\xi b^n\}$, $n \geq 0$, cannot all lie in an interval of length strictly smaller than $1/b$.*

For any integer $b \geq 2$, the example of the rational number $b/(b^2 - 1) = 0 \cdot 101010\ldots$ with purely periodic b-ary expansion shows that it is necessary to assume that ξ is irrational in Theorem 3.1.

PROOF. Let $\sum_{k \geq 1} a_k b^{-k} = 0 \cdot a_1 a_2 \ldots$, with $a_k \in \{0, 1, \ldots, b-1\}$ for $k \geq 1$, be the b-ary expansion of $\{\xi\}$. Let k be a positive integer and observe that $\{\xi b^k\} = 0 \cdot a_{k+1} a_{k+2} \ldots$ Since ξ is irrational, this implies that

$$\frac{a_{k+1}}{b} < \{\xi b^k\} < \frac{a_{k+1}}{b} + \frac{1}{b}.$$

Thus, if there exist $i, j \geq 0$ such that $a_{j+1} - a_{i+1} \geq 2$, then we get

$$\{\xi b^j\} - \{\xi b^i\} > \frac{a_{j+1}}{b} - \frac{a_{i+1}}{b} - \frac{1}{b} \geq \frac{2}{b} - \frac{1}{b} = \frac{1}{b}.$$

Consequently, we can assume without loss of generality that $a_1, a_2, \cdots \in \{\ell, \ell+1\}$ for a certain integer ℓ in $\{0, 1, \ldots, b-2\}$ and we can write ξ under the form $\lfloor \xi \rfloor + \ell/(b-1) + 0 \cdot w_1 w_2 \ldots$, where $w_k = a_k - \ell$ for $k \geq 1$. Since ξ is irrational, the infinite word $\boldsymbol{w} := w_1 w_2 \ldots$, defined on the alphabet $\{0, 1\}$, is not ultimately periodic and its complexity function $p(m, \boldsymbol{w}, \{0, 1\})$ satisfies $p(m, \boldsymbol{w}, \{0, 1\}) \geq m + 1$ for every $m \geq 1$, by Theorem A.3. This implies that, for every $m \geq 1$, there exists (at least) one block W_m of m letters such that both $0W_m$ and $1W_m$ are subblocks of $\boldsymbol{w}$. In other words, there exist integers u_m, v_m and infinite words $\boldsymbol{w}', \boldsymbol{w}''$ such that $\{\xi b^{u_m}\} - \ell/(b-1) = 0 \cdot 0W_m \boldsymbol{w}'$ and $\{\xi b^{v_m}\} - \ell/(b-1) = 0 \cdot 1W_m \boldsymbol{w}''$. Hence $\{\xi b^{v_m}\} - \{\xi b^{u_m}\} > b^{-1} - b^{-m}$. Since m can be taken arbitrarily large, we conclude that no interval of length strictly smaller than $1/b$ can contain all the $\{\xi b^n\}$ with $n \geq 0$. □

It is worth noting that Theorem 3.1 is best possible, since, for any integer $b \geq 2$, there exist irrational real numbers ξ such that all the real numbers $\{\xi b^n\}$, $n \geq 0$, are lying in a semi-open interval of length $1/b$. For instance, for any irrational number θ with $0 < \theta < 1$ and any real number ρ, the real number $\sum_{k\geq 1} s_k b^{-k}$, where $(s_k)_{k\geq 1}$ is the sequence $\boldsymbol{s}_{\theta,\rho}$ defined in Theorem A.4, has this property. Theorem 2.1 from [161] describes the irrational numbers ξ such that all the fractional parts $\{\xi b^n\}$ with $n \geq 0$ are contained in an open or semi-open interval of length $1/b$.

3.2 Mahler's Z-numbers

According to Mahler [468], the problem whether Z-numbers do exist was proposed to him by a Japanese colleague.

DEFINITION 3.2. A positive real number ξ is a *Z-number* if the inequalities $0 \leq \{\xi(3/2)^n\} < 1/2$ hold for all integers $n \geq 0$. More generally, for a given real number α with $\alpha > 1$ and real numbers s, t satisfying $0 \leq s < s + t \leq 1$, we define the set $Z_\alpha(s, s+t)$ as

$$Z_\alpha(s, s+t) = \{\xi \neq 0 : s \leq \{\xi\alpha^n\} < s+t, \quad \text{for all } n \geq 0\}.$$

Mahler has not succeeded in solving this problem. He established various partial results, including that for any non-negative integer m, the real interval $(m, m+1)$ contains at most one Z-number; see Exercise 3.1. Consequently, there are at most countably many Z-numbers. Mahler showed that the cardinality of the set of Z-numbers less than x is $O(x^\delta)$, with $\delta = \log_2((1+\sqrt{5})/2)$, a result improved in 1992 by Flatto [307] who proved that one can take $\delta = \log_2(3/2)$.

While Mahler's method is rather elementary, Flatto's approach relies on notions from symbolic dynamics and β-transformations (see Section 9.3), and it covers the more general case of sequences $(\{\xi(p/q)^n\})_{n\geq 0}$ for coprime integers p and q such that $p > q \geq 2$. A first step towards the proof (or disproof) of the existence of Z-numbers was made in 1995 by Flatto, Lagarias and Pollington [308], who established the next result (for positive ξ).

THEOREM 3.3. *Let p and q be coprime integers satisfying $p > q \geq 2$. Then, the inequality*

$$\limsup_{n\to\infty} \left\{\xi\left(\frac{p}{q}\right)^n\right\} - \liminf_{n\to\infty} \left\{\xi\left(\frac{p}{q}\right)^n\right\} \geq \frac{1}{p}$$

holds for all non-zero real numbers ξ.

According to Theorem 3.1, Theorem 3.3 remains true for $q = 1$ when ξ is irrational, and it is best possible in that case. For coprime integers p and q with $p > q \geq 2$, Pollington [568] was the first to establish that the set $Z_{p/q}(0, 1/p) \cap \mathbb{R}_{>0}$ is empty, a result reproved by Flatto [307].

The original proof of Theorem 3.3 given in [308] is quite intricate. An alternative proof was found by Dubickas [238], which is more natural and can be extended in various directions; see Theorem 3.5. Here, we reproduce the even simpler proof from [251].

PROOF. Set $\alpha = p/q$. Let n be a positive integer and define

$$x_n = \lfloor \xi\alpha^n \rfloor, \quad y_n = \{\xi\alpha^n\}. \tag{3.1}$$

Since

$$\xi\alpha^{n+1} = x_{n+1} + y_{n+1} = \frac{p}{q}\xi\alpha^n = \frac{p}{q}(x_n + y_n),$$

the real number

$$s_n := qx_{n+1} - px_n = -qy_{n+1} + py_n \tag{3.2}$$

is a rational integer. Furthermore, as y_n and y_{n+1} are lying in $[0, 1)$, we deduce that

$$s_n \in \{-q+1, -q+2, \ldots, p-2, p-1\}.$$

First, we prove that $(s_n)_{n\geq 1}$ is not ultimately periodic. We argue by contradiction and assume that there is a positive integer ℓ such that $s_n = s_{n+\ell}$ for every sufficiently large integer n. Iterating ℓ times the recursion $y_n = (py_{n-1} - s_{n-1})/q$, we then deduce that the sequence

$$\left(y_{t\ell} - (p/q)^{\ell} y_{(t-1)\ell}\right)_{t\geq 2}$$

is ultimately constant and rational. This implies the existence of an integer t_0 and of rational numbers $r = (p/q)^{\ell}$ and r' such that $y_{t\ell} = ry_{(t-1)\ell} - r'$ for every $t \geq t_0 + 1$ and, consequently,

$$y_{t\ell} = r^{t-t_0} y_{t_0\ell} - (r^{t-t_0-1} + \cdots + r + 1)r', \quad \text{for } t \geq t_0 + 1.$$

This can be rewritten as

$$y_{t\ell} - \frac{r'}{r-1} = r^{t-t_0}\left(y_{t_0\ell} - \frac{r'}{r-1}\right).$$

Since r exceeds 1 and the quantities $y_{t\ell} - r'/(r-1)$ and $y_{t_0\ell} - r'/(r-1)$ are both bounded independently of t, we deduce that $y_{t\ell} = r'/(r-1)$ for every $t \geq t_0$. Therefore, the rational numbers $\xi(p/q)^{t_0\ell}$ and $\xi(p/q)^{(t_0+1)\ell}$

have the same fractional part. Hence, ξ must be rational. Write $\xi = a/b$ and set

$$c_t := (a/b)(p/q)^{t\ell} - (a/b)(p/q)^{t_0\ell}, \quad t \geq t_0.$$

Since

$$ap^{t\ell} = q^{(t-t_0)\ell}(bq^{t_0\ell}c_t + ap^{t_0\ell}),$$

and c_t is a rational integer for $t \geq t_0$, we get that every integral power of q^ℓ divides a. This is a contradiction since $q \geq 2$. Consequently, the sequence $(s_n)_{n\geq 1}$ is not ultimately periodic.

Let m be a positive integer. Since $(s_n)_{n\geq 1}$ is not ultimately periodic, it follows from Corollary A.4 that there exist integers $z_0, z_0', z_1, \ldots, z_m$ such that $z_0 > z_0'$ and the finite words $z_0z_1\ldots z_m$ and $z_0'z_1\ldots z_m$ occur infinitely often in the infinite word $s_1s_2s_3\ldots$ This implies that there are a rational number r_m, infinitely many integers n such that

$$y_{n+m+1} = \alpha^{m+1}y_n - \frac{z_0'\alpha^m}{q} - r_m,$$

and infinitely many integers n such that

$$y_{n+m+1} = \alpha^{m+1}y_n - \frac{z_0\alpha^m}{q} - r_m.$$

Setting $\lambda = \liminf_{n\geq 1} y_n$ and $\mu = \limsup_{n\geq 1} y_n$, we get

$$\mu - \alpha^{m+1}\lambda \geq -\frac{z_0'\alpha^m}{q} - r_m$$

and

$$\alpha^{m+1}\mu - \lambda \geq \frac{z_0\alpha^m}{q} + r_m.$$

Since $z_0 - z_0' \geq 1$, adding the last two inequalities gives

$$\mu - \lambda \geq \frac{\alpha^m}{q(1+\alpha^{m+1})} \geq \frac{1}{p} - \left(\frac{q}{p}\right)^{m+1}.$$

As m can be taken arbitrarily large, this shows that $\mu - \lambda \geq 1/p$ and that all the $\{\xi\alpha^n\}$, $n \geq 0$, cannot be contained in an interval of length smaller than $1/p$. □

Besides Theorem 3.3, Flatto, Lagarias and Pollington [308] established various results on the sets $Z_{p/q}(s, s+t)$ and derived several conditions for p, q, s and t under which $Z_{p/q}(s, s+t)$ contains at most one element in the interval $(m, m+1)$, for each non-negative integer m. This extends Mahler's result on $Z_{3/2}(0, 1/2)$ and complements Theorem 2.14.

3.3 On the fractional parts of powers of algebraic numbers

The proof of Theorem 3.3 presented in Section 3.2 can be extended to the study of the sequence $(\{\xi\alpha^n\})_{n\geq 1}$, for a given algebraic number α greater than 1.

We begin with the definitions of the length and the reduced length of an algebraic number. The latter notion was introduced by Dubickas [238].

DEFINITION 3.4. The *length* of a complex polynomial is the sum of the moduli of its coefficients. Let α be an algebraic number and denote by $P(X)$ its minimal defining polynomial over the integers. The *length* of α, denoted by $L(\alpha)$, is equal to the length of $P(X)$. The *reduced length* of α, denoted by $\ell(\alpha)$, is equal to the infimum of the lengths of the polynomials $P(X)G(X)$, where $G(X)$ runs through the set of real polynomials whose leading coefficient or whose constant coefficient is equal to 1.

Theorem 3.5 is Theorem 2 from Dubickas [240]. The special case $\eta = 0$ was proved earlier in [238].

THEOREM 3.5. *Let $\alpha > 1$ be a real algebraic number and η be a real number. Let ξ be a non-zero real number that lies outside the field generated by α if α is a Pisot or a Salem number. Then,*

$$\limsup_{n\to+\infty} \{\xi\alpha^n + \eta\} - \liminf_{n\to+\infty} \{\xi\alpha^n + \eta\} \geq 1/\ell(\alpha).$$

It is worth noting that Theorem 3.5 extends Theorems 3.1 and 3.3. Indeed, according to Exercise 3.2, for any coprime integers p, q with $p > q \geq 1$, the reduced length of the rational number p/q is equal to p.

The additional assumption on ξ when α is a Pisot or a Salem number is necessary; see Section 3.4.

Before proving Theorem 3.5, we state an auxiliary lemma, which will be used several times in the sequel.

LEMMA 3.6. *Let $P(X) = a_d X^d + \cdots + a_1 X + a_0$ be a complex polynomial of degree $d \geq 1$ with distinct roots $\alpha_1, \ldots, \alpha_d$. Let $X_1, \ldots, X_d$ and $Z_0, \ldots, Z_{d-1}$ be such that*

$$X_1\alpha_1^k + \cdots + X_d\alpha_d^k = Z_k, \quad k = 0, 1, \ldots, d-1. \tag{3.3}$$

Then we have $X_j P'(\alpha_j) = \sum_{k=0}^{d-1} \beta_{j,k} Z_k$ for $j = 1, \ldots, d$, where we have set $\beta_{j,k} = \sum_{h=k+1}^{d} a_h \alpha_j^{h-k-1}$ for $j = 1, \ldots, d$ and $k = 0, \ldots, d-1$.

PROOF. The linear system (3.3) has a unique solution, since its Vandermonde determinant is non-zero. Let $j = 1, \ldots, d$. Setting

$$\begin{aligned} P_j(X) = \frac{P(X) - P(\alpha_j)}{X - \alpha_j} &= \sum_{0 \le k \le d} a_k \frac{X^k - \alpha_j^k}{X - \alpha_j} \\ &= \sum_{1 \le k \le d} a_k (X^{k-1} + \alpha_j X^{k-2} + \cdots + \alpha_j^{k-1}) \\ &= \sum_{k=0}^{d-1} \beta_{j,k} X^k, \end{aligned}$$

we have $P_j(\alpha_j) = P'(\alpha_j)$ and $P_j(\alpha_k) = 0$ for $k = 1, \ldots, j-1, j+1, \ldots, d$. Using (3.3), this shows that

$$\sum_{k=0}^{d-1} \beta_{j,k} Z_k = X_1 P_j(\alpha_1) + \cdots + X_d P_j(\alpha_d) = X_j P'(\alpha_j),$$

as asserted. □

PROOF OF THEOREM 3.5. Write $Q(X) = q_d X^d + \cdots + q_1 X + q_0$ for the minimal defining polynomial of α over the rationals. Let n be a positive integer and define

$$x_n = \lfloor \xi \alpha^n + \eta \rfloor, \quad y_n = \{\xi \alpha^n + \eta\}.$$

Set

$$\begin{aligned} s_n &= q_d x_{n+d} + \cdots + q_1 x_{n+1} + q_0 x_n \\ &= -(q_d y_{n+d} + \cdots + q_1 y_{n+1} + q_0 y_n) + \eta Q(1). \end{aligned}$$

The crucial point is again the proof that the sequence $(s_n)_{n \ge 1}$ is not ultimately periodic. We argue by contradiction and assume that there are positive integers n_0 and ℓ such that $s_n = s_{n+\ell}$ for every positive integer $n \ge n_0$. The sequence $(u_n)_{n \ge 1}$ defined for $n \ge 1$ by $u_n = x_{n+\ell} - x_n$ satisfies the linear recurrence

$$q_0 u_n + \cdots + q_d u_{n+d} = 0, \quad \text{for } n \ge n_0.$$

Denote by $\alpha_1 = \alpha, \alpha_2, \ldots, \alpha_d$ the Galois conjugates of α. By Theorem F.1, there exist algebraic numbers $\zeta_1, \cdots, \zeta_d$ such that

$$u_n = \zeta_1 \alpha_1^n + \cdots + \zeta_d \alpha_d^n,$$

for $n \ge n_0$. Let $m \ge n_0$ be an integer. Applying Lemma 3.6 to the linear system

$$X_1 \alpha_1^{n-m} + \cdots + X_d \alpha_d^{n-m} = u_n, \quad n = m, m+1, \ldots, m+d-1,$$

we see that there is an integer polynomial $G_m(X)$ of degree at most $d-1$ such that

$$Q'(\alpha_j)\zeta_j\alpha_j^m = G_m(\alpha_j), \quad j = 1, \ldots, d.$$

Then, ζ_j is in the field generated by α_j, and the algebraic numbers $\zeta_1, \ldots, \zeta_d$ are conjugates over $\mathbb{Q}$, hence, ζ_1 is non-zero. Let a be a positive integer such that

$$a/(Q'(\alpha)\zeta_1), a\alpha/(Q'(\alpha)\zeta_1), \ldots, a\alpha^{d-1}/(Q'(\alpha)\zeta_1)$$

are algebraic integers. Then, $a\alpha^m = aG_m(\alpha)/(Q'(\alpha)\zeta_1)$ is also an algebraic integer. Since m can be taken arbitrarily large, this can only be possible if α is an algebraic integer.

For $n \geq n_0$, set

$$\begin{aligned}\delta_n &= y_{n+\ell} - y_n = \xi(\alpha^\ell - 1)\alpha^n - u_n \\ &= (\xi(\alpha^\ell - 1) - \zeta_1)\alpha_1^n - \zeta_2\alpha_2^n - \cdots - \zeta_d\alpha_d^n.\end{aligned}$$

Applying Lemma 3.6 to the linear system

$$X_1\alpha_1^{n-m} + \cdots + X_d\alpha_d^{n-m} = \delta_n, \quad n = m, m+1, \ldots, m+d-1,$$

we see that $X_1, \ldots, X_d$ are in absolute value bounded by a constant depending only on α and independent of m, as $|\delta_n|$ is at most equal to 1 for $n \geq n_0$. Since

$$X_1 = (\xi(\alpha^\ell - 1) - \zeta_1)\alpha_1^m, X_2 = -\zeta_2\alpha_2^m, \ldots, X_d = -\zeta_d\alpha_d^m,$$

this implies that $\zeta_1 = \xi(\alpha^\ell - 1)$ and that $\alpha_2, \ldots, \alpha_d$ are all in the closed unit disc. Since α is an algebraic integer, this proves that it is a Pisot or a Salem number. Moreover, ξ lies in the field $\mathbb{Q}(\alpha)$ because ζ_1 is in $\mathbb{Q}(\alpha)$. This contradicts our assumption on ξ. Consequently, the sequence $(s_n)_{n\geq 1}$ cannot be ultimately periodic.

We continue similarly as in the proof of Theorem 3.3. Let m be a positive integer and $b_1, \ldots, b_m$ be real numbers. Set

$$P(X) = Q(X) \times (1 + b_1X + \cdots + b_mX^m) =: c_0 + c_1X + \cdots + c_{m+d}X^{m+d},$$

and let n be a positive integer. Since $\xi\alpha^nP(\alpha) = 0$, we get

$$\begin{aligned}v_n &:= c_{m+d}x_{n+m+d} + \ldots + c_0x_n \\ &= -(c_{m+d}y_{n+m+d} + \cdots + c_0y_n) + \eta P(1)\end{aligned} \tag{3.4}$$

and we check that

$$v_n = b_ms_{n+m} + \cdots + b_1s_{n+1} + s_n. \tag{3.5}$$

Let us view $(s_n)_{n\geq 1}$ as an infinite word $\boldsymbol{s}$. Since it is not ultimately periodic, we get from Corollary A.4 that there exist integers $z_0, z_0', z_1, \ldots, z_m$ such that the blocks $z_0 z_1 \ldots z_m$ and $z_0' z_1 \ldots z_m$ occur infinitely often in $\boldsymbol{s}$ and $z_0 > z_0'$.

Set $\lambda = \liminf_{n\geq 1} y_n$ and $\mu = \limsup_{n\geq 1} y_n$. Then, for any positive real number ε, we have

$$\lambda - \varepsilon \leq y_n \leq \mu + \varepsilon$$

for every sufficiently large integer n.

We infer from (3.4) and (3.5) that there are arbitrarily large integers n such that

$$\begin{aligned} v_n &= z_0 + z_1 b_1 + \cdots + z_m b_m \\ &< \eta P(1) + (\mu + \varepsilon)\Big(\sum_{j:c_j<0} (-c_j)\Big) - (\lambda - \varepsilon)\Big(\sum_{j:c_j>0} c_j\Big), \end{aligned}$$

and also arbitrarily large integers n with

$$\begin{aligned} -v_n' &= -z_0' - z_1 b_1 - \cdots - z_m b_m \\ &< -\eta P(1) + (\mu + \varepsilon)\Big(\sum_{j:c_j>0} c_j\Big) - (\lambda - \varepsilon)\Big(\sum_{j:c_j<0} (-c_j)\Big). \end{aligned}$$

Adding the last two inequalities gives

$$1 \leq z_0 - z_0' < (\mu - \lambda + 2\varepsilon) L(P).$$

Since ε is arbitrary, it follows that $\mu - \lambda \geq 1/L(P)$. Furthermore, the same proof goes with polynomials $P(X)$ obtained by multiplying $Q(X)$ by a monic polynomial. Consequently, we have shown that $\mu - \lambda \geq 1/\ell(\alpha)$, as asserted. □

3.4 On the fractional parts of powers of Pisot and Salem numbers

If the algebraic number α is a Pisot or a Salem number, then the extra assumption that the non-zero real number ξ does not belong to the number field generated by α is required in the statement of Theorem 3.5. The purpose of this section is to discuss the necessity of this assumption.

Dubickas [241] considered the sequence of fractional parts $(\{\xi\alpha^n\})_{n\geq 1}$ where $\alpha > 1$ is a Pisot number and ξ is a positive element of $\mathbb{Q}(\alpha)$. He found the set of limit points of this sequence and described all cases when it has a unique limit point. We refer the reader to his paper and choose to mainly devote this section to the case of Salem numbers.

Vijayarhagavan [708] established that the sequence $(\alpha^n)_{n\geq 0}$ is dense modulo one, for α being the real root greater than 1 of the reciprocal polynomial $X^4 - X^3 - X^2 - X + 1$. This is a particular case of the following general result of Pisot and Salem [567].

THEOREM 3.7. *For any Salem number α, the sequence $(\alpha^n)_{n\geq 1}$ is dense modulo one, but is not uniformly distributed modulo one.*

Before proving Theorem 3.7 we state an auxiliary lemma.

LEMMA 3.8. *Let $\varepsilon > 0$ and α be a Salem number of degree d. Let $\alpha_3, \ldots, \alpha_d$ be the Galois conjugates of α lying on the unit circle, ordered in such a way that α_{2j-1} and α_{2j} are complex conjugates for $j = 2, \ldots, d/2$. Let $\eta_3, \ldots, \eta_d$ be complex numbers lying on the unit circle such that η_{2j-1} and η_{2j} are complex conjugates for $j = 2, \ldots, d/2$. Then, there are arbitrarily large integers n such that*

$$|\alpha_k^n - \eta_k| < \varepsilon,$$

for $k = 3, \ldots, d$.

PROOF. For $j = 3, \ldots, d-1$ and j odd, let ω_j be such that $\alpha_j = \mathrm{e}^{2i\pi\omega_j}$. Assume that there are rational integers $a_1, a_3, \ldots, a_{d-1}$ such that $a_1 + a_3\omega_3 + \cdots + a_{d-1}\omega_{d-1} = 0$. Then, we have

$$\prod_{3\leq j\leq d-1,\ j \text{ odd}} (\alpha_j)^{a_j} = 1. \tag{3.6}$$

For $h = 3, \ldots, d-1$ and h odd, let σ_h be the complex embedding defined by $\sigma_h(\alpha_h) = \alpha$. Applying σ_h to (3.6), we get

$$\alpha^{a_h} \cdot \prod_{3\leq j\leq d-1, j\neq h,\ j \text{ odd}} (\sigma_h(\alpha_j))^{a_j} = 1,$$

which implies that $a_h = 0$, since $\sigma_h(\alpha_j)$ is on the unit circle for $j \geq 3$ odd and $j \neq h$. This shows that $1, \omega_3, \ldots, \omega_{d-1}$ are linearly independent over the integers.

For $j = 3, \ldots, d-1$ and j odd, let β_j be such that $\eta_j = \mathrm{e}^{2i\pi\beta_j}$. For any positive real number δ, it follows from Theorem 1.18 that there are arbitrarily large integers n such that $||n\omega_j - \beta_j|| < \delta$ for $j = 3, \ldots, d-1$ and j odd. This concludes the proof of the lemma. □

PROOF OF THEOREM 3.7. Let d be the degree of α and denote by α, $1/\alpha, \alpha_3, 1/\alpha_3, \ldots, \alpha_{d-1}, 1/\alpha_{d-1}$ its Galois conjugates. For $j = 3, \ldots, d-1$

and j odd, let ω_j be such that $\alpha_j = \mathrm{e}^{2i\pi\omega_j}$. For any positive integer n, the number $\alpha^n + \alpha^{-n} + \alpha_3^n + \alpha_3^{-n} + \cdots + \alpha_{d-1}^{-n}$ is a rational integer. Note also that α^{-n} tends to zero when n tends to infinity.

Let ρ be in $(0,1)$ and η_3 be a complex number on the unit circle whose real part equals $\rho/2$. Put $\eta_5 = \cdots = \eta_{d-1} = i$. Let ε be a positive real number. By Lemma 3.8, there are arbitrarily large integers n such that $\rho - d\varepsilon \leq \alpha^{-n} + \alpha_3^n + \alpha_3^{-n} + \cdots + \alpha_{d-1}^{-n} \leq \rho + d\varepsilon$. Consequently, there are arbitrarily large integers n for which there exists an integer r_n with $\rho - d\varepsilon \leq r_n - \alpha^n \leq \rho + d\varepsilon$. This implies that the sequence $(\alpha^n)_{n\geq 1}$ is dense modulo one.

Since the function $x \mapsto 2\cos(2\pi x)$ is 1-periodic and continuous and since, for every integer h whose absolute value is sufficiently large, the integral

$$\int_0^1 \exp\{4i\pi h \cos 2\pi t\} \,\mathrm{d}t = J_0(4\pi h),$$

where J_0 is the Bessel function of order 0 (see e.g. [126, p. 13]), is non-zero, we deduce from Theorem 1.19 that the sequence

$$(2\cos(2n\pi\omega_3) + \cdots + 2\cos(2n\pi\omega_{d-1}))_{n\geq 1}$$

is not uniformly distributed modulo one. This implies that the sequence $(\alpha^n)_{n\geq 1}$ is not uniformly distributed modulo one. □

THEOREM 3.9. *Let $\alpha > 1$ be an algebraic number. Then, α is a Pisot or a Salem number if, and only if, for every positive real number ε, there exists a non-zero algebraic number ξ in $\mathbb{Q}(\alpha)$ such that $\|\xi\alpha^n\| < \varepsilon$ for every integer $n \geq 0$.*

PROOF. We proceed with the 'only if' part and leave the 'if' part as Exercise 3.4. Let α be a Pisot or a Salem number of degree d. Similarly as in the proof of Theorem 2.10 and by Lemma 2.11, there exists a Pisot number θ such that $\mathbb{Q}(\theta) = \mathbb{Q}(\alpha)$. Let δ denote the maximum of the absolute values of the Galois conjugates of θ (other than θ). Let ε be a positive real number and h be a positive integer satisfying $\delta^h < \varepsilon/d$. Then, for every integer $n \geq 0$, the trace of $\theta^h\alpha^n$ is a rational integer and

$$|\mathrm{Tr}(\theta^h\alpha^n) - \theta^h\alpha^n| \leq (d-1)\delta^h < \varepsilon.$$

This shows that the algebraic number $\xi := \theta^h$ satisfies the required conclusion. □

It follows from Theorem 3.9 that the dependence on ξ in the upper bound (2.2) of Lemma 2.2 cannot be removed.

We conclude this section by a result of Zaïmi [742].

THEOREM 3.10. *Let α be a Salem number. For any non-zero real number ξ in the field $\mathbb{Q}(\alpha)$, we have*

$$\limsup_{n\to+\infty}\{\xi\alpha^n\} - \liminf_{n\to+\infty}\{\xi\alpha^n\} > 0$$

and even

$$\limsup_{n\to+\infty}\{\xi\alpha^n\} - \liminf_{n\to+\infty}\{\xi\alpha^n\} \ge \frac{1}{L(\alpha)},$$

if $\alpha - 1$ is a unit. Furthermore, if $\alpha - 1$ is not a unit, then

$$\inf_{\xi\neq 0}\left(\limsup_{n\to+\infty}\{\xi\alpha^n\} - \liminf_{n\to+\infty}\{\xi\alpha^n\}\right) = 0.$$

PROOF. We follow the proof of Theorem 3.5 with $\eta = 0$. Let $Q(X) = q_dX^d + \cdots + q_1X + q_0$ be the minimal defining polynomial of α. For $n \ge 1$, put $x_n = \lfloor \xi\alpha^n \rfloor$, $y_n = \{\xi\alpha^n\}$, and

$$s_n = -(q_d y_{n+d} + \cdots + q_0 y_n).$$

Set $\lambda = \liminf_{n\ge 1} y_n$ and $\mu = \limsup_{n\ge 1} y_n$. Let ε be a positive real number. Then we have

$$\lambda - \varepsilon \le y_n \le \mu + \varepsilon,$$

for every sufficiently large integer n. This shows that

$$\begin{aligned}(\lambda-\varepsilon)\Big(\sum_{q_i>0} q_i\Big) + (\mu+\varepsilon)\Big(\sum_{q_i<0} q_i\Big) &\le -s_n \\ &\le (\lambda-\varepsilon)\Big(\sum_{q_i<0} q_i\Big) + (\mu+\varepsilon)\Big(\sum_{q_i>0} q_i\Big)\end{aligned}$$

holds for every sufficiently large integer n. If the sequence $(s_n)_{n\ge 1}$ takes infinitely many times at least two distinct values, then there are arbitrarily large integers n such that $s_{n+1} - s_n \ge 1$, hence,

$$(\mu-\lambda+2\varepsilon)L(\alpha) = (\mu-\lambda+2\varepsilon)\Big(\sum_{q_i>0} q_i\Big) + (\lambda-\mu-2\varepsilon)\Big(\sum_{q_i<0} q_i\Big) \ge 1.$$

Since ε can be taken arbitrarily small we get

$$\limsup_{n\to+\infty}\{\xi\alpha^n\} - \liminf_{n\to+\infty}\{\xi\alpha^n\} = \mu - \lambda \ge \frac{1}{L(\alpha)} > 0. \tag{3.7}$$

This proves the first statement of the theorem. Assume now that there are integers n_0 and s such that $s_n = s$ for all $n \ge n_0$. We then have

$$q_0(y_{n+1} - y_n) + q_1(y_{n+2} - y_{n+1}) + \cdots + q_d(y_{n+d-1} - y_{n+d}) = 0, \tag{3.8}$$

for $n \geq n_0$. Let $\alpha_1 = \alpha, \alpha_2 = 1/\alpha, \alpha_3, \ldots, \alpha_d$ be the Galois conjugates of α. Since the sequence $(y_{n+1} - y_n)_{n \geq n_0}$ satisfies the recursion (3.8), Theorem F.1 implies that there exist complex numbers $\gamma_1, \ldots, \gamma_d$ such that

$$y_{n+1} - y_n = \gamma_1 \alpha_1^n + \cdots + \gamma_d \alpha_d^n \tag{3.9}$$

for all $n \geq n_0$. Since $|y_{n+1} - y_n| \leq 1$ for $n \geq n_0$, we get $\gamma_1 = 0$.

Let $\xi_1 = \xi, \xi_2, \ldots, \xi_d$ be the Galois conjugates of ξ, where $\xi_i = \sigma_i(\xi)$ and σ_i is the complex embedding of $\mathbb{Q}(\alpha)$ sending α to α_i. For $n \geq n_0$, set

$$z_n = \xi_1 \alpha_1^n + \xi_2 \alpha_2^n + \cdots + \xi_d \alpha_d^n,$$

and note that z_n is a rational number satisfying

$$y_n = z_n - x_n - (\xi_2 \alpha_2^n + \cdots + \xi_d \alpha_d^n).$$

Consequently, for $n \geq n_0$, we have

$$y_{n+1} - y_n = (z_{n+1} - x_{n+1}) - (z_n - x_n) - \xi_2(\alpha_2 - 1)\alpha_2^n - \cdots - \xi_2(\alpha_2 - 1)\alpha_2^n,$$

and it follows from (3.9) that

$$(\gamma_2 + \xi_2(\alpha_2 - 1))\alpha_2^n + \cdots + (\gamma_d + \xi_d(\alpha_d - 1))\alpha_d^n = (z_{n+1} - x_{n+1}) - (z_n - x_n)$$

is a rational number. Let $n_1 \geq n_0$ be an integer. Applying Lemma 3.6 to the system

$$\begin{aligned} 0\alpha_1^{n-n_1} &+ (\gamma_2 + \xi_2(\alpha_2 - 1))\alpha_2^{n_1}\alpha_2^{n-n_1} + \cdots \\ &+ (\gamma_d + \xi_d(\alpha_d - 1))\alpha_d^{n_1}\alpha_d^{n-n_1} \\ &= (z_{n+1} - x_{n+1}) - (z_n - x_n), \quad n = n_1, \cdots, n_1 + d - 1, \end{aligned}$$

we get that

$$\sum_{k=0}^{d-1} \Bigl(\sum_{h=k+1}^{d} q_h \alpha^{h-k-1} \Bigr) \bigl((z_{n_1+k+1} - x_{n_1+k+1}) - (z_{n_1+k} - x_{n_1+k}) \bigr) = 0.$$

Since q_d is non-zero and $(z_{n_1+k+1} - x_{n_1+k+1}) - (z_{n_1+k} - x_{n_1+k})$ is rational for $k = 0, \ldots, d-1$, we deduce that $(z_{n_1+k+1} - x_{n_1+k+1}) - (z_{n_1+k} - x_{n_1+k}) = 0$ for $k = 0, \ldots, d-1$. Consequently, $z_n - x_n = z_{n_0} - x_{n_0}$ for $n \geq n_0$ and there is a rational number c such that

$$y_n = c - (\xi_2 \alpha_2^n + \cdots + \xi_d \alpha_d^n) \tag{3.10}$$

for $n \geq n_0$. Using that $0 < \alpha_2 < 1$ and $|\alpha_3| = \cdots = |\alpha_d| = 1$, it follows from Lemma 3.8 that

$$\mu = c + |\xi_3| + \cdots + |\xi_d| \quad \text{and} \quad \lambda = c - |\xi_3| - \cdots - |\xi_d|.$$

This gives

$$\limsup_{n\to+\infty} \{\xi\alpha^n\} - \liminf_{n\to+\infty} \{\xi\alpha^n\} = \mu - \lambda \geq 2(|\xi_3| + \cdots + |\xi_d|) > 0.$$

Since ξ is non-zero, we get $0 \leq \lambda < \mu \leq 1$ and $0 < c < 1$. It follows from (3.10) that

$$s_n = -(q_d y_{n+d} + \cdots + q_0 y_n) = -c(Q(1)) = s,$$

for $n \geq n_0$. Since s is an integer and $0 < c < 1$, the number $Q(1)$ cannot be ± 1, thus $\alpha - 1$ cannot be a unit. This means that, when $\alpha - 1$ is a unit, the sequence $(s_n)_{n\geq 1}$ takes infinitely many times at least two distinct values and (3.7) holds. This proves the second statement of the theorem. For the proof of the last statement we refer the reader to [742]. □

3.5 The sequence $(\|\xi\alpha^n\|)_{n\geq 1}$

As observed by Dubickas [240], under the assumption of Theorem 3.5, we can deduce a lower bound for the largest limit point of the sequence $(\|\xi\alpha^n\|)_{n\geq 1}$. Recall that the quantities $L(\alpha)$ and $\ell(\alpha)$ have been introduced in Definition 3.4.

THEOREM 3.11. *Let $\alpha > 1$ be a real algebraic number. Let ξ be a non-zero real number that lies outside the field generated by α if α is a Pisot or a Salem number. Then, the largest limit point of the sequence $(\|\xi\alpha^n\|)_{n\geq 1}$ is at least equal to $1/\min\{L(\alpha), 2\ell(\alpha)\}$.*

PROOF. If the largest limit point μ of the sequence $(\|\xi\alpha^n\|)_{n\geq 1}$ is smaller than $1/2\ell(\alpha)$, then the limit points of the sequence $(\{\xi\alpha^n + 1/2\})_{n\geq 1}$ all belong to the open interval of endpoints $1/2-\mu$ and $1/2+\mu$, whose length is smaller than $1/\ell(\alpha)$. This contradicts Theorem 3.5 applied with $\eta = 1/2$. In view of Theorem 2.7, this proves the theorem. □

When $\alpha > 1$ is a rational number, Theorem 3.11 was improved by Dubickas [240]. Before stating his result, we introduce the generating function

$$T(z) := \prod_{m=0}^{+\infty} (1 - z^{2^m}) = 1 - z - z^2 + z^3 + \cdots$$

of the Thue–Morse word $(t_k)_{k\geq 0}$ on $\{-1, 1\}$ defined by $t_0 = 1$ and $t_{2k+1} = -t_{2k} = -t_k$ for $k \geq 0$. Furthermore, we put

$$E(z) := \frac{1 - (1-z)T(z)}{2z}. \tag{3.11}$$

THEOREM 3.12. *Let ξ be an irrational, real number. Let b be an integer with $b \ge 2$. Then the sequence $(\|\xi b^n\|)_{n\ge1}$ has a limit point greater than or equal to*

$$\xi_b := \frac{1-(1-b^{-1})\prod_{m=0}^{+\infty}(1-b^{-2^m})}{2} = \frac{E(1/b)}{b}$$

and a limit point smaller than or equal to

$$\xi_b' := \frac{e(b)\prod_{m=0}^{+\infty}(1-b^{-2^m})}{2} = \frac{e(b)\,T(1/b)}{2},$$

where $e(b) = 1$ if b is even and $e(b) = 1 - 1/b$ if b is odd. Furthermore, both bounds are best possible.

For $b = 2$ a version of Theorem 3.12 can be found in [41, 45]. For $b \ge 2$, the real number $T(1/b)$ is transcendental, thus ξ_b and ξ_b' are transcendental; see Chapter 8.

PRELIMINARIES TO THE PROOF OF THEOREM 3.12. We begin with some notation. To each finite or infinite word $\boldsymbol{v} = v_1v_2v_3\dots$ on the alphabet $\{1,2\}$ and to each real number r with $0 < r < 1$, we associate the real number

$$E(\boldsymbol{v},r) = 1 - r^{v_1} + r^{v_1+v_2} - r^{v_1+v_2+v_3} + \cdots = \sum_{k\ge0}(-1)^k r^{v_1+\cdots+v_k}.$$

At various steps we use that $E(\boldsymbol{v},r)$ is given by an alternate sum. Let

$$\boldsymbol{z} = z_1z_2\dots = 2112221121121122211\dots$$

be the fixed point of the morphism σ defined by $\sigma(1) = 2$ and $\sigma(2) = 211$. For r in $(0,1)$, put

$$E(\boldsymbol{z},r) = 1 - r^2 + r^3 - r^4 + r^6 - r^8 + \cdots = \sum_{k\ge0}(-1)^k r^{a_k}, \tag{3.12}$$

where $a_0 = 0$ and $a_k = z_1 + \cdots + z_k$ for $k \ge 1$.

It follows from Theorem A.16 that the generating function of the Thue–Morse sequence and $E(\boldsymbol{z},r)$ are connected by

$$2rE(\boldsymbol{z},r) = 1-(1-r)(t_0+t_1r+t_2r^2+\cdots),$$

which, by (3.11), gives

$$E(\boldsymbol{z},r) = (1-(1-r)T(r))/(2r) = E(r). \tag{3.13}$$

For later use, we check that

$$E(r) < \min\left\{\frac{1}{2r}, \frac{2-r}{1+r}, \frac{1}{1+r^3}, \frac{1+r}{1+r^2}\right\}, \quad \text{for } 0<r<1. \tag{3.14}$$

As in Appendix A, for a word $\boldsymbol{v}$ and a non-negative integer ℓ, we denote by $\boldsymbol{v}_\ell$ the word $\boldsymbol{v}$ deprived of its first ℓ letters. It follows from Lemma A.15 that

$$\begin{aligned} 1 - r + r^3 - r^4 + r^5 - r^7 &< E(\boldsymbol{z}_i, r) \\ &< 1 - r^2 + r^3 - r^4 + r^6, \quad \text{for } 0 < r < 1 \text{ and } i \geq 0. \end{aligned} \tag{3.15}$$

Since $r(1-r^2+r^3-r^4+r^6) < 1-r+r^3-r^4+r^5-r^7$ for $0 < r \leq 0.6775$, we deduce from (3.15) that

$$rE(\boldsymbol{z}_j, r) < E(\boldsymbol{z}_i, r), \quad \text{for } 0 < r \leq 0.6775 \text{ and } i, j \geq 0. \tag{3.16}$$

We use the ordering $\succ$, defined in Appendix A, for the (finite and infinite) words over the alphabet $\{1, 2\}$. Let r with $0 < r \leq 0.6775$ and $i \geq 1$ be an integer. By Theorem A.13, we have $\boldsymbol{z} \succ \boldsymbol{z}_i$, so there is a smallest index k such that the kth letters of $\boldsymbol{z}$ and of $\boldsymbol{z}_i$ are different. If k is odd, then these letters are 2 and 1, respectively, and we get from (3.12) that

$$E(\boldsymbol{z}, r) - E(\boldsymbol{z}_i, r) = -r^{a_k} E(\boldsymbol{z}_k, r) + r^{a_k - 1} E(\boldsymbol{z}_{i+k}, r),$$

which is positive for $0 < r \leq 0.6775$ by (3.16). If k is even, then these letters are 1 and 2, respectively, and we get from (3.12) that

$$E(\boldsymbol{z}, r) - E(\boldsymbol{z}_i, r) = r^{a_k} E(\boldsymbol{z}_k, r) - r^{a_k + 1} E(\boldsymbol{z}_{i+k}, r),$$

which is positive for $0 < r \leq 0.6775$ by (3.16). Consequently, we have established that

$$E(\boldsymbol{z}_i, r) < E(\boldsymbol{z}, r), \quad \text{for } 0 < r \leq 0.6775 \text{ and } i > 0. \tag{3.17}$$

We are now armed to establish a key auxiliary result for the proof of Theorem 3.12.

LEMMA 3.13. *Let r be a real number with $0 < r \leq 0.6775$. Let $\boldsymbol{v}$ be an infinite word which is not ultimately periodic. For any positive real number ε, there are infinitely many positive integers ℓ such that $E(\boldsymbol{v}_\ell, r) > E(r) - \varepsilon$.*

PROOF. Let ε be a positive real number. For $m \geq 0$, put $A_m = \sigma^m(2)$, where σ is the morphism defined at the beginning of the preliminaries. Fix m sufficiently large such that $r^{a_k+1} < \varepsilon$, where $k = f_m$ is the length of the word A_m. Since f_m is odd, for any word $\boldsymbol{v}'$ which begins with A_m, we have $E(A_m, r) < E(\boldsymbol{v}', r) < E(A_m, r) + \varepsilon$. As $\boldsymbol{z}$ begins with A_m, this gives $E(\boldsymbol{v}', r) > E(\boldsymbol{z}, r) - \varepsilon$. This proves the above assumption if A_m occurs infinitely often in $\boldsymbol{v}$. Otherwise, by Theorem A.14, there is a finite word U with $U \succ \boldsymbol{z}$ having infinitely many occurrences in $\boldsymbol{v}$. Let k be the length of U. Without loss of generality, we can assume that the first $k-1$ letters of U and $\boldsymbol{z}$ coincide and that their kth letters are 2 and 1 if k is odd and 1 and 2 if k is even. Let ℓ be an integer such that U starts at the ℓth place of $\boldsymbol{v}$. Write U' for the word U deprived of its kth letter. If k is odd, then we have

$$E(U', r) = E(\boldsymbol{z}, r) + r^{a_k} E(\boldsymbol{z}_k, r) = E(\boldsymbol{v}_{\ell-1}, r) + r^{a_k+1} E(\boldsymbol{v}_{\ell+k-1}, r).$$

In this case, if $E(\boldsymbol{v}_{\ell-1}, r)$ and $E(\boldsymbol{v}_{\ell+k-1}, r)$ are both at most equal to $E(\boldsymbol{z}, r)$, we get that $E(\boldsymbol{z}_k, r) \le rE(\boldsymbol{z}, r)$, a contradiction with (3.16). If k is even, then we have

$$E(U', r) = E(\boldsymbol{z}, r) - r^{a_k} E(\boldsymbol{z}_k, r) = E(\boldsymbol{v}_{\ell-1}, r) - r^{a_k-1} + r^{a_k+\delta} E(\boldsymbol{v}_{\ell+k}, r),$$

where $\delta = 0$ or 1, whence

$$E(\boldsymbol{v}_{\ell-1}, r) - E(\boldsymbol{z}, r) = r^{a_k-1}\big(1 - r^{1+\delta} E(\boldsymbol{v}_{\ell+k}, r) - rE(\boldsymbol{z}_k, r)\big).$$

If $E(\boldsymbol{v}_{\ell+k}, r) \le E(\boldsymbol{z}, r)$, then, by (3.14) and (3.17), we have

$$r^{1+\delta} E(\boldsymbol{v}_{\ell+k}, r) + rE(\boldsymbol{z}_k, r) \le 2rE(\boldsymbol{z}, r) < 1,$$

whence $E(\boldsymbol{v}_{\ell-1}, r) > E(\boldsymbol{z}, r)$. Consequently, for any k, at least one among the numbers $E(\boldsymbol{v}_{\ell-1}, r)$, $E(\boldsymbol{v}_{\ell+k-1}, r)$ and $E(\boldsymbol{v}_{\ell+k}, r)$ is greater than $E(\boldsymbol{z}, r)$. □

Completion of the proof of Theorem 3.12. We start with coprime integers p and q satisfying $p > q \ge 1$. We prove a more general result than the theorem enounced. Namely, we establish a lower bound for the largest limit point and an upper bound for the smallest limit point of the sequence $(\|\xi(p/q)^n\|)_{n\ge1}$, provided that $q/p \le 0.6775$. The result follows by taking $p = b$ and $q = 1$.

Set $\alpha = p/q$ and $r := q/p$. For $n \ge 1$, set $x_n = \lfloor \xi\alpha^n + 1/2 \rfloor$, $y_n = \{\xi\alpha^n + 1/2\}$, and

$$s_n = qx_{n+1} - px_n.$$

Observe that

$$\|\xi\alpha^n\| = |y_n - 1/2| = |\{\xi\alpha^n + 1/2\} - 1/2|$$

and

$$s_n = -q(y_{n+1} - 1/2) + p(y_n - 1/2).$$

Thus,

$$y_n - 1/2 = \frac{s_n}{p} + r(y_{n+1} - 1/2) = \frac{s_n}{p} + r\Big(\frac{s_{n+1}}{p} + \cdots\Big),$$

and, by iteration, we get

$$\|\xi\alpha^n\| = |y_n - 1/2| = |(1/p)(s_n + s_{n+1}r + s_{n+2}r^2 + \cdots)|. \qquad (3.18)$$

The sequence $(s_n)_{n\geq 1}$ takes integer values and it was shown in the proof of Theorem 3.5 that it is not ultimately periodic.

For $n \geq 1$, set

$$U_n(r) := s_n + s_{n+1}r + s_{n+2}r^2 + \cdots$$

Let n be such that $|s_n| \geq 2$. From

$$2 \leq |s_n| = |U_n(r) - rU_{n+1}(r)| \leq |U_n(r)| + r|U_{n+1}(r)|,$$

we get that at least one number among $|U_n(r)|$ and $r|U_{n+1}(r)|$ is greater than or equal to $2/(r+1)$, and thus exceeds $E(r)$, by (3.14).

Let n be such that $s_n = \pm 1$, $s_{n+1} = s_{n+2} = 0$. From

$$1 = |s_n| = |U_n(r) - r^3U_{n+3}(r)| \leq |U_n(r)| + r^3|U_{n+3}(r)|,$$

we get that at least one of the numbers $|U_n(r)|$ and $r^3|U_{n+3}(r)|$ is greater than or equal to $1/(1+r^3)$, and thus exceeds $E(r)$, by (3.14).

Let n be such that $s_n = 1$, $s_{n+1} = 0$ and $s_{n+2} = 1$. Then, we have $U_n(r) - r^3U_{n+3}(r) = 1 + r^2$ and at least one number among $|U_n(r)|$ and $r^3|U_{n+3}(r)|$ is greater than or equal to $(1+r^2)/(1+r^3)$, and thus exceeds $E(r)$, by (3.14). The same conclusion holds if $s_n = -1$, $s_{n+1} = 0$ and $s_{n+2} = -1$.

Let n be such that $s_n = s_{n+1} = 1$. Then $U_n(r) - r^2U_{n+2}(r) = 1 + r$ and at least one number among $|U_n(r)|$ and $r^2|U_{n+2}(r)|$ is greater than or equal to $(1+r)/(1+r^2)$, and thus exceeds $E(r)$. The same conclusion holds if $s_n = s_{n+1} = -1$.

In all of these cases, it follows from (3.18) that $\|\xi\alpha^n\| \geq E(r)/p$. Assume now that $\boldsymbol{s}$ is a word on $\{-1, 0, 1\}$ in which the words 100, -100, 101, $-10-1$, 11 and $-1-1$ occur only finitely often. Consequently, there exists an n_0 such that the non-periodic sequence $s_{n_0}, s_{n_0+1}, \ldots$ begins with $s_{n_0} = 1$ and has the form $1, \hat{0}, -1, \hat{0}, 1, \hat{0}, -1, \ldots$, where $\hat{0}$ means that the letter 0 may or may not occur. We associate to $(s_n)_{n\geq n_0}$ a word $\boldsymbol{v} = (v_n)_{n\geq 0}$ on $\{1, 2\}$ by writing 2 if two consecutive units are separated by the symbol 0 and by writing 1 otherwise. For instance, we associate to $10-11-1010-1\ldots$ the word $21122\ldots$ Observe that $\boldsymbol{v}$ is not ultimately periodic. Let ℓ be an integer greater than n_0 and let $t = t(\ell)$ be the number of 0's among $s_{n_0}, \ldots, s_{\ell-1}$. If $s_\ell = \pm 1$, then $|U_\ell(r)|$ is equal to $E(\boldsymbol{v}_{\ell-n_0-t}, r)$ and, otherwise,

$$|U_{\ell+1}(r)| = \frac{|U_\ell(r)|}{r} = E(\boldsymbol{v}_{\ell-n_0-t(\ell)}, r) = E(\boldsymbol{v}_{\ell+1-n_0-t(\ell+1)}, r),$$

since $s_{\ell+1}$ is non-zero. By Lemma 3.13, for any positive real number ε, there are arbitrarily large integers ℓ for which $E(\boldsymbol{v}_{\ell-n_0-t(\ell)}, r)$ exceeds $E(\boldsymbol{z}, r) - \varepsilon$, since $\boldsymbol{v}$ is not ultimately periodic. Combined with (3.18), this shows that the smallest limit point of the sequence $(\|\xi\alpha^n\|)_{n\geq 1}$ is greater than or equal to $E(\boldsymbol{z}, r)/p$. For $p = b$ and $q = 1$, the latter quantity is equal to $E(1/b)/b$. This proves the first statement of the theorem.

The proof of the second statement is similar. For $n \geq 1$, write the fractional part $\{\xi\alpha^n\}$ under the form $1/2 + g_n$, where $-1/2 \leq g_n < 1/2$. Observe that

$$\begin{aligned} p\lfloor \xi(p/q)^n \rfloor - q\lfloor \xi(p/q)^{n+1} \rfloor &= -p(1/2 + g_n) + q(1/2 + g_{n+1}) \\ &= (q-p)/2 - pg_n + qg_{n+1}, \end{aligned}$$

and set

$$s_n := -qg_{n+1} + pg_n.$$

The sequence $(s_n)_{n\geq 1}$ is not ultimately periodic and takes integer values if $p+q$ is even, and values in $1/2 + \mathbb{Z}$ if $p+q$ is odd. As above, we get by iteration that

$$g_n = (1/p)(s_n + s_{n+1}r + s_{n+2}r^2 + \cdots). \tag{3.19}$$

The case with $p+q$ even corresponds to the case already studied in the first statement of the theorem, and we deduce that the sequence $(\|\xi\alpha^n\|)_{n\geq 1}$ has a limit point smaller than or equal to $1/2 - E(r)/p$. For $q = 1$ and $p = b \geq 3$ odd, the latter quantity is equal to $1/2 - E(1/b)/b$. By (3.13), this gives the result stated.

Suppose now that $p+q$ is odd. Then, multiplying both sides of (3.19) by 2, we get

$$2g_n = (1/p)(2s_n + 2s_{n+1}r + 2s_{n+2}r^2 + \cdots), \tag{3.20}$$

where $2s_n, 2s_{n+1}, \ldots$ are odd integers. For $n \geq 1$, set

$$V_n(r) := 2s_n + 2s_{n+1}r + 2s_{n+2}r^2 + \cdots$$

Let n be such that $|2s_n| \geq 3$. From

$$3 \leq |2s_n| = |V_n(r) - rV_{n+1}(r)| \leq |V_n(r)| + r|V_{n+1}(r)|,$$

we get that at least one number among $|V_n(r)|$ and $r|V_{n+1}(r)|$ is greater than or equal to $3/(1+r)$. Similarly, if $2s_n = 2s_{n+1} = 2s_{n+2} = \pm 1$, then upon writing $1 + r + r^2 = |V_n(r) + r^3 V_{n+3}(r)|$, we deduce that at least one number among $|V_n(r)|$ and $r^3|V_{n+3}(r)|$ exceeds $(1+r+r^2)/(1+r^3)$. By (3.14), we can then assume that, starting from a certain place n_0, the sequence $(s_n)_{n\geq n_0}$ takes only the values ± 1, with at most two equal values in a row. Denote by h the map sending a series $H(r)$ to $(1 + (1 - r)H(r))/2$. Then, h sends the tail of the sequence $(s_n)_{n\geq 1}$ to a sequence of the form $1, \hat{0}, -1, \hat{0}, 1, \hat{0}, -1, \ldots$, where $\hat{0}$ means that the letter 0 may or may not occur. Since, by (3.11), the map h sends $(1-T(r))/r$ to $E(r)$, we deduce from the proof of the first statement that, for any positive real number ε, there are arbitrarily large integers n such that $|2g_n|$ exceeds $(1 - T(r))/(rp) - \varepsilon$. Consequently, the sequence $(|g_n|)_{n\geq 1}$ has a limit point at least equal to $(1-T(r))/(2q)$ and the sequence $(\|\xi\alpha^n\|)_{n\geq 1}$ has a limit point smaller than or equal to $1/2 - (1 - T(r))/(2q)$. For $q = 1$ and $p = b \geq 2$ even, the latter quantity is equal to $T(1/b)/2$, as stated in the theorem.

Finally, the combination of Theorem A.13, (3.18) and (3.20) shows that $\|\xi_b b^n\| < \xi_b$ and $\|\xi'_b b^n\| < \xi'_b$ for every integer $n \geq 1$ and $b \geq 2$. This establishes the last assertion of the theorem. □

The proof of Theorem 3.12 covers powers of rational numbers greater than or equal to 10/6.775 and can be extended to powers of rational numbers greater than 1, but some extra work is needed to prove that Lemma 3.13 still holds for every real number r between 0 and 1. This would introduce further complication, and we prefer to refer the reader to [240]. We display below the special case $\alpha = p/2$, with $p \geq 3$ an odd integer. Unlike Theorem 3.12, the next result is presumably not best possible.

THEOREM 3.14. *Let $p \geq 3$ be an odd integer. Let ξ be a non-zero real number. Then, the sequence $(\|\xi(p/2)^n\|)_{n\geq 1}$ has a limit point greater than or equal to $(p - (p-2)T(2/p))/(4p)$ and a limit point smaller than or equal to $(1 + T(2/p))/4$. In particular, the sequence $(\|\xi(3/2)^n\|)_{n\geq 1}$ has a limit point greater than or equal to $(3 - T(2/3))/12 = 0.238117\ldots$ and a limit point smaller than or equal to $(1 + T(2/3))/4 = 0.285647\ldots$*

Let ξ be a non-zero real number. It follows from Theorem 3.3 that every closed sub-interval of $[0, 1]$ of length $2/3$ contains a limit point of the sequence $(\{\xi(3/2)^n\})_{n\geq 1}$. The first part of Theorem 3.14 asserts that the interval $[0.238117\ldots, 0.761882\ldots]$ of length $0.523764\ldots$ contains a limit point of that sequence. At present, we have no example of an interval of length smaller than 1/2 with the same property. However, Dubickas [248] established that the union of three intervals $[0, 8/39] \cup [18/39, 21/39] \cup [31/39, 1]$, of total length 19/39, always contains a limit point of $(\{\xi(3/2)^n\})_{n\geq 1}$, see Exercise 3.5.

3.6 Constructions of Pollington and of Dubickas

Let p and q be coprime integers satisfying $p > q \geq 2$. We gather in this section several constructions of positive real numbers ξ such that all the fractional parts $\{\xi(p/q)^n\}$, $n \geq 0$, or all the numbers $||\xi(p/q)^n||$, $n \geq 0$, belong to a small interval.

We begin with a method introduced by Pollington [572, 574] for constructing real numbers ξ such that $||\xi(3/2)^n||$ is large for every positive integer n.

THEOREM 3.15. *There exists a real number ξ such that $||\xi(3/2)^n|| \geq 4/65$ holds for every positive integer n.*

PROOF. Let ε be a real number with $0 < \varepsilon < 1/2$ and n be a positive integer. Assume that there is an integer $m = m(n)$, depending on n, such that the interval $[(3/2)^m(n + \varepsilon), (3/2)^m(n + 1 - \varepsilon)]$ contains an interval $[\ell + \varepsilon, \ell + 1 - \varepsilon]$, where $\ell = \ell(n)$ is an integer and

$$[(2/3)^i(\ell + \varepsilon), (2/3)^i(\ell + 1 - \varepsilon)]$$

does not contain an integer for $i = 1, \ldots, m$. Observe that $m(n) = m(n + a2^m)$ for every non-negative integer a, since $(3/2)^m a2^m$ is an integer. Put then

$$\delta(n) = \min_{0\leq i\leq m} \min\{||(\ell + \varepsilon)(2/3)^i||, ||(\ell + 1 - \varepsilon)(2/3)^i||\}.$$

Observe that for x in $[n+\varepsilon, n+1-\varepsilon] \cap [(2/3)^m(\ell+\varepsilon), (2/3)^m(\ell+1-\varepsilon)]$ we then have

$$||x(3/2)^i|| \geq \delta(n), \quad \text{for } i = 1, 2, \ldots, m.$$

We illustrate this for $n = 1$ and $\varepsilon = 2/5$. The interval $[(9/4)(1 + \varepsilon), (9/4)(2 - \varepsilon)]$ contains an interval $[3 + \varepsilon, 4 - \varepsilon]$ such that there is no integer in $[(2/3)(3 + \varepsilon), (2/3)(4 - \varepsilon)] \cup [(4/9)(3 + \varepsilon), (4/9)(4 - \varepsilon)]$. Thus, we can put $m(1) = 2$, $\ell(1) = 3$, and $\delta(1) = 4/15$.

Let $n_1 \bmod 2^{m_1}$, …, $n_k \bmod 2^{m_k}$ be a finite system of covering congruences, in the sense that for every positive integer n, there exists j with

$1 \le j \le k$ and $n \equiv n_j \bmod 2^{m_j}$. Assume furthermore that $m_j = m_j(n_j)$ for $j = 1, \ldots, k$. Then, there exists ξ such that

$$\|\xi(3/2)^i\| \ge \min_{1\le j\le k} \delta(n_j), \tag{3.21}$$

for $i \ge 1$. To see this, put $I_1 = [1, 2]$. Let $j \ge 1$ be an integer and assume that $I_j = [n, n+1]$. Let $h = h(n)$ be in $\{1, \ldots, k\}$ and a_h be such that $n = n_h + a_h 2^{m_h}$. Put then $I_{j+1} = [\ell_h + a_h 3^{m_h}, \ell_h + a_h 3^{m_h} + 1]$ and

$$J_j = \left(\frac{2}{3}\right)^{m_{h(t_1)}+\cdots+m_{h(t_j)}} I_{j+1},$$

where t_i denotes the left endpoint of I_i for $i = 1, \ldots, j$. Then, the sequence $(J_j)_{j\ge1}$ is a decreasing sequence of nested closed intervals. Its intersection is reduced to one point ξ, which satisfies (3.21) for every $i \ge 1$. The theorem then follows by choosing for $(n_1, \ell_1, m_1), \ldots, (n_{12}, \ell_{12}, m_{12})$ the triples

$$(1,3,2),(2,5,2),(3,12,3),(4,14,3),(7,36,4),(8,44,4),(15,117,5),$$
$$(16,125,5),(31,360,6),(32,368,6),(0,8,6),(63,720,6),$$

and by selecting ε suitably. Indeed, it can be checked that this choice gives $\min_{1\le j\le 12} \delta(n_j) = 4/65$. □

Pollington [572] announced that the constant 4/65 in Theorem 3.15 can be replaced by 0.088. Using a totally different method, Dubickas [248] established that every interval $(m, m+1)$ with m a positive integer contains a real number ξ such that $\|\xi(3/2)^n\| > 5/48 = 0.1041\ldots$ for every integer $n \ge 0$.

We continue with several results of Dubickas [248], which, in the opposite direction, show that for rational numbers p/q, there are non-zero real numbers ξ such that $\|\xi(p/q)^n\|$ is quite small for every $n \ge 0$.

THEOREM 3.16. *For every odd integer $p \ge 3$ there exists a non-zero real number ξ such that*

$$\|\xi(p/2)^n\| < 1/p, \quad \textit{for every } n \ge 0.$$

Let p and q be integers with $p > 2q \ge 4$. If m is an integer, put $\delta_m = 1$ when m is odd and $\delta_m = 0$ otherwise. Then, there exist non-zero real numbers ξ_1 and ξ_2 such that

$$\|\xi_1(p/q)^n\| < \frac{q - \delta_q}{2(p-q)}, \quad \textit{for every } n \ge 0,$$

and

$$||\xi_2(p/q)^n|| > \frac{p-2q+\delta_p}{2(p-q)}, \quad \textit{for every } n \geq 0.$$

The first assertion of Theorem 3.16 with $p = 3$ was first proved by Akiyama, Frougny and Sakarovitch [33], who established more precisely that the set of positive real numbers ξ such that $||\xi(3/2)^n|| < 1/3$ for all positive integers n is infinite and countable. Their approach rests on the representation of real numbers in the base composed of $1/p, q/p^2, q^2/p^3, \ldots$; see also [32] for subsequent results. This shows that the constant 0.238117 in Theorem 3.14 cannot be replaced by a constant greater than 1/3. For large p, the bound $1/p$ in Theorem 3.16 is close to best possible, since, by Theorem 3.14, it cannot be replaced by $1/p - 4/p^3$.

With a similar method it is proved in [248] that for coprime integers p and q with $q > 1$ and $(\sqrt{2}+1)(q-1) < p < q^2 - q$, there exists a non-zero real number ξ_3 such that $||\xi_3(p/q)^n|| < (pq-p-q)/(p^2-q^2)$ for every $n \geq 0$. In some cases, for instance when $p/q = 11/4$, this gives a better bound than Theorem 3.16.

A slight improvement of the second assertion of Theorem 3.16 has been obtained by Dubickas [259] when q is even and $p > 2q$. Namely, he established the existence of a non-zero real number ξ_4 such that

$$||\xi_4(p/q)^n|| < \frac{q(p-2)}{2p(p-q)}, \quad \text{for every } n \geq 0.$$

Before proceeding with the proof of Theorem 3.16, we state an auxiliary result which allows us to get strict inequalities in Theorems 3.16 and 3.18.

LEMMA 3.17. *Let p, q be relatively prime integers with $p > q > 1$. For every fixed real number ξ, t with $\xi \neq 0$ and $0 \leq t < 1$, there exist only finitely many integers n such that $\{\xi(p/q)^n\} = t$.*

PROOF. This is left as Exercise 3.6. □

PROOF OF THEOREM 3.16. Let $p > q \geq 2$ be integers and $\mathcal{A}$ be a set of integers containing at least q consecutive integers. Let x_0 be a positive integer. We construct two sequences $(x_n)_{n\geq 0}$ and $(s_n)_{n\geq 0}$ of integers and a real number ξ as follows. Let n be a positive integer such that x_{n-1} is known. Choose s_{n-1} in $\mathcal{A}$ such that q divides $px_{n-1} + s_{n-1}$. This is always possible since $\mathcal{A}$ contains q consecutive integers. Set then $x_n := (px_{n-1} + s_{n-1})/q$. Observe that there may be several possible choices for s_{n-1}. Put

$$\xi := x_0 + \frac{1}{p} \sum_{j \geq 0} s_j \Big(\frac{q}{p}\Big)^j = x_0 + \frac{1}{p}\Big(s_0 + s_1\Big(\frac{q}{p}\Big) + \cdots\Big)$$

and, for every $n \geq 0$,

$$y_n := \frac{1}{p}\Big(s_n + s_{n+1}\Big(\frac{q}{p}\Big) + \cdots\Big).$$

An immediate induction shows that

$$\xi(p/q)^n = x_n + y_n, \qquad \text{for every } n \geq 0.$$

For the first assertion of the theorem, choose $q = 2$ and $\mathcal{A} = \{-1, 0, 1\}$. Since $x_n := (px_{n-1} + s_{n-1})/2$, we must take $s_{n-1} = 0$ if x_{n-1} is even, but we have two choices for s_{n-1} if x_{n-1} is odd. So we can choose the sequence $(s_n)_{n \geq 0}$ of the form $0^{k_1}10^{k_2}(-1)0^{k_3}1\ldots$, where $k_1, k_2, \ldots$ are non-negative integers. It follows that all the y_n belong to the interval $[-1/p, 1/p]$. Consequently, $|y_n| = ||\xi(p/2)^n||$ for $n \geq 0$. Furthermore, if $y_{n_0} = \pm 1/p$ for some integer n_0, then $y_n = y_{n_0}$ for $n \geq n_0$, a contradiction to Lemma 3.17. It follows that $y_n \in (-1/p, 1/p)$ for $n \geq 0$.

For the second statement, choose $\mathcal{A} = \{-(q-1)/2, -(q-1)/2 + 1, \ldots, (q-1)/2\}$ if q is odd and $\mathcal{A} = \{-q/2+1, -q/2+2, \ldots, q/2\}$ if q is even, and take s_n in $\mathcal{A}$ for $n \geq 0$. Assume that q is odd. Then, every y_n belongs to the interval $[-(q-1)/(2(p-q)), (q-1)/(2(p-q))]$. Since $q - 1 < p - q$, we get that $|y_n| = ||\xi(p/q)^n||$. Moreover, we have $y_n = \pm(q-1)/(2(p-q))$ if, and only if, $s_n = s_{n+1} = \cdots = \pm(q-1)/2$. Thus, if $y_{n_0} = \pm(q-1)/(2(p-q))$ for some integer n_0, then $y_n = y_{n_0}$ for $n \geq n_0$, a contradiction to Lemma 3.17. We argue similarly when q is even.

The last statement of the theorem is left as Exercise 3.7. □

There are further results if one replaces the function distance to the nearest integer by the fractional part function. The next statement was essentially proved in [308], see also [248]. Its proof is left as Exercise 3.8.

THEOREM 3.18. *Let p and q be coprime integers with $q \geq 2$ and $p > 2q$. Let t be an integer with $0 \leq t \leq p - 2q$. Then there exists a non-zero real number ξ such that $t/(p-q) < \{\xi(p/q)^n\} < (t+q-1)/(p-q)$ for $n \geq 0$. In particular, for every odd integer $p \geq 5$, there exists a non-zero real number ξ such that $\{\xi(p/2)^n\} < 1/(p-2)$ for $n \geq 0$.*

The last statement of Theorem 3.18 (with $\leq$ instead of $<$) was already proved by Tijdeman [687] with a method close to that used in the proof of Theorem 2.14; see Exercise 2.7. For large p, the bound $1/(p-2)$ is close to best possible, since, by Theorem 3.3, it cannot be replaced by a number smaller than $1/p$.

3.7 Waring's problem

The content of this section is slightly aside the main theme of the book, however, we feel that Waring's problem cannot be omitted when dealing with fractional parts of powers of rational numbers. The reader is directed to the excellent survey of Vaughan and Wooley [702].

Let $n \geq 2$ be an integer. Hilbert [345] proved that there exists an integer $s(n)$ such that every positive integer can be expressed as the sum of at most $s(n)$ positive integers, all of which are nth powers. Denote by $g(n)$ the smallest integer with this property. Observe that the integer $2^n \lfloor (3/2)^n \rfloor - 1$ can only be represented by the powers 1^n and 2^n because it is less than 3^n. Since

$$2^n \lfloor (3/2)^n \rfloor - 1 = (\lfloor (3/2)^n \rfloor - 1) 2^n + (2^n - 1) 1^n,$$

we have

$$g(n) \geq 2^n + \lfloor (3/2)^n \rfloor - 2. \tag{3.22}$$

It is known (see e.g. [334, Chapter XXI]) that equality holds in (3.22) when

$$2^n \{(3/2)^n\} + \lfloor (3/2)^n \rfloor \leq 2^n. \tag{3.23}$$

If (3.23) fails, then

$$g(n) = 2^n + \lfloor (3/2)^n \rfloor + \lfloor (4/3)^n \rfloor - \theta,$$

where θ is 2 or 3, according as

$$\lfloor (4/3)^n \rfloor \cdot \lfloor (3/2)^n \rfloor + \lfloor (4/3)^n \rfloor + \lfloor (3/2)^n \rfloor$$

is equal to 2^n or is larger than 2^n. A quick check shows that, to prove that (3.23) holds for n, it is sufficient to establish that

$$\| (3/2)^n \| \geq (3/4)^{n-1}. \tag{3.24}$$

Mahler showed that (3.24) holds for every sufficiently large n (see Exercise 3.9), but his method does not enable us to get an explicit upper bound for the possible values of n for which (3.24) does not hold. By means of other techniques, Beukers [94] proved that $\| (3/2)^n \| \geq 2^{-9n/10}$ for $n > 5000$. This estimate has been refined by Dubickas [233], Habsieger [331] and Zudilin [751], who established that $\| (3/2)^n \| \geq 0.5803^n$ for every n greater than an effectively computable constant. Note that the condition (3.23) is known to hold for n up to $471,600,000$.

3.8 On the integer parts of powers of algebraic numbers

Let $\alpha > 1$ be a real number. The problem of deciding whether or not there are infinitely many composite numbers of the form $\lfloor \alpha^n \rfloor$ is closely related to the distribution of $(\{\alpha^n/2\})_{n\geq 1}$, since we have $\lfloor \alpha^n \rfloor = 2\lfloor \alpha^n/2 \rfloor$ whenever $\{\alpha^n/2\} < 1/2$. If the latter happens for abitrarily large integers n, then infinitely many elements of the sequence $(\lfloor \alpha^n \rfloor)_{n\geq 1}$ are even and composite.

The first result on this problem was established in 1967 by Forman and Shapiro [311], who showed that there are infinitely many composite numbers in the sequence $(\lfloor \alpha^n \rfloor)_{n\geq 1}$ for $\alpha = 3/2$ and $\alpha = 4/3$. The same conclusion holds for $\alpha = 5/4$ (see [265]), for α a quadratic unit [178] and, more generally, for all Pisot and Salem numbers [235, 236]. Further examples of algebraic and transcendental numbers with this property are given in [238] and in [37], respectively.

Dubickas [242] introduced the set $\mathcal{Z}$ of real numbers $\alpha > 1$ for which there is a non-zero real number ξ such that the integer parts $\lfloor \xi\alpha^n \rfloor$ are all even numbers for $n \geq 1$. He obtained various results on $\mathcal{Z}$ and on its complement, including examples of real numbers lying in $(1, +\infty) \setminus \mathcal{Z}$; see also [238].

It is proved in [252] that for real numbers ξ and ν with $\xi \neq 0$, both sequences $(\lfloor \xi 2^n + \nu \rfloor)_{n\geq 1}$ and $(\lfloor \xi(-2)^n + \nu \rfloor)_{n\geq 1}$ contain infinitely many composite elements. Moreover, if ξ is irrational, then infinitely many elements of each of these sequences are divisible by 2 or 3.

3.9 Exercises

EXERCISE 3.1 (*cf.* [468]). Establish that, for any non-negative integer m, the real interval $(m, m+1)$ contains at most one Z-number. [Hint. Assume that ξ is a Z-number. For an integer $n \geq 0$, write $x_n = \lfloor \xi(3/2)^n \rfloor$ and put $\varepsilon_n = 0$ if x_n is even and $\varepsilon_n = 1$ otherwise. Prove that $3\{\xi\} = \varepsilon_0 + (2/3)\varepsilon_1 + (2/3)^2\varepsilon_2 + \cdots$ Conclude.]

EXERCISE 3.2. Let p and q be coprime integers with $p > q \geq 1$. Compute the reduced length of $qX - p$. Find assumptions on the real polynomial $P(X)$ under which $\ell(P)$ can be estimated.

EXERCISE 3.3. Let $\alpha > 1$ be a real algebraic number and $q_d X^d + \cdots + q_1 X + q_0$ its minimal defining polynomial over $\mathbb{Z}$. Put $L_+(\alpha) = \sum_{i=0}^d \max\{0, q_i\}$ and $L_-(\alpha) = \sum_{i=0}^d \max\{0, -q_i\}$. Let ξ be a positive real number not in $\mathbb{Q}(\alpha)$ if α is a Pisot or a Salem number. Prove that

$$\limsup_{n\to+\infty}\{\xi\alpha^n\} \geq \min\Big\{\frac{1}{L_+(\alpha)}, \frac{1}{L_-(\alpha)}\Big\}.$$

EXERCISE 3.4. By a suitable adaptation of the proof of Theorem 2.5, establish the 'if' part of Theorem 3.9.

EXERCISE 3.5 (*cf.* [248]). Prove that the union of three intervals $[0, 8/39] \cup [18/39, 21/39] \cup [31/39, 1]$ always contains a limit point of $(\{\xi(3/2)^n\})_{n\geq 1}$. Let η be in $[1/5, 8/39]$. Let ξ be a real number such that $\{\xi(3/2)^n\}$ is in $(\eta, 9\eta/4) \cup (1 - 9\eta/4, 1 - \eta)$ for every $n \geq 0$. Define x_n, y_n and s_n as in (3.1) and (3.2) with $p/q = 3/2$. Prove that $s_n = 0$ or 1 for every $n \geq 0$ and, more precisely, $s_n = 0$ if $y_n \in (\eta, 9\eta/4)$ and $s_n = 1$ if $y_n \in (1 - 9\eta/4, 1 - \eta)$. Let m be the smallest integer such that $s_m = 1$ and one among $s_0, s_1, \ldots, s_{m-1}$ is equal to 0. Recall that $(s_n)_{n\geq 0}$ is not ultimately periodic. Prove that $y_n \in (1-27\eta/8, 9\eta/4) \cup (1-9\eta/4, 27\eta/8)$ for every $n \geq m+1$. For $n \geq m+1$, prove that $s_{n+1} = 0$ if, and only if, $s_n = 1$. Conclude.

EXERCISE 3.6. Prove Lemma 3.17.

EXERCISE 3.7. Prove the last assertion of Theorem 3.16.

EXERCISE 3.8. Prove Theorem 3.18.

EXERCISE 3.9 (*cf.* [465]). Let p and q be coprime integers with $p > q \geq 2$. Apply Ridout's Theorem E.8 to show that, for every positive real number ε, there exists an integer $n_0(\varepsilon, p, q)$ such that $||(p/q)^n|| > 2^{-\varepsilon n}$ for every $n > n_0(\varepsilon, p, q)$.

3.10 Notes

▷ Allouche and Glen [43, 44] pointed out that, for $b = 2$, Theorem 2.1 from [161] was already proved by Veerman [703, 704] and subsequently rediscovered by several authors. In [43] they gave a complete description of the minimal intervals containing all the fractional parts $\{\xi 2^n\}$ for some positive real number ξ and all $n \geq 0$; see also [415].

▷ Vijayarhagavan [706] established that, for every integer $b \geq 2$ and every irrational number ξ, the set of limit points of the sequence $(\{\xi b^n\})_{n\geq 1}$ is infinite. If, in addition, some integral power of ξ is equal to b, he deduced that the set of limit points of the fractional parts of the powers of ξ is infinite.

▷ Let p and q be coprime integers with $p > q \geq 2$. It was shown in [308] that the set of s in $[0, 1 - 1/p]$ for which $Z_{p/q}(s, s + 1/p)$ is empty is everywhere dense in $[0, 1 - 1/p]$. Bugeaud [145] proved that this set has full Lebesgue measure. Subsequently, for $p < q^2$, Dubickas [254] established that it is the whole interval $[0, 1 - 1/p]$. More precisely, he proved that, if p and q are coprime integers with $2 \leq q < p < q^2$ and I is a closed interval of length $1/p$ of the torus $\mathbb{T}$, then for each non-zero real number ξ, there are infinitely many positive integers n such that $\{\xi(p/q)^n\}$ does not belong to I.

▷ Dubickas and Mossinghoff [264] developed algorithms to search for Z-numbers. They proved that there are no Z-numbers up to 2^{57}.

▷ In a series of notes [188–194], Choquet stated various results (mostly without proofs) on the distribution of the sequence $(\{\xi\alpha^n\})_{n\geq 1}$, mainly in the case $\alpha = 3/2$.

▷ Strauch [676], Adhikari, Rath and Saradha [22], and Rath [600] studied the distribution functions of the sequence $(\{\xi(p/q)^n\})_{n\geq 1}$ for coprime integers p and q with $p > q \geq 2$.

▷ Schinzel [622, 623] studied the reduced length of polynomials with real coefficients.

▷ As shown by Dubickas [244], Theorem 3.5 can be extended to linear recurrent sequences. For example, for any coprime integers p and q with $p > q \geq 2$ and for any non-negative integer ℓ, we have

$$\limsup_{n\to+\infty} \{n^\ell (p/q)^n\} - \liminf_{n\to+\infty} \{n^\ell (p/q)^n\} \geq 1/p^{\ell+1}.$$

See also [382] for other results on the distribution of $(\{\xi u_n\})_{n\geq 1}$, where $(u_n)_{n\geq 1}$ is a recurrence sequence of rational integers.

▷ The distribution of the sequence $(\{\xi(-p/q)^n\})_{n\geq 1}$, for a non-zero real number ξ and coprime integers p and q with $p > q \geq 1$, has been studied by Dubickas [245]. Among other results, he proved that, for any real irrational number ξ and any integer $b \geq 2$, we have

$$\liminf_{n\to+\infty} \{\xi(-b)^n\} \leq \prod_{k=1}^{+\infty} \left(1 - b^{-(2^k + (-1)^{k-1})/3}\right),$$

and that the above inequality cannot be replaced by a strict one. He also established that, for any non-zero real number ξ, the sequence $(\{\xi(-3/2)^n\})_{n\geq 1}$ has a limit point smaller than 0.533547 and a limit point greater than 0.466452. The word $\boldsymbol{z}$ (Definition A.12) plays a key role in [245].

▷ Let $\alpha > 1$ be an algebraic number and ξ be a positive real number. Kaneko [365] established a lower bound for the number of integers n such that $0 \le n < N$ and $\{\xi\alpha^n\} > \min\{1/L_+(\alpha), 1/L_-(\alpha)\}$, where $L_+(\alpha)$ and $L_-(\alpha)$ are defined in Exercise 3.3.

▷ Theorem 3.5 has been extended to powers of complex non-real algebraic numbers in [260].

▷ Kaneko [363] extended Theorem 3.12 by giving explicit lower bounds for the largest limit point of $(||\xi\alpha^n||)_{n\ge1}$ for a class of algebraic numbers α of degree at least two. He [364] further established various results on the greatest limit point of $(\{\xi\alpha^n\})_{n\ge1}$ and on the difference between the greatest and the smallest limit point of this sequence, when α is an algebraic number of degree at least two satisfying some additional assumptions. In addition, he showed that if α is a quadratic Pisot number whose Galois conjugate α_2 lies in $(0, 2-\sqrt{2})$, then, for any real number ξ not in $\mathbb{Q}(\alpha)$, the greatest limit point of $(\{\xi\alpha^n\})_{n\ge1}$ is at least equal to $1/(\alpha-\alpha_2)$, with equality when ξ equals $(\sum_{m\ge1}\alpha^{-m!})/(\alpha-\alpha_2)$.

▷ Independently, Dubickas [234] and Luca [451] proved that if α is a Pisot number such that $\{\alpha^n\}$ tends to 0 as n tends to infinity, then α is a rational integer.

▷ Akiyama and Tanigawa [34] established that, if α is a Salem number of large degree, then the sequence $(\{\alpha^n\})_{n\ge1}$ is not far from being uniformly distributed modulo one; see also [227]. Additional results on fractional parts of powers of Salem numbers have been proved by Zaïmi [741, 742].

▷ Y. Meyer [511] proved that a real number $\alpha > 1$ is a Pisot or a Salem number if, and only if, for every $\varepsilon > 0$, there exists a positive real number L such that every real interval of length L has a non-empty intersection with the set of real numbers ξ such that $||\xi\alpha^n|| < \varepsilon$ for every $n \ge 1$. Further results on Pisot and Salem numbers can be found in [511].

▷ Dubickas [248] and Akiyama [32] proved independently that for coprime integers p and q with $p \ge q^2+1$, there are non-zero real numbers ξ such that the closure of the set $\{\xi(p/q)^n\}$, $n \ge 1$, is of Lebesgue measure zero and even (see [32]) of Hausdorff dimension at most $(\log q)/\log(p/q)$, which is less than 1.

▷ Let $(F_n)_{n \geq 0}$ denote the Fibonacci sequence defined by $F_0 = 0$, $F_1 = 1$ and $F_{n+2} = F_{n+1} + F_n$ for $n \geq 0$. Among other results, Dubickas [250] proved that, for any real number ξ not in the field $\mathbb{Q}(\sqrt{5})$, there are infinitely many positive integers n such that $\{\xi F_n\} > 2/3$.

▷ Kamae [361] established that if $(m_n)_{n \geq 1}$ is a non-decreasing sequence of positive integers tending to infinity and such that $2^{m_n}/n$ is bounded, then the sequence $(3^n/2^{m_n})_{n \geq 1}$ is uniformly distributed modulo one.

▷ Chowla and De Leon [195] proved that, if 0 is a limit point of the sequence $(||\sqrt{6}(\sqrt{2}+1)^n/4||)_{n \geq 1}$, then the pair $(\sqrt{2}, \sqrt{3})$ satisfies the Littlewood conjecture, that is, for any positive ε, there exists a positive integer q such that $q \cdot ||q\sqrt{2}|| \cdot ||q\sqrt{3}|| < \varepsilon$.

4
Normal numbers

Throughout this chapter, b denotes an integer at least equal to 2 and ξ is a real number given by its b-ary expansion, that is,

$$\xi = \lfloor \xi \rfloor + \sum_{k \geq 1} \frac{a_k}{b^k} = \lfloor \xi \rfloor + 0 \cdot a_1 a_2 \dots, \tag{4.1}$$

where the digits $a_1, a_2, \dots$ are integers from $\{0, 1, \dots, b-1\}$ and an infinity of the a_k are not equal to $b-1$.

For a positive integer N and a digit d in $\{0, 1, \dots, b-1\}$, set

$$A_b(d, N, \xi) := \mathrm{Card}\{j : 1 \leq j \leq N, a_j = d\}.$$

More generally, for a block $D_k = d_1 \dots d_k$ of k digits from $\{0, 1, \dots, b-1\}$, set

$$A_b(D_k, N, \xi) := \mathrm{Card}\{j : 0 \leq j \leq N - k, a_{j+1} = d_1, \dots, a_{j+k} = d_k\}.$$

Often, we identify the block D_k with the integer $d_1 b^{k-1} + \dots + d_{k-1} b + d_k$. We keep this notation throughout this chapter.

The notion of normal numbers was introduced by Émile Borel [114] in a seminal paper published in 1909; see also [115, pp. 194–196]. We reproduce his original definition.

Definition 4.1. Let $b \geq 2$ be an integer. The *frequency* of a digit d in the b-ary expansion of a real number ξ is equal to the limit of the sequence $(A_b(d, N, \xi)/N)_{N \geq 1}$ if this sequence converges, and does not exist otherwise. A real number ξ is called *simply normal* to base b if every digit $0, 1, \dots, b-1$ occurs in its b-ary expansion with the same frequency $1/b$, that is, if

$$\lim_{N \to +\infty} \frac{A_b(d, N, \xi)}{N} = \frac{1}{b}, \quad \text{for } d = 0, 1, \dots, b-1. \tag{4.2}$$

It is called *normal* to base b if each of $\xi, b\xi, b^2\xi, \ldots$ is simply normal to every base $b, b^2, b^3, \ldots$

There are several equivalent definitions of a normal number, which are discussed in Sections 4.1 and 4.3. Theorem 4.14 shows the close relationship between the notion of normality of a real number ξ to a given base b and the distribution modulo one of the sequence $(\xi b^n)_{n\geq 1}$. Explicit constructions of normal numbers are given in Section 4.2. The notion of 'richness', which is much weaker than that of normality, is defined and briefly discussed in Section 4.4. Finally, we study in Section 4.5 the rational approximation to a subclass of the normal numbers constructed in Section 4.2.

4.1 Equivalent definitions of normality

In the sequel, we often use that (4.2) holds if, and only if, there are positive integers $\ell_0, \ldots, \ell_{b-1}$ such that

$$\lim_{N\to+\infty} \frac{A_b(d, \ell_d N, \xi)}{\ell_d N} = \frac{1}{b}, \quad \text{for } d = 0, 1, \ldots, b-1.$$

This fact is easy to check.

Pillai [562] showed that the conditions imposed by Borel are redundant in part and established the following result.

THEOREM 4.2. *Let $b \geq 2$ be an integer. A real number ξ is normal to base b if, and only if, ξ is simply normal to every base $b, b^2, b^3, \ldots$*

Theorem 4.2 was used without proof by Sierpiński [654].

PROOF. The 'only if' part is immediate from Definition 4.1, but some work is needed to establish the 'if' part. Let ξ be a real number simply normal to every base $b, b^2, b^3, \ldots$ It is sufficient to show that $b\xi$ is simply normal to every base $b, b^2, b^3, \ldots$ Since

$$|A_b(d, N, \xi) - A_b(d, N, b\xi)| \leq 1,$$

for $d = 0, 1, \ldots, b-1$ and $N \geq 1$, it is immediate that $b\xi$ is simply normal to base b. Let $j \geq 2$ and $r \geq 2$ be integers. Let $d = 0, \ldots, b^j - 1$. Given an integer $D = 0, \ldots, b^{jr} - 1$, write $D = d'_1 b^{jr-1} + \cdots + d'_{jr-1} b + d'_{jr}$ to base b. Then, express $bD - d'_1 b^{jr} = d'_2 b^{jr-1} + \cdots + d'_{jr-1} b^2 + d'_{jr} b$ as

$$bD - d'_1 b^{jr} = d_1 b^{j(r-1)} + \cdots + d_{r-1} b^j + d_r$$

to base b^j. The cardinality of the set $\mathcal{D}'_b(j,r,k)$ composed of the integers D in $\{0,1,\ldots,b^{jr}-1\}$ such that the digit d occurs exactly k times among $d_1,\ldots,d_{r-1}$ is equal to

$$\binom{r-1}{k} b^j (b^j-1)^{r-1-k}, \tag{4.3}$$

since the values of the j digits d'_1 and $d'_{jr-j+2},\ldots,d'_{jr}$ do not affect the values of $d_1,\ldots,d_{r-1}$. It follows from the definition of $\mathcal{D}'_b(j,r,k)$ that

$$A_{b^j}(d,Nr,b\xi) \ge \sum_{k=1}^{r-1} k \sum_{D\in\mathcal{D}'_b(j,r,k)} A_{b^{jr}}(D,N,\xi).$$

Let ε be a positive real number. Since ξ is simply normal to base b^{jr}, we infer from (4.3) and the binomial identity that, for sufficiently large N, we have

$$\begin{aligned}\frac{A_{b^j}(d,Nr,b\xi)}{Nr} &\ge \sum_{k=1}^{r-1} \frac{k}{r}\binom{r-1}{k} b^j(b^j-1)^{r-1-k}(b^{-jr}-\varepsilon)\\ &= (1-r^{-1})b^j(b^{-jr}-\varepsilon)\sum_{k=1}^{r-1}\binom{r-2}{k-1}(b^j-1)^{r-2-(k-1)}\\ &= (1-r^{-1})b^{-j}(1-\varepsilon b^{jr}),\end{aligned}$$

thus

$$\liminf_{N\to+\infty} \frac{A_{b^j}(d,N,b\xi)}{N} \ge \frac{1-r^{-1}}{b^j}. \tag{4.4}$$

Since (4.4) holds for every digit $d=0,\ldots,b^j-1$ and

$$\sum_{d=0}^{b^j-1} \frac{A_{b^j}(d,N,b\xi)}{N} = 1,$$

we deduce that

$$\limsup_{N\to+\infty} \frac{A_{b^j}(d,N,b\xi)}{N} \le \frac{1+b^j r^{-1}}{b^j}, \tag{4.5}$$

for $d=0,\ldots,b^j-1$. Choosing r arbitrarily large, it follows from (4.4) and (4.5) that $b\xi$ is simply normal to base b^j. Since $j\ge 2$ is arbitrary, we have established the theorem. □

We continue with an easy lemma.

LEMMA 4.3. *Let b and r be integers greater than or equal to 2. If a real number is simply normal to base b^r, then it is simply normal to base b.*

PROOF. Let k be an integer with $0 \le k \le r$. Let $d = 0, \dots, b-1$. Denote by $\mathcal{D}_b(r,d,k)$ the set of integers D from $\{0,1,\dots,b^r-1\}$ such that in the representation $D = d_1 b^{r-1} + \cdots + d_{r-1} b + d_r$, with $d_1, \dots, d_r$ in $\{0,1,\dots,b-1\}$, exactly k digits among $d_1, \dots, d_r$ are equal to d. Observe that the cardinality $p_b(r,k)$ of $\mathcal{D}_b(r,d,k)$ is given by

$$p_b(r,k) = \binom{r}{k} (b-1)^{r-k},$$

and that

$$\sum_{k=0}^{r} \frac{k}{r} \, p_b(r,k) = b^{r-1}, \tag{4.6}$$

by the binomial identity.

Let ξ be a real number simply normal to base b^r. Let ε be a positive real number and N_0 be an integer such that

$$\frac{A_{b^r}(D,N,\xi)}{N} = \frac{1}{b^r} + \varepsilon_{D,N} \quad \text{with} \quad |\varepsilon_{D,N}| \le \frac{\varepsilon}{b^{r-1}},$$

for every $N \ge N_0$ and for $D = 0,1,\dots,b^r-1$. Let $N \ge N_0$ be given. Then, we deduce from (4.6) that

$$\begin{aligned} \left| \frac{A_b(d,rN,\xi)}{rN} - \frac{1}{b} \right| &= \left| \Bigl(\sum_{k=0}^{r} \frac{k}{r} \sum_{D \in \mathcal{D}_b(r,d,k)} \frac{A_{b^r}(D,N,\xi)}{N} \Bigr) - \frac{1}{b} \right| \\ &= \left| \sum_{k=0}^{r} \frac{k}{r} \frac{p_b(r,k)}{b^r} - \frac{1}{b} + \sum_{k=0}^{r} \frac{k}{r} \sum_{D \in \mathcal{D}_b(r,d,k)} \varepsilon_{D,N} \right| \\ &\le b^{r-1} \max_{D \in \{0,1,\dots,b^r-1\}} |\varepsilon_{D,N}| \le \varepsilon. \end{aligned}$$

This proves that ξ is simply normal to base b. □

The converse of Lemma 4.3 does not hold. Indeed, it is easy to construct explicit examples of real numbers simply normal to a given base $b \ge 2$, which are not simply normal to every base b^r with $r \ge 2$. Take for instance the rational number whose b-ary expansion is purely periodic of period $012\dots(b-1)$.

We derive from Lemma 4.3 and Theorem 4.2 a theorem of Maxfield [494].

THEOREM 4.4. *Let $b \ge 2$ be an integer, and r, s be positive integers. A real number is normal to base b^s if, and only if, it is normal to base b^r.*

PROOF. Clearly, it is sufficient to establish the result for $s = 1$. The 'only if' part follows directly from Theorem 4.2, so it remains for us to establish the 'if' part. Let $r \geq 2$ be an integer and ξ be a real number normal to base b^r. Let t be a positive integer. By Definition 4.1, ξ is simply normal to base $b^{rt} = (b^t)^r$, and then to base b^t, by Lemma 4.3. Consequently, ξ is normal to base b, by Theorem 4.2. □

We have already used several times the one-to-one correspondence between digits in base b^k (that is, elements of $\{0, 1, \ldots, b^k - 1\}$) and blocks of k digits in base b (that is, words of length k on $\{0, 1, \ldots, b-1\}$). Thus, looking at the expansion of a number to base b^k should be essentially the same as looking at blocks of k digits in its expansion to base b. This point of view allows us to formulate an equivalent definition of normality. In his seminal paper [114] (see also [115, p. 195]) Borel also wrote that the characteristic property of a normal number is the following:

Un groupement quelconque de p chiffres consécutifs étant considéré, si l'on désigne par c_n le nombre de fois que se rencontre ce groupement dans les n premiers chiffres décimaux, on a

$$\lim_{n \to +\infty} \frac{c_n}{n} = \frac{1}{10^p}.$$

This is precisely the content of Theorem 4.5.

THEOREM 4.5. *Let $b \geq 2$ be an integer. A real number ξ is normal to base b if, and only if, for every $k \geq 1$, every block of k digits from $\{0, 1, \ldots, b-1\}$ occurs in the b-ary expansion of ξ with the same frequency $1/b^k$, that is, if and only if,*

$$\lim_{N \to +\infty} \frac{A_b(D_k, N, \xi)}{N} = \frac{1}{b^k}, \quad \text{for every } k \geq 1 \text{ and every block } D_k \text{ of } k \text{ digits from } \{0, 1, \ldots, b-1\}. \tag{4.7}$$

Several authors, including Champernowne [184], Koksma [389] and Copeland and Erdős [203], have taken property (4.7) as the definition of a normal number. Hardy and Wright [334] stated that property (4.7) is equivalent to Definition 4.1, but gave no proof.

A complete proof of Theorem 4.5, independent of Theorem 4.2, was given in 1951 by Niven and Zuckerman [544], and simplified one year later by Cassels [180]. Theorem 4.5 was applied by Maxfield [493] to give a short proof of Theorem 4.2.

The 'if' part of Theorem 4.5 is an immediate consequence of the next result, established independently by Piatetski-Shapiro [559] again in 1951, and which is called the 'hot spot lemma' by Borwein and Bailey [120].

THEOREM 4.6. *A real number ξ is normal to base b if, and only if, there exists a positive number C such that*

$$\limsup_{N\to+\infty} \frac{A_b(D_k,N,\xi)}{N} \le \frac{C}{b^k}, \quad \text{for every } k\ge 1 \text{ and every block } D_k \text{ of } k \text{ digits from } \{0,1,\ldots,b-1\}. \tag{4.8}$$

To prove Theorem 4.6, we need the following lemma. Recall that $p_b(n,k)$ denotes the number of blocks of n digits from $\{0,1,\ldots,b-1\}$ containing exactly k times a given digit.

LEMMA 4.7. *Let $b\ge 2$ and $n\ge 1$ be integers. For every integer j such that $2\le j\le (b-1)n$ or $-n\le j\le -2$, we have*

$$p_b(bn,n+j) < b^{bn}\mathrm{e}^{-j^2/(4bn)}.$$

If $n\ge b^{15}$, then, for every real number ε with $n^{-1/3}\le\varepsilon\le 1$, we have

$$\sum_{-n\le j\le -\lceil\varepsilon n\rceil} p_b(bn,n+j) + \sum_{\lceil\varepsilon n\rceil\le j\le (b-1)n} p_b(bn,n+j) \le 2^{14}b^{bn}\mathrm{e}^{-\varepsilon^2 n/(10b)}.$$

PROOF. If $j\ge 2$, then we have

$$\begin{aligned}
\frac{p_b(bn,n+j)}{p_b(bn,n)} &= \frac{(bn-n)(bn-n-1)\ldots(bn-n-j+1)}{(n+1)(n+2)\ldots(n+j)(b-1)^j}\\
&= \frac{n}{n+1}\cdot\frac{n}{n+2}\cdots\frac{n}{n+j}\cdot\frac{(b-1)(b-1-1/n)\ldots(b-1-(j-1)/n)}{(b-1)^j}\\
&< \Big(1-\frac{1}{n(b-1)}\Big)\Big(1-\frac{2}{n(b-1)}\Big)\cdots\Big(1-\frac{j-1}{n(b-1)}\Big)\\
&< \exp\Big\{-\frac{1}{n(b-1)}-\frac{2}{n(b-1)}-\cdots-\frac{j-1}{n(b-1)}\Big\}\\
&= \exp\{-j(j-1)/2n(b-1)\} < \exp\{-j^2/(4bn)\},
\end{aligned}$$

as well as, in the same way,

$$\begin{aligned}
\frac{p_b(bn,n-j)}{p_b(bn,n)} &= \Big(1-\frac{1}{n}\Big)\Big(1-\frac{2}{n}\Big)\cdots\Big(1-\frac{j-1}{n}\Big)\\
&\quad\times\frac{(b-1)^j}{(b-1+1/n)\ldots(b-1+j/n)}\\
&< \exp\{-j(j-1)/2n\} < \exp\{-j^2/(4bn)\}.
\end{aligned}$$

Combined with the obvious estimate

$$p_b(bn,n) < b^{bn},$$

this gives the first assertion of the lemma. As for the second one, it is sufficient to note that we estimate a sum of at most $2bn$ terms, all smaller than $b^{bn}\mathrm{e}^{-\varepsilon^2 n/(4b)}$, and that

$$2bn\mathrm{e}^{-\varepsilon^2 n/(4b)} = 2bn\mathrm{e}^{-3\varepsilon^2 n/(20b)}\mathrm{e}^{-\varepsilon^2 n/(10b)}.$$

From our assumptions $n \geq b^{15}$ and $\varepsilon \geq n^{-1/3}$, it then follows that

$$2bn\mathrm{e}^{-3\varepsilon^2 n/(20b)} \leq 2bn\mathrm{e}^{-3n^{1/3}/(20b)} \leq 2bn\mathrm{e}^{-3(bn)^{1/4}/20} \leq 2^{14}.$$

This concludes the proof. □

Proof of Theorems 4.5 and 4.6. We begin with the 'only if' part of Theorem 4.5. Express ξ as in (4.1). Assume that ξ is normal to base b and let D_k be given. Write D_k as a single digit, say d, to base b^k. For integers $j = 0, 1, \ldots, k-1$ and $r \geq 0$, the occurrence of the block D_k as $a_{kr+j+1} \cdots a_{kr+j+k}$ corresponds to the digit d being in the $(r+1)$th position in the expansion of $\{b^j\xi\}$ to base b^k. Observe that

$$\frac{1}{k}\sum_{j=0}^{k-1} \frac{A_{b^k}(d, N, b^j\xi)}{N} - \frac{1}{N} \leq \frac{A_b(D_k, Nk, \xi)}{Nk} \leq \frac{1}{k}\sum_{j=0}^{k-1} \frac{A_{b^k}(d, N, b^j\xi)}{N}.$$

Since, by Definition 4.1, the real number $b^j\xi$ is simply normal to base b^k for $j = 0, \ldots, k-1$, we get that

$$\lim_{N\to+\infty} \frac{A_b(D_k, N, \xi)}{N} = \frac{1}{b^k},$$

as required.

We establish now the 'if' part of Theorem 4.6. Let ξ, b and C be such that (4.8) holds. Let $r \geq 1$ and $\ell \geq b^{5r}$ be integers. Combine the digits of ξ to base b^r into blocks of $b^r\ell^3$ digits. Let d be in $\{0, 1, \ldots, b^r - 1\}$. A block of $b^r\ell^3$ digits on $\{0, 1, \ldots, b^r - 1\}$ is called a *good* block if the digit d occurs in it no more than $\ell^3 + b^r\ell^2$ times, and no less than $\ell^3 - b^r\ell^2$ times. Otherwise, the block is called a *bad* block. It follows from the second assertion of Lemma 4.7 applied with $n = \ell^3$ and $\varepsilon = b^r/\ell$ that there are at most

$$2^{14}b^{rb^r\ell^3}\mathrm{e}^{-b^r\ell/10}$$

different bad blocks of $b^r\ell^3$ digits on $\{0, 1, \ldots, b^r - 1\}$.

Consider a bad block as a block of $rb^r\ell^3$ digits to base b. By assumption, for sufficiently large N, it occurs no more than $2CNb^{-rb^r\ell^3}$ times in the prefix of length N of the b-ary expansion of ξ. Consequently, viewed as a block of $b^r\ell^3$ digits to base b^r, it occurs no more than $2CNr(b^r)^{-b^r\ell^3}$

times in the prefix of length N of the b^r-ary expansion of ξ, for sufficiently large N. Consequently, if N is large enough, then the number of occurrences $A_{b^r}(d, N, \xi)$ of the digit d among the first N digits in the b^r-ary expansion of ξ satisfies

$$A_{b^r}(d, N, \xi) \le \frac{N}{b^r \ell^3} \cdot (\ell^3 + b^r \ell^2) + \left(2^{15} C N r \mathrm{e}^{-b^r \ell/10}\right) b^r \ell^3 + b^r \ell^3. \quad (4.9)$$

Let ε be a given positive real number. Take ℓ sufficiently large in order that the right-hand side of (4.9) is less than $N b^{-r}(1+\varepsilon) + b^r \ell^3$. This implies that

$$\limsup_{N \to +\infty} \frac{A_{b^r}(d, N, \xi)}{N} \le \frac{1}{b^r}.$$

Since the same estimate holds for every digit in $\{0, 1, \ldots, b^r - 1\}$ and

$$\sum_{d'=0}^{b^r-1} A_{b^r}(d', N, \xi) = N,$$

we deduce that ξ is simply normal to base b^r. It then follows from Theorem 4.2 that ξ is normal to base b. □

We establish now a result given by Borel [114] in his fundamental paper.

THEOREM 4.8. *Almost all real numbers are normal to all integer bases.*

PROOF. By Theorem 4.2, and since a countable union of null sets is a null set, it is sufficient to prove that almost all numbers are simply normal to a given base. Let $b \ge 2$ and d be integers with $0 \le d \le b-1$. For an integer j and a positive integer n, let $S_{d,n}(j)$ denote the set of real numbers ξ in $(0,1)$ such that there are exactly $n+j$ digits equal to d among the first bn digits of the b-ary expansion of ξ. In view of Lemma 4.7, if $-n \le j \le -2$ or $2 \le j \le (b-1)n$, then the set $S_{d,n}(j)$ is contained in a union of at most $p_b(bn, n+j)$ intervals of the form $[\xi_s, \xi_s + 1/b^{bn}]$ of total length at most equal to

$$b^{-bn} p_b(bn, n+j) \le \mathrm{e}^{-j^2/(4bn)}.$$

Let T and n be positive integers with $n \ge \max\{b^{15}, T^3\}$. Let $S'_{d,n}(T)$ denote the union of the sets $S_{d,n}(j)$ with $-n \le j \le -\lceil n/T \rceil$ or $\lceil n/T \rceil \le j \le (b-1)n$. By Lemma 4.7 applied with $\varepsilon = 1/T$, the Lebesgue measure of $S'_{d,n}(T)$ is at most $2^{14} \mathrm{e}^{-n/(10bT^2)}$. Since the series $\sum_{n \ge 1} \mathrm{e}^{-n/(10bT^2)}$ converges, Lemma C.1 implies that, for any given positive integer T and any d in $\{0, 1, \ldots, b-1\}$, almost all real numbers ξ belong to only finitely many sets $S'_{d,n}(T)$, $n \geq 1$.

Let now ξ be a real number that is not normal to base b. Then, there exists $d = 0, \ldots, b-1$ such that

$$\lim_{N \to +\infty} \frac{A_b(d, bN, \xi)}{bN} = \frac{1}{b}$$

does not hold. This means that there exist a positive real number τ and infinitely many positive integers N such that

$$\left| \frac{A_b(d, bN, \xi)}{bN} - \frac{1}{b} \right| \geq \frac{\tau}{b} = \frac{1}{bN} \cdot \tau N.$$

Setting $T = \lceil 2/\tau \rceil$, this proves that ξ belongs to infinitely many sets $S'_{d,n}(T)$, $n \geq 1$. Combined with the conclusion of the previous paragraph, we have established that almost all real numbers are normal to base b. This implies the theorem. □

In [114] Borel asked for the construction of a real number normal to all integer bases. We discuss this question in Section 5.2.

4.2 The Champernowne number

The first explicit example of a real number normal to a given base was given by Champernowne [184] in 1933.

THEOREM 4.9. *The real number*

$$\xi_{\mathbf{c}} = 0.12345678910111213 14\ldots,$$

whose sequence of decimals is the increasing sequence of all positive integers, is normal to base ten.

The real number $\xi_{\mathbf{c}}$ is often called the Champernowne number. At the end of his paper, Champernowne stated (without proof) further related results, including that the real number 0.46891012141516... whose sequence of decimals is the increasing sequence of all composite numbers is normal to base ten, as is the number $0 \cdot x_1 x_2 \ldots x_r \ldots$, where x_r denotes the integer part of $r \log r$, for any positive integer r. He concluded by 'It would be reasonable to suppose that the decimal formed by the sequence of prime numbers is also normal to the scale of ten, but of this I have no proof'.

The latter problem was solved in 1946 by Copeland and Erdős [203], who established the particular case $m = 1$ of the next statement. It contains Theorem 4.9 and all the results stated in [184].

For a positive integer c and an integer $b \geq 2$, we denote by $(c)_b$ the word on $\{0, 1, \ldots, b-1\}$ representing c in base b, and we let $\ell_b(c)$ denote its length. With this convention, we check that $19 = (19)_{10} = (10011)_2$ and $\ell_2(19) = 5$.

THEOREM 4.10. *Let $b \geq 2$ be an integer. Let $(c_j)_{j \geq 1}$ be an increasing sequence of positive integers such that, for every $\theta > 1$, we have $c_N < N^\theta$ for every sufficiently large integer N. Let m be a real number with $m \geq 1$. Then the real number*

$$0 \cdot (c_1)_b \ldots (c_1)_b (c_2)_b \ldots (c_2)_b (c_3)_b \ldots,$$

where the block of digits $(c_j)_b$ is repeated $\lfloor m^{\ell_b(c_j)} \rfloor$ times for every $j \geq 1$, is normal to base b.

Theorem 4.10 for $m > 1$ has not been published previously. Since the nth prime number is at most equal to a constant times $n \log 3n$, the next statement, established in [203], is an immediate consequence of Theorem 4.10.

COROLLARY 4.11. *The real number*

$$0.23571113171923293 1 \ldots,$$

whose sequence of decimals is the increasing sequence of all prime numbers, is normal to base ten.

The proof of Theorem 4.10, and of most of the results of the same flavour, rests on the concept of (ε, k)-normality, introduced by Besicovitch [93].

DEFINITION 4.12. Let $b \geq 2$, $k \geq 1$ and $\ell \geq 1$ be integers, and let ε be a positive real number. A finite word W of length ℓ on the alphabet $\{0, 1, \ldots, b-1\}$ is *(ε, k)-normal* to base b if the total number of occurrences in W of any combination of k digits is comprised between $(b^{-k} - \varepsilon)\ell$ and $(b^{-k} + \varepsilon)\ell$. A positive integer c is *(ε, k)-normal* to base b if the word $(c)_b$ is (ε, k)-normal to base b.

We begin with a lemma asserting that there are not too many integers which fail to be (ε, k)-normal to a given base.

LEMMA 4.13. *Let $b \geq 2$, $k \geq 1$ be integers, and let ε be a positive real number with $\varepsilon < 1/2$. There exists a positive real number $\delta = \delta(b, k)$,* *depending only on b and k, such that the number of integers up to N which are not (ε, k)-normal to base b is less than $N^{1-\delta\varepsilon^2}$, provided that N is sufficiently large.*

PROOF. Let n be an integer with $n \geq b^{15k}$ and $n \geq \varepsilon^{-3}$. It follows from Lemma 4.7 that the number of blocks of $b^k n$ letters on $\{0, 1, \ldots, b^k - 1\}$ containing more than $n(1+\varepsilon)$ or less than $n(1-\varepsilon)$ occurrences of a given digit is at most equal to

$$2^{14}(b^k)^{b^k n} \mathrm{e}^{-\varepsilon^2 n/(10 b^k)}.$$

Consequently, there exists a positive real number δ, depending only on b and k such that, for n sufficiently large, the number of blocks of length $kb^k n$ on $\{0, 1, \ldots, b-1\}$ which are not (ε, k)-normal to base b is less than $(b^k)^{(1-\delta\varepsilon^2)b^k n}$. Thus, for n sufficiently large, the number of integers between $(b^k)^{b^k n}$ and $(b^k)^{b^k(n+1)} - 1$ which are not $(2\varepsilon, k)$-normal to base b is less than $(b^k)^{b^k(n+1)(1-\delta\varepsilon^2)}$. This implies that, if N is a sufficiently large integer of the form $(b^k)^{b^k n}$, then there are at most $N^{1-(\delta\varepsilon^2/2)}$ integers up to N which fail to be $(2\varepsilon, k)$-normal to base b. The same holds for every sufficiently large integer N if δ is replaced by $\delta/2$. This finishes the proof of the lemma. □

PROOF OF THEOREM 4.10. Fix a positive integer k and an integer N, which is assumed to be sufficiently large. Set $\nu = \ell_b(c_N)$. Let ε be a positive real number with $\varepsilon < 1/2$. For integers h, j with $0 \leq h \leq \lfloor m^\nu \rfloor - 1$ and $0 \leq j \leq \nu - 1$, consider the block

$$X_N = X_{N,h,j} := (c_1)_b \ldots (c_1)_b \ldots (c_N)_b \ldots (c_N)_b (c_N)_{b,j}$$

on base b, where the block $(c_N)_b$ is repeated h times and $(c_N)_{b,j}$ is the prefix of $(c_N)_b$ of length j. Let t_N denote the total number of blocks in X_N and u_N denote the total number of distinct blocks in X_N.

Set $\theta = 1 + \varepsilon^4$ and assume that N is sufficiently large in order that $c_N < N^\theta$. By Lemma 4.13, there exists $\delta > 0$, depending only on b and k, such that the number of integers j such that $1 \leq j \leq N$ and the block $(c_j)_b$ is not (ε, k)-normal to base b is at most $N^{(1-\delta\varepsilon^2)\theta}$.

There are at most $b^{\lceil \nu(1-\varepsilon^3) \rceil}$ positive integers j such that the representation of c_j to base b has length less than or equal to $\lceil \nu(1-\varepsilon^3) \rceil$. Thus, there are at least $N - b^{\lceil \nu(1-\varepsilon^3) \rceil}$ integers j such that $1 \leq j \leq N$ and $\ell_b(c_j) \geq \lceil \nu(1-\varepsilon^3) \rceil$. Consequently, the length L_N of the block X_N satisfies

$$L_N \geq \nu(1-\varepsilon^3) \lfloor m^{\nu(1-\varepsilon^3)} \rfloor (N - 1 - b^{\lceil \nu(1-\varepsilon^3) \rceil}),$$

and the same argument also gives that

$$L_N \geq \nu(1-\varepsilon^3)\big(t_N - (bm)^{\lceil \nu(1-\varepsilon^3)\rceil}\big).$$

Observe also that, since $b^{\nu-1} \leq c_N \leq b^\nu$, we have

$$b^{(\nu-1)/\theta} \leq N \leq c_N \leq b^\nu. \tag{4.10}$$

In particular, we get $N > 2b^{\lceil \nu(1-\varepsilon^3)\rceil}$ and $t_N > 2(bm)^{\lceil \nu(1-\varepsilon^3)\rceil}$.

Let D_k be a block of k digits from $\{0,1,\ldots,b-1\}$. Taking into account the overlaps, we deduce that

$$\begin{aligned}\frac{A_b(D_k, L_N, \xi)}{L_N} &< \frac{1}{b^k} + \varepsilon + \frac{k t_N}{\nu(1-\varepsilon^3)(t_N - (bm)^{\lceil \nu(1-\varepsilon^3)\rceil})} \\ &\quad + \frac{\nu m^\nu N^{(1-\delta\varepsilon^2)\theta}}{\nu(1-\varepsilon^3)\lfloor m^{\nu(1-\varepsilon^3)}\rfloor\big(N-1-b^{\lceil \nu(1-\varepsilon^3)\rceil}\big)} \\ &< \frac{1}{b^k} + \varepsilon + \frac{2k}{\nu(1-\varepsilon^3)} + \frac{2m^{\varepsilon^3\nu} N^{(1-\delta\varepsilon^2)\theta-1}}{1-\varepsilon^3},\end{aligned} \tag{4.11}$$

for N sufficiently large. It follows from (4.10) that $\nu \geq (\log N)/(\log b)$ and

$$m^{\varepsilon^3\nu} N^{(1-\delta\varepsilon^2)\theta-1} \leq m^{\varepsilon^3} N^{(\varepsilon^3\theta(\log m)/(\log b))+(1-\delta\varepsilon^2)(1+\varepsilon^4)-1}. \tag{4.12}$$

If ε is sufficiently small, then the right-hand side of (4.12) is at most equal to $mN^{-\delta\varepsilon^2/2}$. For N large enough in terms of ε and k, we deduce that the left-hand side of (4.11) is less than $b^{-k} + 2\varepsilon$. We conclude by applying Theorem 4.6. □

4.3 Normality and uniform distribution

Unsurprisingly, the notion of normality to base b for a real number ξ is deeply related to the distribution modulo one of the sequence $(\xi b^n)_{n\geq 1}$. The next theorem was proved by Wall [725] in his Ph.D. thesis.

THEOREM 4.14. *Let $b \geq 2$ be an integer. The real number ξ is normal to base b if, and only if, the sequence $(\xi b^n)_{n\geq 1}$ is uniformly distributed modulo one.*

PROOF. Assume first that the sequence $(\xi b^n)_{n\geq 1}$ is uniformly distributed modulo one. Let $D_k = d_1 \ldots d_k$ be a block of k digits on $\{0, 1, \ldots, b-1\}$ and put

$$u = \frac{d_1}{b} + \frac{d_2}{b^2} + \cdots + \frac{d_k}{b^k}, \qquad v = \frac{d_1}{b} + \frac{d_2}{b^2} + \cdots + \frac{d_k+1}{b^k}.$$

Let $\xi = \lfloor \xi \rfloor + 0 \cdot a_1 a_2 \ldots$ be the b-ary expansion of ξ given by (4.1). Observe that, for $n \geq 1$, the block $a_n a_{n+1} \ldots a_{n+k-1}$ is identical with the block $d_1 \ldots d_k$ if, and only if, $\{b^{n-1}\xi\}$ is in $[u, v)$. Consequently, for $N \geq k$, we have

$$A_b(D_k, N, \xi) = \mathrm{Card}\{n : 1 \leq n \leq N - k + 1, u \leq \{\xi b^{n-1}\} < v\}. \quad (4.13)$$

Combined with our assumption

$$\lim_{N \to +\infty} \frac{\mathrm{Card}\{n : 1 \leq n \leq N, u \leq \{\xi b^n\} < v\}}{N} = v - u = \frac{1}{b^k},$$

it then follows from Theorem 4.5 that ξ is normal to base b.

Conversely, let ξ be a real number normal to base b. Let u, v be real numbers with $0 \leq u < v \leq 1$. Observe that, in view of Theorem 4.5 and (4.13),

$$\lim_{N \to +\infty} \frac{\mathrm{Card}\{n : 1 \leq n \leq N, u \leq \{\xi b^n\} < v\}}{N} = v - u$$

holds if u and v are rational numbers whose denominators are powers of b. Let ε be a positive real number. There exist rational numbers u_1, v_1, u_2, v_2, whose denominators are powers of b, such that $u_1 \leq u \leq u_2 < v_2 \leq v \leq v_1$ and $u_2 - u_1 < \varepsilon$, $v_1 - v_2 < \varepsilon$. Since, for every positive integer N, we have

$$\begin{aligned} \mathrm{Card}\{n : 1 \leq n \leq N, u_1 \leq \{\xi b^n\} < v_1\} & \\ \leq \mathrm{Card}\{n : 1 \leq n \leq N, u \leq \{\xi b^n\} < v\} & \\ \leq \mathrm{Card}\{n : 1 \leq n \leq N, u_2 \leq \{\xi b^n\} < v_2\}, & \end{aligned}$$

we deduce that, if N is large enough, then

$$v - u - 2\varepsilon \leq \frac{\mathrm{Card}\{n : 1 \leq n \leq N, u \leq \{\xi b^n\} < v\}}{N} \leq v - u + 2\varepsilon$$

holds. As ε is arbitrary, this proves that the sequence $(\xi b^n)_{n \geq 1}$ is uniformly distributed modulo one. □

This point of view allows us to define normality in a non-integer base.

DEFINITION 4.15. Let α be a real number with $|\alpha| > 1$. The real number ξ is said to be *normal* to base α if the sequence $(\xi \alpha^n)_{n \geq 1}$ is uniformly distributed modulo one.

We conclude this section by a restatement of Corollary 1.9 and Exercise 1.2.

THEOREM 4.16. *Let α be a real number with $|\alpha| > 1$. Then, almost every real number ξ is normal to base α.*

In the special case where $\alpha \geq 2$ is an integer, this gives an alternative proof of Theorem 4.8.

4.4 Block complexity and richness

To prove that a given real number is normal to some integer base is often much too difficult. We introduce in this section the weaker notion of richness (sometimes also called disjunctivity). It seems that this notion first appeared in theoretical computer science and not in number theory; see [202]. Note that the terminology 'rich' has been used since 2008 in combinatorics on words with a different meaning.

Let $b \geq 2$ be an integer. A natural way to measure the *complexity* of a real number ξ whose b-ary expansion is given by (4.1) is to count the number of distinct blocks of given length in the infinite word $\boldsymbol{a} = a_1a_2a_3\ldots$ As in Section A.1, for an infinite word $\boldsymbol{w}$ on the alphabet $\{0, 1, \ldots, b-1\}$ and for any positive integer n, we denote by $p(n, \boldsymbol{w}, b)$ the number of distinct blocks of n letters occurring in $\boldsymbol{w}$. Furthermore, we set $p(n, \xi, b) = p(n, \boldsymbol{a}, b)$ with $\boldsymbol{a}$ as above. Clearly, we have

$$p(n, \xi, b) = \mathrm{Card}\{a_k a_{k+1} \ldots a_{k+n-1} : k \geq 1\}$$

and

$$1 \leq p(n, \xi, b) \leq b^n, \tag{4.14}$$

where both inequalities are sharp (take for example the analogue in base b of the Champernowne number $\xi_{\mathbf{c}}$ to see that the right-hand inequality is an equality for certain real numbers ξ).

THEOREM 4.17. *Let $b \geq 2$ be an integer and ξ be a real number. If ξ is irrational, then we have $p(n, \xi, b) \geq n+1$ for every positive integer n. If ξ is rational, then there exists a positive constant C, depending only on ξ and b, such that $p(n, \xi, b) \leq C$ for every $n \geq 1$.*

PROOF. Recalling that ξ is rational if, and only if, its b-ary expansion is ultimately periodic, the theorem is a reformulation of Theorem A.3. □

For every integer $b \geq 2$, there exist uncountably many real numbers ξ such that $p(n, \xi, b) = n + 1$ for every positive n; see Theorem A.5.

For an infinite word $\boldsymbol{w}$ on $\{0, 1, \ldots, b-1\}$ and for positive integers n, n' we have

$$p(n + n', \boldsymbol{w}, b) \leq p(n, \boldsymbol{w}, b) \cdot p(n', \boldsymbol{w}, b).$$

This implies that the sequence $(\log p(n, \boldsymbol{w}, b))_{n\geq 1}$ is subadditive, thus, $((\log p(n, \boldsymbol{w}, b))/n)_{n\geq 1}$ converges. We then set

$$E(\boldsymbol{w}, b) := \lim_{n\to+\infty} \frac{\log p(n, \boldsymbol{w}, b)}{n}$$

and observe that $0 \leq E(\boldsymbol{w}, b) \leq \log b$, by (4.14).

DEFINITION 4.18. Let $b \geq 2$ be an integer. The *entropy* to base b of a real number ξ is the quantity

$$E(\xi, b) = \lim_{n\to+\infty} \frac{\log p(n, \xi, b)}{n}.$$

The set of real numbers ξ such that $E(\xi, b) = 0$ for some $b \geq 2$ has zero Hausdorff dimension; see Exercise 4.5.

DEFINITION 4.19. Let $b \geq 2$ be an integer. The real number ξ is *rich* (or *disjunctive*) to base b if $p(n, \xi, b) = b^n$ for every $n \geq 1$.

Clearly, every real number normal to base b is also rich to base b, but the converse does not hold. The next result, whose proof is left as Exercise 4.6, was established in [278]; see also [342].

THEOREM 4.20. *Let $b \geq 2$ be an integer, and r, s be positive integers. A real number is rich to base b^s if, and only if, it is rich to base b^r.*

By arguing as in the proof of Theorem 4.14, it is easy to show that the real number ξ is rich to base b if, and only if, the sequence $(\xi b^n)_{n\geq 1}$ is dense modulo one.

4.5 Rational approximation to Champernowne-type numbers

The aim of the present section is to investigate the quality of rational approximation to certain real numbers which have been proved to be normal to a given base in Theorem 4.10. The first result in this direction was proved in 1937 by Mahler [462, 463], who established that the Champernowne number $\xi_{\mathbf{c}}$ is transcendental and is not a Liouville number; see [472, p. 136] for an anecdote on Mahler's life. This is a particular instance of a more general result; further extensions of which have been given in [50, 462, 463, 470, 748, 749].

DEFINITION 4.21. Let $b \geq 2$ be an integer. Let a and n be positive integers. We denote by $W_{b,n,a}$ the integer, written in base b, whose digits are the concatenation, in non-decreasing order, of a copies of every positive integer having exactly n digits in base b.

Note that $W_{10,1,2} = 112233445566778899$. The special case $c = 1$ of Theorem 4.22 was established by Mahler [462, 463].

THEOREM 4.22. *Let $b \geq 2$ be an integer and $c \geq 1$ be a real number. Then, the real number $\xi_{b,c}$ whose b-ary expansion is given by*

$$\xi_{b,c} := 0 \cdot W_{b,1,\lfloor c \rfloor} W_{b,2,\lfloor c^2 \rfloor} \ldots W_{b,n,\lfloor c^n \rfloor} \ldots$$

is normal to base b, transcendental, and is not a Liouville number.

Observe that $\xi_{10,1}$ is the Champernowne number $\xi_{\mathbf{c}}$.

The key lemma for Mahler's proof of Theorem 4.22 is the following result.

LEMMA 4.23. *Let $b \geq 2$ be an integer. Let a and n be positive integers. Then, the integer $W_{b,n,a}$ has exactly*

$$D_{b,n,a} := nab^{n-1}(b-1)$$

digits, and

$$W_{b,n,a} = \frac{b^{(a+1)n-1} - b^{n-1} + 1}{(b^n - 1)(b^{an} - 1)} b^{D_{b,n,a}} - \frac{b^{(a+1)n} - b^n + 1}{(b^n - 1)(b^{an} - 1)}. \tag{4.15}$$

PROOF. We follow the proof of Mahler [470]. Since exactly $b^n - b^{n-1}$ integers have n digits in base b, the first assertion is clear. For the second one, observe that

$$\begin{aligned}
W_{b,n,a} &= b^{D_{b,n,a}} \sum_{k=b^{n-1}}^{b^n-1} k(b^{-n} + b^{-2n} + \cdots + b^{-an})\, b^{-an(k-b^{n-1})} \\
&= b^{D_{b,n,a}} \frac{1-b^{-an}}{b^n-1} \sum_{k=0}^{b^n-b^{n-1}-1} (b^{n-1}+k)\, b^{-ank} \\
&= b^{D_{b,n,a}} \frac{1-b^{-an}}{b^n-1} \Bigg(b^{n-1} \frac{b^{-an(b^n-b^{n-1})}-1}{b^{-an}-1} + \\
&\quad \frac{(b^n-b^{n-1}-1)b^{-an(b^n-b^{n-1}+1)}(b^{-an}-1) - b^{-an(b^n-b^{n-1}-1)} + 1}{b^{an}(1-b^{-an})^2} \Bigg) \\
&= b^{D_{b,n,a}} \frac{1-b^{-an}}{b^n-1} \left(\frac{-b^{n-1}}{b^{-an}-1} + \frac{b^{-an}}{(b^{-an}-1)^2} \right) + \\
&\quad \frac{1-b^{-an}}{b^n-1} \left(\frac{b^{n-1}}{b^{-an}-1} + \frac{(b^n-b^{n-1}-1)(b^{-an}-1)-1}{(b^{-an}-1)^2} \right).
\end{aligned}$$

This proves the lemma. □

The important feature of Lemma 4.23 is that all the integers occurring in (4.15), except $b^{D_{b,n,a}}$, are much smaller than $b^{D_{b,n,a}}$. Consequently, the Champernowne number $\xi_{\mathbf{c}}$ can be expressed as a lacunary sum

$$\sum_{n\geq 1} \frac{c_n}{10^{f(n)}},$$

where f is a rapidly increasing integer-valued function and c_n is a *rational number* whose denominator and numerator are small compared to $10^{f(n)}$. As will be clear in the proof of Theorem 4.22, this is the key point for establishing that $\xi_{\mathbf{c}}$ is transcendental.

PROOF OF THEOREM 4.22. It follows from Theorem 4.10 that $\xi_{b,c}$ is normal to base b for $b \geq 2$ and $c \geq 1$. Thus, it only remains to prove that $\xi_{b,c}$ is transcendental and not a Liouville number. For a positive integer n, put

$$r_n = \frac{b^{\lfloor c^{n+1}\rfloor(n+1)+n} - b^n + 1}{(b^{n+1}-1)(b^{\lfloor c^{n+1}\rfloor(n+1)}-1)} - \frac{b^{\lfloor c^n\rfloor n+n} - b^n + 1}{(b^n-1)(b^{\lfloor c^n\rfloor n}-1)}$$

and

$$d_n = (b-1)\sum_{j=1}^{n} j b^{j-1}\lfloor c^j\rfloor. \tag{4.16}$$

We deduce from Lemma 4.26 that

$$\xi_{b,c} = \lim_{N\to+\infty} \xi_{b,c,N},$$

where, for any positive integer N, we have set

$$\xi_{b,c,N} = \frac{b^{\lfloor c\rfloor}}{(b-1)(b^{\lfloor c\rfloor}-1)} + \sum_{n=1}^{N} r_n\, b^{-d_n}.$$

Let $N \geq 2$ be an integer. Define

$$q_N = b^{d_N} \prod_{n=1}^{N+1} (b^n-1)(b^{\lfloor c^n\rfloor n}-1). \tag{4.17}$$

Then, $p_N := q_N\xi_{b,c,N}$ is an integer which may not be coprime with q_N. Observe that the height $H(\xi_{b,c,N})$ of $\xi_{b,c,N}$ (see Definition E.4) is at most equal to q_N. Note that

$$\frac{b-1}{b}\sum_{j=1}^{N} j(bc)^j = \frac{b-1}{c}\,\frac{N(bc)^{N+1}-(N+1)(bc)^N+1}{(bc-1)^2} \tag{4.18}$$

and

$$\sum_{n=1}^{N+1} (n+1)\lfloor c^n \rfloor \le (N+1)(N+2)c^{N+1}. \qquad (4.19)$$

Let ε be a positive real number. The combination of (4.16), (4.17), (4.18) and (4.19) shows that, for N large enough, we have

$$q_N^{bc-\varepsilon} < q_{N+1} < q_N^{bc+\varepsilon}, \quad b^{d_N} < q_N < b^{d_N(1+\varepsilon)}, \qquad (4.20)$$

and

$$\begin{aligned} 0 < |\xi_{b,c} - \xi_{b,c,N}| &\le 2b^{-d_{N+1}} \\ &\le 2q_N^{-(bc-\varepsilon)/(1+\varepsilon)} \le 2H(\xi_{b,c,N})^{-(bc-\varepsilon)/(1+\varepsilon)}. \end{aligned}$$

Except when $b = 2$ and $c = 1$, by taking ε sufficiently small, we deduce from the Roth Theorem E.7 that $\xi_{b,c}$ is transcendental. To get the same conclusion if $b = 2$ and $c = 1$, we apply Ridout's Theorem E.8; see Exercise 4.7.

It only remains for us to prove that $\xi_{b,c}$ is not a Liouville number. We show that, since $\xi_{b,c}$ has many good rational approximations, it cannot be too well approximated by rational numbers. First, note that

$$|\xi_{b,c} - p_N/q_N| \le q_N^{-3/2}$$

and

$$|\xi_{b,c} - p_N/q_N| \ge b^{-d_{N+1}-2} \ge q_N^{-bc-1}, \qquad (4.21)$$

for every sufficiently large integer N. Let p/q be a rational number with q positive and large. Let N be such that $q_{N-1} < (2q)^2 \le q_N$. If $p/q \ne p_N/q_N$, then we have

$$\begin{aligned} |\xi_{b,c} - p/q| &\ge |p/q - p_N/q_N| - |\xi_{b,c} - p_N/q_N| \\ &\ge 1/(qq_N) - q_N^{-3/2} \\ &\ge 1/(2qq_N) \ge q^{-1}q_{N-1}^{-bc-1} \ge q^{-2(bc+2)}. \end{aligned}$$

Otherwise, we get from (4.20) and (4.21) that

$$|\xi_{b,c} - p/q| = |\xi_{b,c} - p_N/q_N| \ge q_N^{-bc-1} \ge q_{N-1}^{-(bc+1)^2} \ge q^{-2(bc+2)^2}.$$

Consequently, for every rational number p/q with q positive and sufficiently large, we have

$$|\xi_{b,c} - p/q| \ge q^{-2(bc+2)^2}.$$

This shows that the irrationality exponent of $\xi_{b,c}$ is finite, thus, $\xi_{b,c}$ is not a Liouville number. □

4.6 Exercises

EXERCISE 4.1. Let $\alpha > 1$ be a real number and $r \geq 2$ be an integer. Prove that if $\xi, \xi\alpha, \xi\alpha^2, \ldots, \xi\alpha^{r-1}$ are all normal to base α^r, then ξ is normal to base α.

EXERCISE 4.2 (*cf.* [494, 725]). Let $b \geq 2$ be an integer and ξ be a real number normal to base b. Prove that $\xi/(b^\ell - 1)$ is normal to base b for every positive integer ℓ. Prove that $r\xi$ is normal to base b for every non-zero rational number r. Deduce that, for every coprime positive integer m, n, the quadratic number $\sqrt{m/n}$ is normal to base b if, and only if, $\sqrt{n/m}$ (resp., $\sqrt{mn}$) is normal to base b.

EXERCISE 4.3. Let $b \geq 2$ be an integer. Prove that a real number ξ is normal to base b if, and only if, ξ is normal to base $-b$.

EXERCISE 4.4. Prove that a real number ξ is rich to an integer base $b \geq 2$ if, and only if, $E(\xi, b) = \log b$.

EXERCISE 4.5. Prove that the set of real numbers whose expansion to some integer base has zero entropy has Hausdorff dimension zero.

EXERCISE 4.6. Prove Theorem 4.20.

EXERCISE 4.7. Apply Ridout's Theorem E.8 to prove that the number $\xi_{2,1}$ defined in Theorem 4.22 is transcendental.

4.7 Notes

▷ Long [449] proved that ξ is normal to base b if, and only if, there exist positive integers $m_1 < m_2 < \ldots$ such that ξ is simply normal to all of the bases b^{m_i}, $i \geq 1$. No finite set of m_i suffices.

▷ Let $b \geq 2$ be an integer. Hertling [344] proved that, if r and s are positive integers such that $s \geq 2$ and r does not divide s, then there are uncountably many real numbers which are simply normal to base b^r but not simply normal to base b^s. The set of real numbers having the latter property has full Hausdorff dimension [155].

▷ An alternative proof of Theorem 4.4, using automata and results from symbolic dynamics, was given by Blanchard [104]. He also proved that real numbers which are near normality to base b^r (resp., b) are also near normality to base b (resp., b^r); see [105] for further results.

▷ According to Hanson [332], a real number ξ is quasi-normal to base b if every number derived from the b-ary expansion of ξ by selecting those digits whose positions form an arithmetic progression is simply normal to base b. He proved that every number normal to base b is also quasi-normal to base b, but the converse does not hold.

▷ In his Ph.D. thesis, Wall [725] proved that the product of a non-zero normal number by a non-zero rational number and the sum of a normal number and a rational number are normal. This result was established independently in [185], and later reproved by Doty, Lutz and Nandakumar [229] by means of totally different methods. Maxfield [492] proved that every non-zero real number is the sum (resp. product) of two normal numbers. Consequently, the set of numbers normal to base b is not closed under multiplication.

▷ Let $b \geq 2$ be an integer. Let $\mathcal{N}(b)$ denote the set of real numbers normal to base b. Wall's result can be reformulated as $\mathcal{N}(b) + r = \mathcal{N}(b)$, for every rational number r. It is quite obvious that there exist irrational numbers γ such that $\mathcal{N}(b) + \gamma = \mathcal{N}(b)$ (some Liouville numbers for example). Non-trivial examples of such γ's were given by Spears and Maxfield [659]. A characterization of those γ's, given by Rauzy [604], has been used by Bernay [77] to show that they form a set of zero Hausdorff dimension.

▷ The hot spot lemma has been improved by Bailey and Misiurewicz [57]; see also [210]. Piatetski-Shapiro [560] improved Theorem 4.6; see also [529].

▷ A proof of Theorem 4.9 was also given by Pillai [562] (the arguments in [561] are not correct). At the end of [561], he asked whether the number 0.248163264128... formed of the increasing sequence of powers of 2 is normal to base ten. This question remains open.

▷ Theorem 4.9 was subsequently generalized by several authors. Davenport and Erdős [215] proved that, if $f(X)$ is a polynomial taking positive integer values at positive integers, then the decimal $0 \cdot f(1)f(2)\ldots$ is normal to base ten. For $f(X) = X^2$, this was proved by Besicovitch [93], who in fact established that the squares of almost all integers are (ε, k)-normal. A common generalization of the result of Davenport and Erdős and of Corollary 4.11 was obtained by Nakai and Shiokawa [537]. The result of Davenport and Erdős was extended by Madritsch, Thuswaldner and Tichy [459] to numbers generated by the values of entire functions; see also [534–536, 683] and [436, 437] for higher-dimensional generalizations. Schiffer [621] also generalized the construction of Davenport and Erdős to $0{\cdot}(f(1)+d_1)(f(2)+d_2)\cdots$, with $f(X)$ as above and $(d_n)_{n\geq 1}$ being a bounded sequence of positive rational numbers. Discrepancy estimates are given in [232, 536, 621], where among other results it is shown that there exists a positive constant c such that $D_N((\xi_{\mathbf{c}}10^n)_{n\geq 1}) \geq c/(\log N)$ for every $N \geq 2$.

▷ For any integer $b \geq 2$ and any positive real number $\theta < 1$, Shiokawa [649] constructed an increasing sequence $(c_j)_{j\geq 1}$ of positive integers satisfying $x^\theta < \sum_{j:c_j\leq x} 1 < b^2 x^\theta$, for every sufficiently large x, and such that the real number $0 \cdot (c_1)_b (c_2)_b (c_3)_b \ldots$ is not normal to base b. This shows that the growth condition in Theorem 4.10 is essentially best possible.

▷ Lehrer [423] described a game between two players which can be used to construct normal numbers to a given base.

▷ An alternative proof of the theorem of Copeland and Erdős was given in [330].

▷ Properties of some dyadic Champernowne-type numbers have been studied in [226, 650].

▷ Further constructions of real numbers normal to a given base have been given by De Koninck and Kátai [219–222].

▷ Let $b \geq 2$ be an integer. Gál and Gál [317] proved that the discrepancy $D_N((\xi b^n)_{n\geq 1})$ is $O((N^{-1}\log\log N)^{1/2})$ for almost all real numbers ξ. Levin [433] constructed explicitly real numbers ξ for which $D_N((\xi b^n)_{n\geq 1}) = O(N^{-1}(\log N)^2)$. In view of Theorem 1.13, this is close to best possible. Earlier and related works include [400, 412, 413, 426–428, 431, 435, 586, 616, 617, 737].

▷ For small variations around Theorem 4.14, see [648].

▷ Two criteria of non-uniform distribution modulo one for sequences $(\xi b^n)_{n\geq 1}$, where b is an integer with $|b| \geq 2$ and ξ is a real number, are given in [530].

▷ For $k \geq 1$, let non-negative real numbers $\mu(d_1, \ldots, d_k)$ be defined for each k-tuple $(d_1, \ldots, d_k)$ in $\{0, 1\}^k$. Assume that these numbers satisfy $\mu(0) + \mu(1) = 1$ and

$$\begin{aligned}\mu(d_1, \ldots, d_k) &= \mu(0, d_2, \ldots, d_k) + \mu(1, d_2, \ldots, d_k)\\ &= \mu(d_1, \ldots, d_{k-1}, 0) + \mu(d_1, \ldots, d_{k-1}, 1),\end{aligned}$$

for $k \geq 2$. Ville [711] (see also [584]) has contructed explicitly real numbers ξ whose binary expansion $\xi = \sum_{k\geq 1} a_k/2^k$ is such that

$$\lim_{N\to+\infty} \frac{A_2(d_1 \dots d_k, N, \xi)}{N} = \mu(d_1, \dots, d_k),$$

for each k-tuple $(d_1, \dots, d_k)$ in $\{0,1\}^k$, with $k \geq 1$; see also [516].

▷ For an integer $b \geq 2$ and a real number ξ, denote by $V(\xi, b)$ the collection of all limit points (in the weak-$*$ topology) of the sequence $(\nu_n)_{n\geq 1}$ of probability measures defined by

$$\nu_n(f) = \frac{1}{n} \sum_{j=0}^{n-1} f(b^j\xi), \qquad f \in C^0(\mathbb{T}).$$

The set $V(\xi, b)$ is a non-empty closed and connected subset of the set $I(b)$ of all probability measures on $\mathbb{T}$ invariant under the transformation $T_b : x \mapsto bx$. Conversely, Colebrook [200] established that, given a non-empty closed and connected subset V of $I(b)$, there always exists a number ξ such that $V(\xi, b) = V$. This extends an earlier result of Piatetski-Shapiro [559], who dealt with the case where V is reduced to one probability measure; see also [269]. Volkmann [718] established that $V(m\xi/n, b) = T_n^{-1} T_m V(\xi, b)$ holds for every real number ξ and every positive integer m, n.

▷ Let $b \geq 2$ be an integer. For a real number ξ and an integer $n \geq 1$, let $p_{b,n}(\xi)$ be the point in the simplex

$$H_b := \{0 \leq x_0, x_1, \dots, x_{b-1} \leq 1 : x_0 + x_1 + \dots + x_{b-1} = 1\}$$

with coordinates

$$(A_b(0, n, \xi)/n, A_b(1, n, \xi)/n, \dots, A_b(b-1, n, \xi)/n).$$

Denote by $V_b(\xi)$ the set of limit points of the sequence $(p_{b,n}(\xi))_{n\geq 1}$. Volkmann [714] proved that, given any closed, connected set S included in H_b, there exist real numbers ξ such that $V_b(\xi) = S$.

▷ Pellegrino [555] established that if $0 \cdot a_1 a_2 a_3 \dots$ is normal to a given integer base, then the real number $0 \cdot a_1 a_1 a_2 a_1 a_2 a_3 \dots$, whose sequence of digits is the concatenation of all blocks $a_1 \dots a_n$, has the same property. Various modifications of normal numbers (including by transducers) have been considered in [25, 51, 266, 267, 682, 719].

▷ A subset S of the set of finite words on $\{0,1\}$ is called a selection rule. For a selection rule S and a real number ξ whose binary expansion reads $0 \cdot a_1 a_2 \ldots$, we define $\xi_S := 0 \cdot a_{t_1} a_{t_2} \ldots$, where $\{t_1 < t_2 < \ldots\} = \{i : a_1 a_2 \ldots a_{i-1} \in S\}$. We call the real number ξ_S proper if $\limsup_{j \to +\infty} t_j/j < +\infty$. A selection rule S is said to preserve normality if for any normal number ξ such that ξ_S is proper, ξ_S is also a normal number. Two kinds of selection rules have been considered: oblivious ones, which correspond to increasing sequences $(n_j)_{j \geq 1}$ for which $\xi_S = 0 \cdot a_{n_1} a_{n_2} \ldots$, and selection rules that depend on the input sequence. Kamae [359, 362] (see also [340]) proved that an oblivious selection rule preserves normality if, and only if, $(n_j)_{j \geq 1}$ is completely deterministic in the sense of Weiss (see Chapters 1 and 8 of [730]). Agafonov [24] (see also [638]) established that selection by a finite automaton (or, equivalently, by a regular language) preserves normality, a result later extended by Kamae and Weiss [362]. Merkle and Reimann [507] have shown that Agafonov's result cannot be extended to certain more complicated classes of languages.

▷ Mendès France [496] proved that, for every real number ξ normal to base 2 and whose binary expansion is given by $0 \cdot a_1 a_2 \ldots$, for every finite sequence $\ell_1 < \ldots < \ell_m$ of positive integers, the number $0 \cdot b_1 b_2 \ldots$, with $1 - 2b_k = \prod_{i=1}^{m} (1 - 2a_{k+\ell_i})$ for $k \geq 1$, is also normal to base 2. An extension to any integer base $b \geq 2$ is given in [498]. Some special non-normal numbers are considered in [497, 498]; see also [325].

▷ Let $b \geq 2$ be an integer. Mauduit and Moreira [489, 490] computed the generalized Hausdorff dimensions of sets of real numbers having zero entropy to base b. They subsequently considered the case of positive entropy [491].

▷ Amou [49] studied the approximation to the Champernowne number by algebraic numbers of bounded degree. It follows from a result established in [10] that the real numbers defined in Theorem 4.22 cannot be U-numbers (see Definition E.13).

▷ Let $b \geq 2$ be an integer. Slivka and Severo [657] studied various ways to decompose the set of numbers in $[0,1]$ which are simply normal to base b.

▷ Let I be an interval in $[0,1)$. For ξ in $[0,1)$ and an integer $n \geq 1$, put $d_n(\xi) = 1$ if an odd number of the real numbers $\{\xi\}, \{2\xi\}, \ldots, \{2^{n-1}\xi\}$ are lying in I, and put $d_n(\xi) = 0$ otherwise. We say that ξ is a normal number mod 2 with respect to I if $(d_1(\xi) + \cdots + d_N(\xi))/N$

tends to $1/2$ as N tends to infinity. If I is not the interval $(1/5, 5/6)$, it is proved in [187] that almost every ξ in $[0, 1)$ is a normal number mod 2 with respect to I. Furthermore, if $I = (1/6, 5/6)$, then, for almost all ξ in $[1/3, 2/3]$ (resp. in $[0, 1/3] \cup [2/3, 1]$) the above arithmetic mean tends to $2/3$ (resp. to $1/3$).

▷ Let $b \geq 2$ be an integer. Ki and Linton [381] proved that the set of real numbers simply normal to base b and the set of real numbers normal to base b are $\mathbf{\Pi}^0_3$-complete. It is an open problem to decide whether the set of real numbers which are normal to at least one integer base is $\mathbf{\Sigma}^0_4$-complete; see [376] for an introduction to the Borel hierarchy.

▷ Šalát [618] proved that the set of simply normal numbers and the set of real numbers which are normal to every integer base are of the first Baire category (that is, are meagre sets). His result has been extended in many different directions; see [35, 36, 350, 545, 642] and the references quoted therein. Pushkin and Rakhmatullina [593] proved that, for $b \geq 2$ being given, the set of real numbers ξ such that each digit $0, 1, \ldots, b-1$ occurs infinitely many times in the b-ary expansion of ξ is a set of second Baire category.

▷ Maxfield [494] called the real d-tuple $(\xi_1, \ldots, \xi_d)$ a normal d-tuple to base b if the sequence $((\xi_1 b^n, \ldots, \xi_d b^n))_{n \geq 1}$ is uniformly distributed modulo one in $\mathbb{R}^d$.

▷ Postnikov [582] gave a geometric form to Champernowne's construction and, for a given Gaussian integer $a + \mathrm{i}b$ with $ab \neq 0$, he constructed a complex number $\alpha + \mathrm{i}\beta$ such that the sequence of fractional parts $(\{(\alpha + \mathrm{i}\beta)(a + \mathrm{i}b)^n\})_{n \geq 1}$ is uniformly distributed in $[0, 1]^2$ (it is understood that $\{x + \mathrm{i}y\} = \{x\} + \mathrm{i}\{y\}$ and that $\mathbb{C}$ is identified with $\mathbb{R}^2$); see also Polosuev [577]. Explicit constructions of normal numbers in matrix number systems have been given by Madritsch [457, 458].

▷ A set S of real numbers is called a normal set (*ensemble normal* in French) if there exists a real sequence $(t_n)_{n \geq 1}$ such that $(\xi t_n)_{n \geq 1}$ is uniformly distributed modulo one if, and only if, ξ is in S. By Theorem 4.14, for any integer $b \geq 2$, the set of real numbers which are normal to base b is a normal set. The study of normal sets was originated by Mendès France [499, 500], who proved that the set of transcendental numbers is a normal set, and continued by Y. Meyer [508–510], Rauzy [602, 603], Dress [231], Zame [746], Mauduit [485, 486], J.-P. Borel [117–119], Coquet [204], Watson [727], among others.

5

Further explicit constructions of normal and non-normal numbers

In Section 4.2, we have constructed explicitly real numbers which are normal to a given base. In the first section of this chapter, we describe another class of explicit real numbers with the same property. Then, we discuss the existence of explicit examples of absolutely normal numbers.

DEFINITION 5.1. A real number is called *absolutely normal* if it is normal to every integer base $b \geq 2$. A real number is called *absolutely non-normal* if it is normal to no integer base $b \geq 2$.

We briefly and partially mention in Section 5.2 the point of view of complexity and calculability theory. Then, in Section 5.3, we give an explicit example of an absolutely non-normal irrational number. We end this chapter with some words on a method proposed by Bailey and Crandall to investigate the random character of arithmetical constants.

5.1 Korobov's and Stoneham's normal numbers

In 1946 Good [324] introduced the so-called 'normal recurring decimals'. Integers $b \geq 2$ and $k \geq 1$ being given, he constructed rational numbers ξ whose b-ary expansion has period b^k and is such that every sequence of k digits from $\{0, 1, \ldots, b-1\}$ occurs in the b-ary expansion of ξ with the same frequency b^{-k}. An example with $b = 2$ and $k = 3$ is given by the rational 23/255 with purely periodic binary expansion of period 00010111. A similar result was independently proved by de Bruijn [143] also in 1946. A few years later Korobov [396, 397] (see also [584]) considered the normal recurring decimals of Good from a different point of view and constructed by a different method what he called a 'normal periodic system'. We refer the reader to the addendum of [674] for a discussion showing that the constructions of the normal recurring decimals of Good and the normal periodic systems of Korobov have actually been studied by many authors since a paper of Flye Sainte-Marie [309] published in 1894.

Stoneham [667] considered the b-ary expansion of negative powers p^{-n} of an odd prime p, assuming that the integer $b \geq 2$ is a primitive root modulo p^2, and showed that the distribution of the digits in the recurring part of the period is (ε, k)-normal in the sense of Besicovitch [93] (see Definition 4.12) when n is large. Subsequently [669–675], he extended his results to broad classes of rational numbers and applied them to construct normal numbers. For instance, in 1973, he proved [672] that the real number

$$\xi_{S,2,3} := \sum_{j \geq 1} \frac{1}{3^j 2^{3^j}}$$

is normal to base 2. The aim of this section is to establish this result and several of its extensions.

We follow the method developed by Bailey and Crandall [56] to establish Theorem 4.8 of [56], which we reproduce below.

THEOREM 5.2. *Let b and c be coprime integers, both at least equal to 2. Let $(m_j)_{j\geq 1}$ and $(n_j)_{j\geq 1}$ be increasing sequences of positive integers such that $(n_{j+1} - n_j)_{j\geq 1}$ is non-decreasing and there exist $\gamma > 1/2$ and an integer j_0 with*

$$\frac{m_{j+1} - m_j}{c^{\gamma n_{j+1}}} \geq \frac{m_j - m_{j-1}}{c^{\gamma n_j}} \quad \text{for } j \geq j_0. \tag{5.1}$$

Then, the real number

$$\xi = \sum_{j \geq 1} \frac{1}{c^{n_j} b^{m_j}} \tag{5.2}$$

is normal to base b.

We point out a particular case of Theorem 5.2.

COROLLARY 5.3. *Let b and c be coprime integers, both at least equal to 2. Let $d \geq 2$ be an integer. Then, the Stoneham number*

$$\xi_{S,b,c} := \sum_{j \geq 1} \frac{1}{c^j b^{c^j}} \tag{5.3}$$

and the Korobov number

$$\xi_{K,b,c,d} := \sum_{j \geq 1} \frac{1}{c^{d^j} b^{c^{d^j}}} \tag{5.4}$$

are normal to base b.

To show that Corollary 5.3 follows from Theorem 5.2, it is sufficient to note that for the real numbers defined by (5.3) and (5.4) the inequality (5.1) is satisfied with $\gamma = 2/3$.

Korobov [395] proved that the numbers $\xi_{K,b,c,d}$ defined in Corollary 5.3 are normal to base b. For the sake of simplicity, we content ourselves to establish Theorem 5.2 when c is an odd prime number. Its proof differs greatly from that of Theorem 4.10 and involves exponential sums.

The strategy of the proof is inspired by the dynamical point of view developed by Bailey and Crandall [55] and briefly presented in Section 5.4. With a number of the form (5.2), we associate the sequence of rational numbers $(x_n)_{n\geq 0}$ defined by $x_0 = 0$ and, for $n \geq 1$, by

$$x_n = \{bx_{n-1}\}, \quad \text{if } n \text{ is not an element of } (m_j)_{j\geq 1},$$

and

$$x_{m_j} = \left\{bx_{m_j-1} + \frac{1}{c^{n_j}}\right\}, \quad \text{for } j \geq 1.$$

For $j \geq 1$, setting $a_j := x_{m_j} c^{n_j}$, we see that $a_1 = 1$,

$$\begin{aligned} x_0 = \cdots = x_{m_1-1} = 0, \quad x_{m_1} = \frac{a_1}{c^{n_1}}, x_{m_1+1} = \left\{\frac{ba_1}{c^{n_1}}\right\}, \ldots, \\ x_{m_2-1} = \left\{\frac{b^{m_2-m_1-1}a_1}{c^{n_1}}\right\}, \ldots, x_{m_j} = \frac{a_j}{c^{n_j}}, x_{m_j+1} = \left\{\frac{ba_j}{c^{n_j}}\right\}, \\ \ldots, x_{m_{j+1}-1} = \left\{\frac{b^{m_{j+1}-m_j-1}a_j}{c^{n_j}}\right\}, \ldots, \end{aligned} \tag{5.5}$$

and $a_{j+1} = b^{m_{j+1}-m_j} c^{n_{j+1}-n_j} a_j + 1$, for $j \geq 1$.

LEMMA 5.4. *Let ξ be given by (5.2) and let $(x_n)_{n\geq 0}$ be the sequence associated to ξ as defined above. Then, ξ is normal to base b if, and only if, the sequence $(x_n)_{n\geq 0}$ is uniformly distributed in $[0,1]$.*

PROOF. Put $m_0 = 0$. For $m \geq m_1$, let j be the index defined by $m_j \leq m < m_{j+1}$ and observe that

$$\xi b^m = \sum_{h=1}^{j} \frac{b^{m-m_h}}{c^{n_h}} + \sum_{h\geq j+1} \frac{b^{m-m_h}}{c^{n_h}}$$

and

$$x_m = \sum_{h=1}^{j} \frac{b^{m-m_h}}{c^{n_h}},$$

thus,

$$0 \leq \xi b^m - x_m \leq \frac{1}{c^{n_j}} \left(\frac{1}{b} + \frac{1}{b^2} + \cdots\right) \leq \frac{1}{c^{n_j}}.$$

Since n_j tends to infinity with j, it follows from Exercise 1.1 that the sequence $(\xi b^m)_{m\geq 1}$ is uniformly distributed modulo one if, and only if, the sequence $(x_n)_{n\geq 0}$ is uniformly distributed in $[0,1]$. This proves the lemma. □

The key tool for the proof of Theorem 5.2 is an estimate for exponential sums. We need two auxiliary lemmas.

Let p be an odd prime number and n a positive integer. Let $b\geq 2$ be an integer not divisible by p. The next lemma shows that the rational numbers $1/p^n, b/p^n, b^2/p^n, \ldots, b^{\mathrm{ord}(b,p^n)-1}/p^n$ are quite well distributed modulo one. This is a particular case of [401, Lemma 2]; see also [402].

LEMMA 5.5. *Let p be an odd prime number and $b\geq 2$ be an integer coprime with p. Let h be a non-zero integer and n a positive integer. Set $d=\gcd(h,p^n)$. If $d=1$ or $d<\mathrm{ord}(b,p^n)/\mathrm{ord}(b,p)$, then, for any integer $J=0,1,\ldots,\mathrm{ord}(b,p^n)$, we have*

$$\Bigl|\sum_{j=0}^{J-1} \mathrm{e}^{2i\pi h b^j/p^n}\Bigr| < \sqrt{\frac{p^n}{d}}\Bigl(1+\log\frac{p^n}{d}\Bigr).$$

PROOF. Set $\tau=\mathrm{ord}(b,p^n)$. For an integer f and a positive integer a coprime with p^n, consider the sum

$$\sigma(a,f)=\sum_{j=0}^{\tau-1}\mathrm{e}^{2i\pi(ab^j/p^n+fj/\tau)}.$$

From the equality between fractional parts

$$\Bigl\{\frac{ab^{j+\tau}}{p^n}+\frac{f(j+\tau)}{\tau}\Bigr\}=\Bigl\{\frac{ab^j}{p^n}+\frac{fj}{\tau}\Bigr\},$$

we get that, for any integer ℓ,

$$\sigma(a,f)=\sum_{j=0}^{\tau-1}\mathrm{e}^{2i\pi(ab^{j+\ell}/p^n+f(j+\ell)/\tau)}=\mathrm{e}^{2i\pi f\ell/\tau}\sum_{j=0}^{\tau-1}\mathrm{e}^{2i\pi(ab^{j+\ell}/p^n+fj/\tau)}.$$

Hence,

$$|\sigma(a,f)|=\Bigl|\sum_{j=0}^{\tau-1}\mathrm{e}^{2i\pi(ab^jb^\ell/p^n+fj/\tau)}\Bigr|.$$

Noticing that the integers $ab, ab^2, \dots, ab^\tau$ are pairwise incongruent modulo p^n and coprime with p^n, we deduce that

$$\begin{aligned}\tau|\sigma(a,f)|^2 &= \sum_{\ell=1}^{\tau}\Big|\sum_{j=0}^{\tau-1} \mathrm{e}^{2i\pi(ab^jb^\ell/p^n+fj/\tau)}\Big|^2 \\ &\le \sum_{m=1}^{p^n}\Big|\sum_{j=0}^{\tau-1} \mathrm{e}^{2i\pi(mb^j/p^n+fj/\tau)}\Big|^2, \\ &= \sum_{h,j=0}^{\tau-1}\Big(\mathrm{e}^{2i\pi(f(h-j)/\tau)}\sum_{m=1}^{p^n}\mathrm{e}^{2i\pi(m(b^h-b^j)/p^n)}\Big) = \tau p^n,\end{aligned}$$

whence,

$$|\sigma(a,f)| \le \sqrt{p^n}. \tag{5.6}$$

Let h be a non-zero integer coprime with p^n. For $J = 0, 1, \dots, \tau$, we have

$$\begin{aligned}\Big|\sum_{j=0}^{J-1}\mathrm{e}^{2i\pi hb^j/p^n}\Big| &= \Big|\frac{1}{\tau}\sum_{f=1}^{\tau}\sum_{\ell=0}^{J-1}\mathrm{e}^{-2i\pi f\ell/\tau}\sum_{j=0}^{\tau-1}\mathrm{e}^{2i\pi(hb^j/p^n+fj/\tau)}\Big| \\ &\le \frac{1}{\tau}\sum_{f=1}^{\tau}\Big|\sum_{\ell=0}^{J-1}\mathrm{e}^{-2i\pi f\ell/\tau}\Big| \times \Big|\sum_{j=0}^{\tau-1}\mathrm{e}^{2i\pi(hb^j/p^n+fj/\tau)}\Big|,\end{aligned} \tag{5.7}$$

and, for $f = 1, \dots, \tau - 1$,

$$\Big|\sum_{\ell=0}^{J-1}\mathrm{e}^{-2i\pi f\ell/\tau}\Big| \le \frac{1}{\sin(\pi f/\tau)} \le \frac{1}{2||f/\tau||}. \tag{5.8}$$

Since $\tau \le p^n - 1$, the combination of (5.6), (5.7) and (5.8) gives

$$\begin{aligned}\Big|\sum_{j=0}^{J-1}\mathrm{e}^{2i\pi hb^j/p^n}\Big| &\le \max_{1\le f\le\tau}\Big|\sum_{j=0}^{\tau-1}\mathrm{e}^{2i\pi(hb^j/p^n+fj/\tau)}\Big|(1+\log\tau) \\ &\le \sqrt{p^n}(1+\log p^n).\end{aligned} \tag{5.9}$$

This establishes the lemma when $d = 1$.

Assume now that $d \ge 2$. Set $n' = \min\{n, \mathrm{ord}_p(b^{\mathrm{ord}(b,p)} - 1)\}$ and note that Corollary B.3 asserts that $\mathrm{ord}(b,p^n) = p^{n-n'}\mathrm{ord}(b,p)$. Consequently, our assumption reads $p \le d \le p^{n-n'-1}$ and there exist an integer h_1 coprime with p and an integer m such that $n'+1 \le m \le n-1$ and $h/p^n = h_1/p^m$. Set $\tau_1 = \mathrm{ord}(b, p^{m-1})$ and note that $\tau = \mathrm{ord}(b,p^m) = p\tau_1$. We follow the proof of Theorem 8 on [402, p. 42]. For an arbitrary integer x, set $\delta_{p^m}(x) = 1$ if p^m divides x and $\delta_{p^m}(x) = 0$ otherwise. We have

$$\begin{aligned}\sum_{\ell=1}^{p^m}\Big|\sum_{j=0}^{\tau-1} e^{2i\pi\ell b^j/p^m}\Big|^2 &= \sum_{\ell=1}^{p^m}\sum_{j,k=0}^{\tau-1} e^{2i\pi\ell(b^j-b^k)/p^m}\\ &= p^m\sum_{j,k=0}^{\tau-1}\delta_{p^m}(b^j-b^k) = p^m\tau.\end{aligned}$$

Consequently, we have

$$\begin{aligned}\sum_{\ell=1,(p,\ell)=1}^{p^m}\Big|\sum_{j=0}^{\tau-1} e^{2i\pi\ell b^j/p^m}\Big|^2 &= \sum_{\ell=1}^{p^m}\Big|\sum_{j=0}^{\tau-1} e^{2i\pi\ell b^j/p^m}\Big|^2\\ &\quad - \sum_{\ell=1,(p,\ell)=p}^{p^m}\Big|\sum_{j=0}^{\tau-1} e^{2i\pi\ell b^j/p^m}\Big|^2\\ &= p^m\tau - \sum_{\ell=1}^{p^{m-1}}\Big|\sum_{j=0}^{\tau-1} e^{2i\pi\ell b^j/p^{m-1}}\Big|^2\\ &= p^m\tau - p^2p^{m-1}\tau_1 = 0.\end{aligned}$$

Thus, since p does not divide h_1, we have

$$\sum_{j=0}^{\tau-1} e^{2i\pi h_1 b^j/p^m} = 0,$$

and, setting $\tau_d = \mathrm{ord}(b, p^n/d)$ and recalling that $h/p^n = h_1/p^m$,

$$\sum_{j=0}^{\tau_d-1} e^{2i\pi hb^j/p^n} = \frac{\tau_d}{\tau}\sum_{j=0}^{\tau-1} e^{2i\pi hb^j/p^n} = 0. \tag{5.10}$$

Let J_d be the integer between 1 and τ_d which is congruent to J modulo τ_d. Then, using (5.10) and arguing as in (5.9), we get

$$\Big|\sum_{j=0}^{J-1} e^{2i\pi hb^j/p^n}\Big| = \Big|\sum_{j=0}^{J_d-1} e^{2i\pi hb^j/p^n}\Big| \le \sqrt{\frac{p^n}{d}}\Big(1+\log\frac{p^n}{d}\Big).$$

This completes the proof of the lemma. □

LEMMA 5.6. *Let p be an odd prime and $b \ge 2$ be an integer coprime with p. Let h be a non-zero integer. For any integers $J \ge 1$ and $n \ge 2$, the condition* $\gcd(h, p^n) < b^{-\mathrm{ord}(b,p)}p^n$ *implies*

$$\left| \sum_{j=0}^{J-1} \mathrm{e}^{2i\pi h b^j / p^n} \right| < \left(p^{n/2} + J b^{\mathrm{ord}(b,p)} p^{-n/2} \right) \log p^n. \tag{5.11}$$

PROOF. By Corollary B.3, we have $\mathrm{ord}(b, p^n) \ge \mathrm{ord}(b,p) p^n b^{-\mathrm{ord}(b,p)}$. Thus, Lemma 5.5 implies that the left-hand side of (5.11) is less than $p^{n/2}(1 + \log p^{n/2})$, as long as J does not exceed $\mathrm{ord}(b, p^n)$. But for larger J, we have at most $\lceil J/\mathrm{ord}(b, p^n) \rceil$ copies of the exponential sum, and this ceiling is bounded from above by $1 + J b^{\mathrm{ord}(b,p)} p^{-n}$. Since $1 + \log p^{n/2} \le \log p^n$, this completes the proof of the lemma. □

We now have the material to complete the proof of Theorem 5.2 when c is an odd prime number.

PROOF OF THEOREM 5.2. We assume that c is an odd prime number and leave the general case to the interested reader. Let h be a non-zero integer. Set $\mu_1 = m_1$ and $\mu_j = m_j - m_{j-1}$, for $j \ge 2$. Set $E = \mu_{j_0} c^{-\gamma n_{j_0}}$. Let $k_1 \ge j_0 + 1$ be an integer such that the conclusion of Lemma 5.6 holds for every power c^{n_k} with $k \ge k_1 - 1$. Let N be a sufficiently large integer in order that we can write

$$N = N_0 + \mu_{k_1} + \cdots + \mu_K + J,$$

with $K \ge k_1$, $N_0 = \mu_1 + \cdots + \mu_{k_1 - 1}$ and $1 \le J \le \mu_{K+1}$. We infer from (5.5) and Lemma 5.6 that

$$\begin{aligned} \left| \sum_{j=0}^{N-1} \mathrm{e}^{2i\pi h x_j} \right| \le N_0 + \sum_{k=k_1}^{K} & \left| \sum_{j=0}^{\mu_k - 1} \mathrm{e}^{2i\pi h a_{k-1} b^j / c^{n_{k-1}}} \right| \\ & + \left| \sum_{j=0}^{J-1} \mathrm{e}^{2i\pi h a_K b^j / c^{n_K}} \right| \\ \le N_0 + \sum_{k=k_1}^{K} & \left(c^{n_{k-1}/2} + \mu_k \mathrm{ord}(b,c) c^{-n_{k-1}/2} \right) \log c^{n_{k-1}} \\ & + \left(c^{n_K/2} + J \mathrm{ord}(b,c) c^{-n_K/2} \right) \log c^{n_K} \\ \le N_0 + & \left(K c^{n_{K-1}/2} + \sum_{k=k_1}^{K} \mu_k \mathrm{ord}(b,c) c^{-n_{k-1}/2} \right) \log c^{n_{K-1}} \\ & + \left(c^{n_K/2} + J \mathrm{ord}(b,c) c^{-n_K/2} \right) \log c^{n_K}. \end{aligned}$$

A short calculation shows that (5.1) implies that

$$\sum_{k=1}^{K} \frac{\mu_k}{c^{n_{k-1}/2}} \le \frac{\mu_K}{c^{n_{K-1}/2}} \left(\frac{1}{1 - c^{1/2 - \gamma}} \right).$$

Since, by (5.1) and the definition of E, we have

$$\frac{c^{n_k/2}}{N} \leq \frac{c^{n_k/2}}{\mu_k} \leq \frac{c^{n_k/2} c^{-\gamma(n_k - n_{j_0})}}{\mu_{j_0}} \leq \frac{c^{n_k(-\gamma+1/2)}}{E},$$

for $k = K-1, K$, there exists a positive constant C, independent of N, such that

$$\frac{C}{N} \Big| \sum_{n=0}^{N-1} e^{2i\pi h x_n} \Big| \leq \frac{1}{N} + \big(K c^{n_{K-1}(-\gamma+1/2)} + c^{-n_{K-1}/2}\big) \log c^{n_{K-1}} \\ + \big(c^{n_K/2(-\gamma+1/2)} + c^{-n_K/2}\big) \log c^{n_K}.$$

Since $\gamma > 1/2$ and $(n_j)_{j\geq 1}$ increases to infinity, it follows from Weyl's criterion (Theorem 1.2) that the sequence $(x_n)_{n\geq 1}$ is uniformly distributed in $[0,1]$. Combined with Lemma 5.4, this proves the theorem. □

We present an alternative proof of a particular case of Corollary 5.3, following [57, 120]. This proof shows the strength of Theorem 4.6.

Alternative proof of the normality of $\xi_{S,2,3}$ to base 2. Let us write ξ for $\xi_{S,2,3}$. Then, the sequence $(x_n)_{n\geq 0}$ introduced after Corollary 5.3 is generated by the recursion defined by $x_0 = 0$ and, for $n \geq 1$, by $x_n = \{2x_{n-1} + r_n\}$, where $r_n = 1/n$ if $n = 3^m$ for some integer m and $r_n = 0$ otherwise. In particular, for any positive p, if n is less than 3^{p+1}, then x_n is an integer multiple of $1/3^p$ and, for $j = 0, \ldots, 3^p - 1$, the rational $j/3^p$ occurs exactly three times among the 3^{p+1} numbers $x_0, \ldots, x_{3^{p+1}-1}$.

Observe that, for $n \geq 1$, we have

$$\{\xi 2^n\} = x_n + \sum_{m=\lfloor \log_3 n \rfloor + 1}^{+\infty} \frac{2^{n-3^m}}{3^m}.$$

From

$$0 < \sum_{m=\lfloor \log_3 n \rfloor + 1}^{+\infty} \frac{2^{n-3^m}}{3^m} \leq \frac{1}{2(n+1)} \sum_{m\geq 0} 3^{-m} \leq \frac{3}{4(n+1)},$$

we get that

$$|\{\xi 2^n\} - x_n| \leq \frac{3}{4(n+1)} < \frac{1}{n}. \tag{5.12}$$

Let k be a positive integer and $d_1, \ldots, d_k$ be in $\{0,1\}$. Set $D_k = 2^{k-1}d_1 + \cdots + 2d_{k-1} + d_k$,

$$u = \sum_{j=1}^{k} \frac{d_j}{2^j} = \frac{D_k}{2^k}, \quad \text{and} \quad v = u + \frac{1}{2^k}.$$

Let $N > 4^k$ be an integer. We wish to bound from above the number of indices n between 1 and N such that $\{\xi 2^n\}$ lies in $[u, v)$. In view of (5.12), it is enough to bound from above the number of indices n between 1 and N such that x_n lies in $[u - 1/n, v + 1/n)$.

Define ℓ by the inequalities $3^\ell \le N < 3^{\ell+1}$. Observe that for $n \ge 2^k$ the interval $[u - 1/n, v + 1/n)$ is contained in $[u - 2^{-k}, v + 2^{-k})$. Since the length of this interval is $3 \cdot 2^{-k}$, it contains at most $\lfloor 3^{\ell+1}2^{-k} \rfloor + 1$ integer multiples of $1/3^\ell$. Thus, recalling that x_n is an integer multiple of $1/3^\ell$ if n is less than N, we see that there can be at most three times this many n's less than N and at least equal to 2^k for which x_n lies in $[u - 2^{-k}, v + 2^{-k})$. Consequently, the number of positive integers n less than N such that $\{\xi 2^n\}$ lies in $[u, v)$ is at most equal to

$$2^k + 3(3^{\ell+1}2^{-k} + 1) < 11N2^{-k}.$$

With the notation of Chapter 4, we get that

$$\limsup_{N \to +\infty} \frac{A_2(D_k, N, \xi)}{N} \le \frac{11}{2^k},$$

and we conclude by applying Theorem 4.6. □

As pointed out in [56], the proof of Theorem 5.2 shows that, for coprime integers b and c both ≥ 2, the discrepancy of the sequence $(\xi_{S,b,c} b^n)_{n \ge 1}$ is $O(N^{-1/2}(\log N)^2)$.

We complement Corollary 5.3 by showing that Stoneham numbers are not absolutely normal. Theorem 5.7 extends a result of Bailey reproduced on [120, p. 329]; see also [357, 722].

THEOREM 5.7. *For every integer b and c with $b \ge 2$, $c \ge 2$, the real number $\xi_{S,b,c}$ is not normal to base bc.*

PROOF. The idea consists in showing that the bc-ary expansion of $\xi_{S,b,c}$ has very large blocks of zeros. If $b = c = 2$, then the result is clear. Assume that $bc \ge 6$. Let n be a positive integer. Since $c^{n-j}b^{n-c^j}$ is an integer for $j = 1, \ldots, \lfloor \log_c n \rfloor$, we have

$$\{\xi_{S,b,c}(bc)^n\} = \Big\{ \sum_{j=\lfloor \log_c n \rfloor + 1}^{+\infty} c^{n-j}b^{n-c^j} \Big\}.$$

If $n = c^m$, where $m \geq 1$ is an integer, then we have

$$\{\xi_{S,b,c}(bc)^n\} \leq (cb^{-(c-1)})^{c^m} \left(1 + \frac{1}{c} + \frac{1}{c^2} + \cdots\right) \leq 2(cb^{-(c-1)})^{c^m}.$$

Since $b \geq 3$ or $c \geq 3$, we have $cb^{-(c-1)} < 1$. This implies that, after the c^mth digit in the bc-ary expansion of ξ, there are at least δc^m consecutive digits 0, for some positive real number δ which does not depend on m. With the notation of Chapter 4, this means that $A_{bc}(0, c^m, \xi) + \lfloor \delta c^m \rfloor = A_{bc}(0, c^m + \lfloor \delta c^m \rfloor, \xi)$. This implies that $A_{bc}(0, N, \xi)/N$ cannot have a limit when N tends to infinity. Consequently, $\xi_{S,b,c}$ is not normal to base bc. □

5.2 Absolutely normal numbers

In 1917, Sierpiński [654] gave an alternative proof of Borel's Theorem 4.8 asserting that almost all real numbers are absolutely normal (note that he used, without proof, the fact that a real number is absolutely normal if it is simply normal to every integer base); see also the paper of Lebesgue [421], written in 1909, but published in 1917. Sierpiński and Lebesgue explained how to construct infinite sets of intervals in order to determine effectively one absolutely normal number, defined as the minimum of an uncountable set. This answers a question of Borel, but not in the most satisfactory way, since it would be desirable to show that *un nombre irrationnel connu est absolument normal.*

In 1962 Schmidt [628] gave a complicated effective construction of absolutely normal numbers.

In 1979 Levin [428] constructed explicitly an absolutely normal number ξ whose discrepancy to every integer base $b \geq 2$ satisfies

$$D_N((\xi b^n)_{n\geq 1}) \leq 10^7 N^{-1/2} (\log N)^3,$$

for every N sufficiently large in terms of b. His construction is unfortunately much too complicated to be reproduced here.

Turing [689] formalized the concept of computability in 1936, about 20 years after the publication of the papers of Sierpiński and Lebesgue. The computable numbers are those whose expansion in some integer base can be generated by a mechanical (finitary) method, outputting each of the digits, one after the other. In an unpublished note, Turing [690] gave a computable construction to establish Borel's Theorem 4.8 together with an algorithm for computing absolutely normal numbers; see [67] and [69] for a comprehensive exposition of Turing's ideas. We insist on the fact that the set of computable real numbers is countable, so there is no evidence for the existence of computable absolutely normal numbers.

Becher and Figueira [68] gave a recursive reformulation of Sierpiński's construction which produces a computable absolutely normal number, together with an algorithm to compute it in doubly exponential time complexity.

Various notions of randomness are discussed in [97, 173, 230, 387, 540]. A sequence is random if it passes all conceivable effectively testable properties of stochasticity. Kolmogorov [343, 392] and Martin-Löf [480] formalized this approach. Further equivalent definitions are summarized in [173, Theorem 6.99]. If $b \geq 2$ is an integer and $(a_k)_{k\geq 1}$ is a random sequence on $\{0, 1, \ldots, b-1\}$, then the real number with b-ary expansion $0 \cdot a_1 a_2 \ldots$ is called a random number. According to [173, Theorem 6.61], this random number is normal to base b. The converse, however, does not hold. Since the sequence 123456789101112... is computable, it is not random, thus the Champernowne number $\xi_{\mathbf{c}} :=$ 0.12345678910111213..., which is normal to base 10, is not a random number. Furthermore, Calude and Jürgensen [175] proved that the notion of random number is base invariant; see also Staiger [660].

A further example of an absolutely normal but not computable number is given by a Chaitin's random number Ω, the halting probability of a universal machine. Since it is random, it is a transcendental number. A procedure to compute the exact values of the first 64 binary digits of a Chaitin's Ω number is given in [174].

A simple construction of an absolutely normal number remains a challenging open problem. No simple explicit example of a real number which is normal to two multiplicatively independent bases is known.

5.3 Absolutely non-normal numbers

Maxfield [494] has pointed out that the set of absolutely non-normal numbers is uncountable and dense. Martin [478] was the first to give an explicit example of an absolutely non-normal irrational number. We reproduce his nice construction.

THEOREM 5.8. *Set* $d_2 = 4$ *and*

$$d_j = j^{d_{j-1}/(j-1)}, \quad \text{for } j \geq 3.$$

Then, the real number

$$\xi := \prod_{j=2}^{+\infty} \left(1 - \frac{1}{d_j}\right) \tag{5.13}$$

is a Liouville number which is simply normal to no base.

PROOF. Since $d_j \geq j^2$ for every $j \geq 2$, the sum $\sum_{j\geq 2} 1/d_j$ and the infinite product (5.13) also converge. Let $j \geq 5$ be an integer. We deduce from $j \leq \sqrt{d_j}$ that

$$d_{j+1} > 5^{d_j/j} \geq 5^{\sqrt{d_j}}.$$

Using the fact that $5^x \geq x^5$ for $x \geq 5$, we conclude that

$$d_{j+1} > (\sqrt{d_j})^5 > 2d_j^2. \tag{5.14}$$

For $k \geq 2$, set

$$\xi_k := \prod_{j=2}^{k} \left(1 - \frac{1}{d_j}\right),$$

and observe that $\xi < \xi_k$ and

$$\begin{aligned} \xi = \xi_k \cdot \prod_{j=k+1}^{+\infty} \left(1 - \frac{1}{d_j}\right) &\geq \xi_k \left(1 - \sum_{j=k+1}^{+\infty} \frac{1}{d_j}\right) \\ &> \xi_k \left(1 - \sum_{j=k+1}^{+\infty} \frac{1}{2^{j-k-1} d_{k+1}}\right) \\ &= \xi_k \left(1 - \frac{2}{d_{k+1}}\right) > \xi_k - \frac{2}{d_{k+1}}, \end{aligned} \tag{5.15}$$

since $d_j > 2d_{j-1}$ and $\xi_k < 1$.

Let $k \geq 2$ be an integer and write

$$\begin{aligned} d_{k+1}\xi_{k+1} &= (d_{k+1} - 1)\xi_k \\ &= \left((k+1)^{d_k/k} - 1\right)\xi_k = \frac{(k+1)^{d_k/k} - 1}{d_k} d_k \xi_k. \end{aligned} \tag{5.16}$$

Noticing that d_k is a power of k, it follows from (5.16) and Corollary B.4 that $d_{k+1}\xi_{k+1}$ is an integer multiple of $d_k\xi_k$. Since $d_2\xi_2 = 3$ is an integer, we deduce that

$$d_k\xi_k \text{ is an integer for } k \geq 2. \tag{5.17}$$

We now prove that, for every $b \geq 2$, the real number ξ is slightly less than but very close to the rational number ξ_b, whose denominator is a power of b. Consequently, the b-ary expansion of ξ has a very long string of digits equal to $b-1$.

Let $b \geq 2$ and r be positive integers. By (5.17), the number $d_{b^r}\xi_{b^r}$ is an integer, thus, the b-ary expansion of ξ_{b^r} terminates after at most

$$\log_b d_{b^r} = rd_{b^r-1}/(b^r - 1)$$

digits. Furthermore, ξ is less than ξ_{b^r}, but, by (5.15), the difference is less than $2/d_{b^r+1}$. Observe that

$$\frac{2}{d_{b^r+1}} \leq \frac{2}{b^{rd_{b^r}/b^r}} \leq b^{1-rd_{b^r}/b^r}.$$

Subtracting this small difference from ξ_{b^r}, the resulting b-ary expansion has occurrences of the digit $b-1$ beginning at the $(rd_{b^r-1}/(b^r-1)+1)$th digit at the latest, and continuing through at least the $(rd_{b^r}/b^r - 1)$th digit. With the notation from Chapter 4, we have

$$\begin{aligned} A_b\Big(b-1, \frac{rd_{b^r}}{b^r}, \xi\Big) &\geq \frac{rd_{b^r}}{b^r} - \frac{rd_{b^r-1}}{b^r-1} - 1 \\ &> \frac{rd_{b^r}}{b^r} - \frac{2rd_{b^r-1}}{b^r} > \frac{rd_{b^r}}{b^r}\Big(1 - \frac{2}{\sqrt{d_{b^r}}}\Big), \end{aligned}$$

by (5.14). Consequently,

$$\limsup_{N \to +\infty} \frac{A_b(b-1, N, \xi)}{N} = 1,$$

and ξ is not simply normal to base b. Since $b \geq 2$ is arbitrary, this shows that ξ is absolutely non-normal. Furthermore, we have shown that, for any positive integer w, there are arbitrarily large integers N such that the first N digits of ξ are followed by more than wN digits $b-1$. This shows that the irrationality exponent of ξ is infinite, meaning that ξ is a Liouville number. Liouville's Theorem E.5 then implies that ξ is a transcendental number. □

5.4 On the random character of arithmetical constants

In this section, we briefly discuss a theory developed by Bailey and Crandall [55] to explain random behaviour for the digits in the integer expansions of fundamental mathematical constants. At the core of their approach is the following general hypothesis concerning the distribution of the iterates generated by dynamical maps.

HYPOTHESIS A. *Let $b \geq 2$ be an integer. Let $p(X)$ and $q(X)$ be integer polynomials such that $0 \leq \deg p(X) < \deg q(X)$ and $q(X)$ has no non-negative integer roots. Define the sequence $(y_n)_{n \geq 0}$ by setting $y_0 = 0$ and*

$$y_{n+1} = \left\{ b y_n + \frac{p(n)}{q(n)} \right\}, \quad \textit{for } n \geq 0.$$

Then this sequence either has finitely many limit points or is uniformly distributed in $[0, 1]$.

Bailey and Crandall [55] established that, assuming the validity of Hypothesis A, then the real number

$$\xi := \sum_{n \geq 0} \frac{p(n)}{q(n)} b^{-n} \tag{5.18}$$

either is rational or is normal to base b. In particular, if Hypothesis A is true, then each of the constants π, $\log 2$ and $\zeta(3)$ is normal to base 2, and $\log 2$ is normal to base 3. To be even more precise, if one could establish that the simple iteration given by $y_0 = 0$ and

$$y_{n+1} = \left\{ 2 y_n + \frac{1}{n+1} \right\}, \quad \text{for } n \geq 0,$$

is uniformly distributed in $[0, 1]$, then it would follow that $\log 2$ is normal to base 2.

A real number ξ having an expansion (5.18) is called in [417] a *BBP-number*, and the associated formula (5.18) is then called a BBP-expansion of ξ to base b. This refers to the algorithm discovered by Bailey, Borwein and Plouffe [54] by which one can rapidly calculate individual digits of certain polylogarithmic constants, including π, written in some base.

Proving Hypothesis A appears intractable.

Lagarias [417] showed that the relation between particular orbits of two discrete dynamical systems underlying Hypothesis A also applies to expansions of real numbers of the form $\sum_{n \geq 1} \varepsilon_n / b^n$, with $(\varepsilon_n)_{n \geq 1}$ a sequence of real numbers tending to 0 as n tends to infinity. However, since every real number has such an expansion, Hypothesis A cannot be true in such generality and thus must be restricted to apply only to expansions of some special form. At the end of [417], there is an interesting discussion on Hypothesis A and Furstenberg's conjecture mentioned as Problem 10.26.

Note that Lemma 5.4 is a particular case of a general result given in Exercise 5.2. It is motivated by Hypothesis A.

The interested reader is advised to consult [55, 56, 417] and the monograph [120].

5.5 Exercises

EXERCISE 5.1. Let b and c be coprime integers, both ≥ 2. Let $d \geq 2$ be an integer. Prove that the real numbers $\xi_{S,b,c}$ and $\xi_{K,b,c,d}$ are transcendental. Are they Liouville numbers?

EXERCISE 5.2. Let $b \geq 2$ be an integer. Let $(\varepsilon_n)_{n\geq 0}$ be a converging sequence of real numbers. Set $x_0 = 0$ and $x_{n+1} = \{bx_n + \varepsilon_n\}$ for $n \geq 0$. Prove that $(x_n)_{n\geq 0}$ is uniformly distributed in $[0, 1]$ if, and only if, the real number $\sum_{n\geq 0} \varepsilon_n/b^n$ is normal to base b.

EXERCISE 5.3 (*cf.* [396, 402]). Let $(x_n)_{n\geq 1}$ be a real sequence which is completely uniformly distributed modulo one. Let $b \geq 2$ be an integer and set $a_k = \lfloor \{x_k\} b \rfloor$ for $k \geq 1$. Prove that the real number ξ with b-ary expansion $0 \cdot a_1 a_2 \ldots$ is normal to base b.

EXERCISE 5.4 (*cf.* [155, 342, 667, 672]). Let b be an odd prime number. Let $a \geq 2$ be an integer coprime with b. Set $\ell_b = \mathrm{ord}_b(a^{\mathrm{ord}(a,b)} - 1)$. Prove that, for any positive integer n and any integer c coprime with b, all blocks of digits from $\{0, 1, \ldots, a-1\}$ of length at most $\lfloor (n - \ell_b) \log_a b \rfloor$ occur in the periodic part of the a-ary expansion of c/b^n.

EXERCISE 5.5 (*cf.* [342], Theorem 5). Let b be an odd prime number. Let $a \geq 2$ be an integer coprime with b. Apply Exercise 5.4 to show that, if $(n_j)_{j\geq 1}$ is a strictly increasing sequence of positive integers satisfying

$$n_{j+1} \geq b^{n_j} \log_b a$$

for infinitely many j, then the real number $\sum_{j\geq 1} b^{-n_j}$ is rich to base a.

5.6 Notes

▷ Let $b \geq 2$ be an integer. Ugalde [691] used graphs and de Bruijn words to construct real numbers which are normal to base b, and, moreover, have quite small discrepancy. In [692], he gave a specific construction of a Cantor-like set in $[0, 1]$, all of whose members are normal to base b, and whose Lebesgue measure can be made arbitrarily close to 1.

▷ Further constructions of normal numbers of Korobov type have been given by Kano [370].

▷ Starčenko [661] (see also [583]) established the converse of Korobov's result stated as Exercise 5.3. Namely, if a real number ξ with b-ary expansion $0 \cdot a_1 a_2 \ldots$ is normal to an integer base $b \geq 2$, then there exists a completely uniformly distributed sequence $(x_n)_{n\geq 1}$ such that $a_k = \lfloor \{x_k\} b \rfloor$ for $k \geq 1$.

▷ Let $r \geq 2$ be an integer. Hertling [342] proved that the real number $\sum_{j\geq 1} r^{-j!-j}$ is not rich to base r but is rich to every base s such that r and s are multiplicatively independent. Inspired by this work, Bugeaud [155] gave an explicit construction of a real number which is rich to every integer base.

▷ Exercise 5.4 asserts that every block of suitably small length occurs in the periodic part of the a-ary expansion of c/b^n; however, we cannot predetermine its location.

▷ A vector $\underline{x} = (x_1, \ldots, x_d)$ is said to be absolutely normal in $\mathbb{R}^d$ if the sequence $(\underline{x}A^n)_{n\geq 1}$ is uniformly distributed modulo one in $\mathbb{R}^d$ for arbitrary non-singular integral matrices A whose eigenvalues are not roots of unity. Pushkin [589] considered certain m-dimensional manifolds Γ in $\mathbb{R}^d$ and showed that λ_Γ-almost all points $\underline{x}$ in Γ are absolutely normal in $\mathbb{R}^d$, where λ_Γ is the Lebesgue measure for the parameter space.

6
Normality to different bases

Keeping in mind that almost all real numbers are normal to every integer base, we investigate the following question: Do there exist real numbers which are normal to one base r, but not normal to another base s? By Theorem 4.4 we know already that the answer is negative when r and s are multiplicatively dependent. However, at the end of the 1950s, Cassels and W. M. Schmidt, independently, gave a positive answer to this question when r and s are multiplicatively independent. Section 6.1 is devoted to their result. In the second section, we discuss its extension to non-integer bases. Then, we investigate what can be said on the expansions of a given number to two different bases. The final section is concerned with the study of the analogous question for representations of integers in two different bases.

6.1 Normality to a prescribed set of integer bases

Theorem 4.4, established by Maxfield [494], asserts that if r and s are multiplicatively dependent integers at least equal to 2, then a real number is normal to base r if, and only if, it is normal to base s. However, this result says nothing if r and s are multiplicatively independent. In 'The new Scottish book' (Problem 144), Steinhaus [663] asked whether normality with respect to infinitely many bases implies normality with respect to all other bases. Answers have been given independently by Cassels [182] and W. M. Schmidt [627]. Below is the statement established by Schmidt.

THEOREM 6.1. *Let $r \geq 2$ and $s \geq 2$ be multiplicatively independent integers. The set of real numbers which are normal to base r but not even simply normal to base s is uncountable.*

The idea behind the proof of Theorem 6.1 is to construct a measure μ supported by the set of real numbers that are not normal to base s, and then to apply the Davenport–Erdős–LeVeque Lemma 1.8 to infer that μ-almost all numbers are normal to base r.

For the sake of simplicity, we establish Theorem 6.1 in the particular case $s = 3$, as was done by Cassels [182]. The general case is slightly more technical but it does not require new ideas. We begin with an auxiliary lemma.

LEMMA 6.2. *There exist absolute positive constants A and δ such that*

$$\sum_{n=0}^{N-1} \prod_{j=1}^{+\infty} |\cos(3^{-j} h b^n \pi)| < A N^{1-\delta},$$

for every positive integer N, every non-zero integer h, and every integer $b \geq 2$ which is not a power of 3.

PROOF. Assume first that b is congruent to 1 modulo 3 and that N is a large power of 3, say $N = 3^r$, with $r \geq 3^{16}$. Set $\ell = \mathrm{ord}_3(b-1)$. Let n be an integer with $0 \leq n < 3^r$. Set $s = \mathrm{ord}_3(b^n - 1)$ and $n_0 = n/3^s$. Lemma B.2 implies that $\mathrm{ord}_3(b^n - 1) = s + \mathrm{ord}_3(b^{n_0} - 1)$, and we deduce from Theorem B.1 that $(b^{n_0} - 1)/(b-1)$ is not divisible by 3. We deduce from $s \leq r - 1$ that $\mathrm{ord}_3(b^n - 1) \leq 3^{r+\ell-1}$. This implies that the b^n, with $0 \leq n < 3^r$, run modulo $3^{\ell+r}$ through all residue classes which are congruent to 1 modulo 3^ℓ. Let h be a positive integer and set $m = \mathrm{ord}_3(h)$. Then the integers hb^n, $0 \leq n < 3^r$, run modulo $3^{\ell+m+r}$ through all residue classes which are congruent to h modulo $3^{\ell+m}$. Consequently, if we have the ternary representation

$$hb^n = \sum_{k \geq 0} \varepsilon_k(n) 3^k, \quad \text{where } \varepsilon_k(n) = 0, 1 \text{ or } 2,$$

then the r-tuple

$$\big(\varepsilon_{\ell+m}(n), \varepsilon_{\ell+m+1}(n), \ldots, \varepsilon_{\ell+m+r-1}(n)\big) \tag{6.1}$$

takes precisely once every one of the 3^r possible values as n runs from 0 to $3^r - 1$.

We now divide the 3^r integers n satisfying $0 \leq n \leq 3^r - 1$ into two classes. Class I comprises the integers such that the digit 1 occurs at least $\lceil r/6 \rceil$ times in the r-tuple (6.1). Class II comprises the remaining integers.

Let us first deal with Class I. Observe that the fractional part of $3^{-j} hb^n$ lies between 1/3 and 2/3 for every integer j such that $\varepsilon_{j-1}(n) = 1$.

For every integer n in Class I, this is precisely the case for at least $\lceil r/6 \rceil$ values of j in $\{\ell+m+1,\dots,\ell+m+r\}$, thus

$$\prod_{j=1}^{+\infty} |\cos(3^{-j}hb^n\pi)| \leq (\cos(\pi/3))^{r/6}.$$

Since Class I contains at most 3^r elements, there exists a positive δ_1 such that

$$\sum_{n\in(\mathrm{I})} \prod_{j=1}^{+\infty} |\cos(3^{-j}hb^n\pi)| \leq 3^{(1-\delta_1)r}. \tag{6.2}$$

Let us now deal with Class II. Note that, on average, the digit 1 occurs $r/3$ times in the r-tuple (6.1). Hence, we may hope that the cardinality of Class II is rather small. This is indeed the case, as it follows from Lemma 4.7 that

$$\sum_{k=0}^{\lceil r/6 \rceil} p_3(r,k) \leq \sum_{k=0}^{\lceil r/6 \rceil} p_3(3\lceil r/3 \rceil, k) \leq 2^{14}3^{r+2}\mathrm{e}^{-r/3240}.$$

Consequently, there exist positive real numbers A_2 and δ_2 such that

$$\sum_{n\in(\mathrm{II})} \prod_{j=1}^{+\infty} |\cos(3^{-j}hb^n\pi)| \leq A_2 3^{(1-\delta_2)r}. \tag{6.3}$$

Combining (6.2) and (6.3), we conclude that there exist positive real numbers A_3 and δ_3 such that

$$\sum_{n=0}^{N-1} \prod_{j=1}^{+\infty} |\cos(3^{-j}hb^n\pi)| < A_3 N^{1-\delta_3}, \tag{6.4}$$

for every N that is a power of 3, including the values of N less than 3^{16}.

Now, let N be a positive integer, and write

$$N = \sum_{0\leq r\leq R} \eta_r 3^r, \quad \text{where } \eta_r = 0, 1 \text{ or } 2 \text{ and } \eta_R \neq 0.$$

The range of summation $0 \leq n < N$ may be divided into $0 \leq n < N - 3^R$ and $N - 3^R \leq n < (N - 3^R) + 3^R$. More generally, this range may be subdivided into $\eta_0 + \eta_1 + \cdots + \eta_R$ intervals of the type

$$N_{r,\eta} \leq n < N_{r,\eta} + 3^r,$$

where there are precisely η_r intervals of length 3^r. Since A_3 and δ_3 do not depend on h, we infer from (6.4) that

$$\sum_{N_{r,\eta} \le n < N_{r,\eta}+3^r} \prod_{j=1}^{+\infty} |\cos(3^{-j} h b^n \pi)| < A_3 (3^r)^{1-\delta_3} \le A_3 N^{1-\delta_3}, \quad (6.5)$$

for every $N_{r,\eta}$ as above. Using that

$$\eta_0 + \eta_1 + \cdots + \eta_R \le 2(\log 3N),$$

it follows from (6.5) that

$$\sum_{n=0}^{N-1} \prod_{j=1}^{+\infty} |\cos(3^{-j} h b^n \pi)| < 2(\log 3N) A_3 N^{1-\delta_3} < A_4 N^{1-\delta_4},$$

for some positive real numbers A_4 and δ_4.

It only remains for us to deal with the integers b that are not congruent to 1 modulo 3. If b is congruent to 2 modulo 3, then b^2 is congruent to 1 modulo 3 and we may separate odd and even integers n in the range of summation $0 \le n < N$ before applying our upper bound for the base b^2 and integers h and hb, respectively, to get the required estimate. If $b = 3^\tau b_1$ with $\tau \ge 1$ and b_1 prime to 3, then we simply observe that

$$\prod_{j=1}^{+\infty} |\cos(3^{-j} h b^n \pi)| = \prod_{j=1-n\tau}^{+\infty} |\cos(3^{-j} h b_1^n \pi)| \le \prod_{j=1}^{+\infty} |\cos(3^{-j} h b_1^n \pi)|$$

to conclude the proof of the lemma. □

We are now in a position to establish Theorem 6.1 for $s = 3$.

PROOF OF THEOREM 6.1. Assume that $s = 3$ and let $r = b$ be a positive integer which is not a power of 3. Let μ_K denote the standard measure on the middle third Cantor set (see Section C.3). Observe that, for every real number t, its Fourier transform $\hat{\mu}_K$ satisfies

$$\begin{aligned} \hat{\mu}_K(t) = \int_0^1 e^{2i\pi t\xi}\, d\mu_K(\xi) &= \lim_{J \to +\infty} 2^{-J} \prod_{1 \le j \le J} \left(1 + e^{2i\pi(2 \cdot 3^{-j})t}\right) \\ &= \prod_{j=1}^{+\infty} \frac{1 + e^{4i\pi 3^{-j} t}}{2}, \end{aligned}$$

hence

$$|\hat{\mu}_K(t)| = \left| \int_0^1 e^{2i\pi t\xi}\, d\mu_K(\xi) \right| = \prod_{j=1}^{+\infty} |\cos(2 \cdot 3^{-j} \pi t)|.$$

Consequently, if h is a non-zero integer and N is a positive integer, then it follows from Lemma 6.2 that

$$\begin{aligned}
\int_0^1 \Big| \sum_{0\le n<N} \mathrm{e}^{2i\pi h b^n \xi} \Big|^2 \mathrm{d}\mu_K(\xi) &= \sum_{0\le m<N} \sum_{0\le n<N} \int_0^1 \mathrm{e}^{2i\pi(hb^m - hb^n)\xi} \,\mathrm{d}\mu_K(\xi) \\
&\le \sum_{0\le m<N} \sum_{0\le n<N} \prod_{j=1}^{+\infty} |\cos(2\cdot 3^{-j} h(b^m - b^n)\pi)| \\
&\le N + 2 \sum_{0\le n<N} \sum_{1\le \ell<N-n} \prod_{j=1}^{+\infty} |\cos(2\cdot 3^{-j} h(b^\ell - 1)b^n\pi)| \\
&\le N + 2NAN^{1-\delta} \le N^{2-\delta/2},
\end{aligned}$$

provided that N is sufficiently large. This proves that the series

$$\sum_{N\ge 1} \frac{1}{N} \int_0^1 \Big| \frac{1}{N} \sum_{0\le n<N} \mathrm{e}^{2i\pi h b^n \xi} \Big|^2 \mathrm{d}\mu_K(\xi)$$

converges, thus, by Lemma 1.8,

$$\lim_{N\to+\infty} \frac{1}{N} \sum_{0\le n<N} \mathrm{e}^{2i\pi h b^n \xi} = 0$$

for μ_K-almost all ξ. By Theorem 1.2, the sequence $(\xi b^n)_{n\ge 1}$ is then uniformly distributed modulo one for μ_K-almost all ξ. By Theorem 4.14, this shows that μ_K-almost all ξ are normal to base b. □

Actually, the proof of Theorem 6.1 gives a stronger result than the one enounced, see Theorem 7.14. We end this section by an extension of Theorem 6.1 obtained by Schmidt [628] by means of a specific construction.

THEOREM 6.3. *Let $\mathcal{R} \cup \mathcal{S}$ be a partition of the set of integers greater than or equal to 2 into two classes such that any two multiplicatively dependent integers fall in the same class. The set of real numbers which are normal to every base from $\mathcal{R}$ but not normal to every base from $\mathcal{S}$ is uncountable.*

Schmidt [628] gave a (fairly complicated) constructive proof of Theorem 6.3, while Theorem 6.1 is 'only' an existence statement. On the other hand, the numbers satisfying the conclusion of Theorem 6.3 may not be simply normal to every base from $\mathcal{S}$. Taking for $\mathcal{S}$ the empty set,

Schmidt constructed explicitly real numbers which are normal to every integer base; see also Section 5.2.

6.2 Normality to non-integer bases

In their proofs of Theorem 6.1, Cassels and Schmidt have used the Cantor measure. This is also the case of Pearce and Keane [554], who gave a 'structurally simple proof' [sic] of Theorem 6.1. To this end, they calculated the Fourier coefficients associated with the Cantor measure. Shortly thereafter, Brown, Moran and Pearce [136] replaced the use of the Cantor measure by that of Riesz product measures, which have the great advantage that their Fourier coefficients are easy to compute. This method, which extends well to non-integer bases, has been developed in subsequent works [138–141, 519, 520, 522].

DEFINITION 6.4. For a real number α with $|\alpha| > 1$, set

$$\mathcal{N}(\alpha) := \{\xi \in \mathbb{R} : (\xi\alpha^n)_{n\geq 1} \text{ is uniformly distributed modulo one}\}.$$

In [502], Mendès France asked whether Maxfield's Theorem 4.4 extends to non-integer bases, that is, α and β being real numbers greater than 1, whether we have $\mathcal{N}(\alpha) = \mathcal{N}(\beta)$ if, and only if, $(\log\alpha)/(\log\beta)$ is rational. In this direction, Moran and Pollington [522] established the following result.

THEOREM 6.5. *Let α and β be real numbers greater than* 1. *If $\mathcal{N}(\alpha)$ is included in $\mathcal{N}(\beta)$, then $(\log\alpha)/(\log\beta)$ is rational.*

Bertrand [83] proved that if the positive integer k and the real number $\alpha > 1$ are such that $\alpha^k + \alpha^{-k}$ or $\alpha^k - \alpha^{-k}$ is a rational integer, then $\mathcal{N}(\alpha)$ is contained in $\mathcal{N}(\alpha^r)$ for every positive integer r.

The answer to Mendès France's question is in general negative, as was first shown by Brown, Moran and Pollington [139]. They established a definitive result in [140], which we reproduce below.

THEOREM 6.6. *Let $\alpha > 1$ be a real number and r and s be distinct positive integers. The set $\mathcal{N}(\alpha^s)$ is included in the set $\mathcal{N}(\alpha^r)$ if, and only if, there is a positive integer k such that either*

$$\alpha^k \in \mathbb{Z} \quad \textit{and} \quad \mathbb{Q}(\alpha^r) \subset \mathbb{Q}(\alpha^s)$$

or

$$\alpha^k + \alpha^{-k} \textit{ or } \alpha^k - \alpha^{-k} \textit{ is in } \mathbb{Z} \textit{ and } s \textit{ divides } r.$$

We do not give a complete proof of Theorem 6.6. We content ourselves with establishing a particular case of it, namely Theorem 6.7 below, which forms the main content of the note [139]. Recall that for any non-square positive integer b, it is expected that $\sqrt{b}$ is normal to base b, while, clearly, $\sqrt{b}$ does not belong to $\mathcal{N}(\sqrt{b})$.

THEOREM 6.7. *Let b be a non-square positive integer with $b \geq 10$. For every positive integer k we have*

$$\mathcal{N}(\sqrt{b}) \subset \mathcal{N}((\sqrt{b})^k) \quad \textit{and} \quad \mathcal{N}(\sqrt{b}) = \mathcal{N}(b^k\sqrt{b}),$$

but

$$\mathcal{N}(b) \not\subset \mathcal{N}(\sqrt{b}).$$

The assumption that b is at least equal to 10 is a technical assumption which allows some simplification in the proof. It follows from Theorem 6.6 that the conclusion of Theorem 6.7 remains true for b in $\{2, 3, 5, 6, 7, 8\}$.

The proofs of Theorems 6.5 and 6.7 follow the same lines. The key idea for the proof of Theorem 6.5 is to find, when $(\log \alpha)/(\log \beta)$ is irrational, a probability measure μ (a discriminatory measure) which assigns zero mass to $\mathcal{N}(\beta)$ and full mass to $\mathcal{N}(\alpha)$.

PRELIMINARIES FOR THE PROOFS OF THEOREMS 6.5 & 6.7. Let λ be a real number with $\lambda > 1$. We construct a suitable Riesz product measure μ such that the sequence $(\xi\lambda^n)_{n\geq 1}$ is not uniformly distributed modulo one for μ-almost all real numbers ξ. Let $\mathcal{A}$ be a set of positive integers whose upper density is positive, that is, satisfying

$$\limsup_{N\to+\infty} \frac{\mathrm{Card}\{1 \leq n \leq N : n \in \mathcal{A}\}}{N} > 0, \tag{6.6}$$

and such that

$$\lambda^{n'-n} > 3 \text{ for all integers } n, n' \text{ in } \mathcal{A} \cup \{0\} \text{ with } n' > n. \tag{6.7}$$

For a positive integer N and a real number t, set

$$\eta_N(t) = \prod_{n\in\mathcal{A}, 1\leq n\leq N} (1 + \cos(2\pi\lambda^n t)) \quad \text{and} \quad \rho(t) = \frac{1 - \cos t}{\pi t^2}.$$

The Fourier transform of ρ is the triangle function equal to 1 at 0 and vanishing on and only on $\{t \in \mathbb{R} : |t| \geq 1\}$. Said differently, we have $\hat{\rho}(t) = \max\{1 - |t|, 0\}$ for t in $\mathbb{R}$. Then, define the measure μ_N by

$$\mathrm{d}\mu_N(t) = \eta_N(t)\rho(t)\,\mathrm{d}t.$$

Observe that $\hat{\mu}_N(t) = 0$ unless the real number t satisfies

$$\Big| t - \sum_{n \in \mathcal{A}, 1 \le n \le N} \varepsilon_n \lambda^n \Big| < 1 \tag{6.8}$$

for some $\varepsilon_1, \ldots, \varepsilon_N$ in $\{\pm 1, 0\}$. Here and below, it is understood that $\varepsilon_n = 0$ if $n \notin \mathcal{A}$. Furthermore, the graph of $\hat{\mu}_N$ is the union of segments of the real axis with triangular bumps of width 2 centred at the points $\varepsilon_1 \lambda_1 + \cdots + \varepsilon_N \lambda_N$ with height $2^{-|\varepsilon_1| - \cdots - |\varepsilon_N|}$, where $\varepsilon_1, \ldots, \varepsilon_N \in \{\pm 1, 0\}$. Since the sequence of Fourier transforms $(\hat{\mu}_N)_{N \ge 1}$ converges pointwise towards a function which is continuous at 0, the continuity theorem (see [98, p. 303]) asserts that the weak star limit μ of the sequence of measures $(\mu_N)_{N \ge 1}$ exists and, furthermore, that μ is a continuous probability measure on $\mathbb{R}$ whose Fourier transform at t is 0 unless (6.8) holds for some integer N and some $\varepsilon_1, \ldots, \varepsilon_N$ in $\{\pm 1, 0\}$, in which case we have $0 \le \hat{\mu}(t) \le 2^{-|\varepsilon_1| - \cdots - |\varepsilon_N|}$. The latter inequality becomes an equality if, and only if, $t = \sum_{n=1}^N \varepsilon_n \lambda^n$.

For further use, observe that, by (6.7), the difference between any two distinct elements of the increasing sequence of all positive real numbers of the form $\pm \lambda^{n_1} \pm \cdots \pm \lambda^{n_h}$ exceeds 2, where $h \ge 1$ and $n_1 < \cdots < n_h$ are in $\mathcal{A}$. This implies that, for any given real number t, inequality (6.8) has at most one solution with $\varepsilon_1, \ldots, \varepsilon_N$ in $\{0, \pm 1\}$.

For $n \ge 1$ and a real number t, put

$$X_n(t) = \mathrm{e}^{2i\pi t \lambda^n},$$

and note that

$$\hat{\mu}(\lambda^n) = \int_{-\infty}^{+\infty} \mathrm{e}^{2i\pi t \lambda^n} \, \mathrm{d}\mu(t) = \int_{-\infty}^{+\infty} X_n(t) \, \mathrm{d}\mu(t).$$

It follows from the properties of $\hat{\mu}$ that, for all distinct positive integers m and n, we have

$$\hat{\mu}(\lambda^m \pm \lambda^n) = \hat{\mu}(\lambda^m) \cdot \hat{\mu}(\lambda^n). \tag{6.9}$$

Indeed, (6.9) becomes $0 = 0$ if m or n does not belong to $\mathcal{A}$, and it reduces to $1/4 = 1/4$ if m and n are in $\mathcal{A}$. Let N be a positive integer. Observe that we have

$$\begin{aligned} \int_{-\infty}^{+\infty} \frac{1}{N} \sum_{1 \le n \le N} \mathrm{e}^{2i\pi t \lambda^n} \, \mathrm{d}\mu(t) &= \frac{1}{N} \sum_{1 \le n \le N} \hat{\mu}(\lambda^n) \\ &\ge \frac{\mathrm{Card}\{1 \le n \le N : n \in \mathcal{A}\}}{2N}, \end{aligned} \tag{6.10}$$

since $\hat{\mu}(\lambda^n) = 1/2$ for every positive integer n in $\mathcal{A}$ and $\hat{\mu}(t) \geq 0$ for every real number t.

Using that the functions $|\hat{\mu}|$ and $|X_1|, |X_2|, \ldots$ are all bounded by 1, we deduce from (6.9) that

$$\int_{-\infty}^{+\infty} \left| \frac{1}{N} \sum_{1 \leq n \leq N} (X_n(t) - \hat{\mu}(\lambda^n)) \right|^2 \mathrm{d}\mu(t) \leq \frac{4}{N}.$$

We then infer from Lemma 1.8 that

$$\lim_{N \to +\infty} \frac{1}{N} \sum_{1 \leq n \leq N} (X_n(t) - \hat{\mu}(\lambda^n)) = 0$$

holds for μ-almost all real numbers t. Combined with (6.6) and (6.10), this gives

$$\limsup_{N \to +\infty} \frac{1}{N} \sum_{1 \leq n \leq N} \mathrm{e}^{2i\pi\xi\lambda^n} > 0$$

for μ-almost all real numbers ξ. Thus, by Theorem 1.2, the sequence $(\xi\lambda^n)_{n\geq 1}$ is not uniformly distributed modulo one for μ-almost all real numbers ξ. □

PROOF OF THEOREM 6.7. Let k be a positive integer. Let ξ be in $\mathcal{N}(\sqrt{b})$ and set $u_n = \xi(\sqrt{b})^n$ for $n \geq 1$. For $h \geq 1$ and $n \geq 1$, set

$$v_{h,n} = u_{n+2h} - u_n = (b^h - 1)\xi(\sqrt{b})^n.$$

Since ξ is in $\mathcal{N}(\sqrt{b})$, Theorem 1.2 implies that the sequence $(v_{h,n})_{n\geq 1}$ is uniformly distributed modulo one for every positive integer h. It then follows from Theorem 1.6 that $(u_{kn})_{n\geq 1}$ is uniformly distributed modulo one. By Theorem 4.14, this means that ξ is normal to base $(\sqrt{b})^k$ and proves the first assertion.

Let ξ be in $\mathcal{N}(b^k\sqrt{b})$. For $h = 0, 1, \ldots, 2k$ and $n \geq 1$, we have

$$\begin{aligned} \xi(\sqrt{b})^h(\sqrt{b})^{n(2k+1)} &= \xi(\sqrt{b})^{2h(k+1)-h(2k+1)}(\sqrt{b})^{n(2k+1)} \\ &= \xi b^{h(k+1)}(b^k\sqrt{b})^{n-h}, \end{aligned}$$

and, since ξ is in $\mathcal{N}(b^k\sqrt{b})$, it follows from Theorem 1.2 that $\xi(\sqrt{b})^h$ belongs to $\mathcal{N}(b^k\sqrt{b})$. We then get from Exercise 4.1 that ξ is in $\mathcal{N}(\sqrt{b})$. This completes the proof of the first two assertions of Theorem 6.7.

To establish the last assertion of Theorem 6.7, we define the measure μ as in the preliminaries with $\lambda = \sqrt{b}$ and $\mathcal{A} = \{n \geq 1 : n \text{ odd}\}$. Observe that (6.7) holds since $b \geq 10$. We conclude from the preliminaries that the sequence $(\xi(\sqrt{b})^n)_{n\geq 1}$ is not uniformly distributed for μ-almost all real numbers ξ. It remains to show that μ-almost all real numbers are normal to base b. By Lemma 1.8 and Theorem 1.2, it is sufficient to establish that

$$\sum_{N=1}^{+\infty} \frac{1}{N^3} \sum_{1 \le m < n \le N} \hat{\mu}\big(h(b^n - b^m)\big) < +\infty, \tag{6.11}$$

for every non-zero integer h.

Fix a non-zero integer h. Let m and n be distinct positive integers. It follows from the preliminaries that the inequality

$$\Big| h(b^n - b^m) - \sum_{i \ge 1} \varepsilon_i b^i \sqrt{b} \Big| < 1 \tag{6.12}$$

has at most one solution $(\varepsilon_i)_{i \ge 1}$ such that $\varepsilon_i \in \{0, \pm 1\}$ for $i \ge 1$. If it has no solution, then set $r(n, m) = +\infty$; otherwise, let $r(n, m)$ be the number of non-zero elements of the sequence $(\varepsilon_i)_{i \ge 1}$. Then, we have

$$|\hat{\mu}\big(h(b^n - b^m)\big)| \le \frac{1}{2^{r(n,m)}}. \tag{6.13}$$

For $n \ge 1$, set $t(n) = \max\{1, \lfloor n/(\log 3n)^2 \rfloor\}$. Since, for any integer $N \ge 3$, we have

$$\begin{aligned} \sum_{n - t(n) \le m < n \le N} \frac{1}{2^{r(n,m)}} &\le \sum_{n=1}^{N} \sum_{m = n - t(n)}^{n-1} \frac{1}{2} \\ &\le \sum_{n=1}^{N} \frac{t(n)}{2} \le N t(N) \le \frac{N^2}{(\log 3N)^2}, \end{aligned} \tag{6.14}$$

it follows from (6.11) and (6.13) that it remains for us to prove that the sum

$$\sum_{N=1}^{+\infty} \frac{1}{N^3} \sum_{1 \le m < n \le N, m < n - t(n)} \frac{1}{2^{r(n,m)}} \tag{6.15}$$

converges. Without any loss of generality, we may assume that there are infinitely many integers m, n with $m < n - t(n)$ for which (6.12) has a solution $(\varepsilon_i)_{i \ge 1}$ such that $\varepsilon_i \in \{0, \pm 1\}$ for $i \ge 1$. Since $t(n)$ tends to infinity with n, it then follows that there exist an integer h_0 and a sequence $(\delta_i)_{i \ge -h_0}$ taking its values in $\{0, \pm 1\}$ such that

$$h\sqrt{b} = \sum_{i = -h_0}^{+\infty} \delta_i b^{-i}. \tag{6.16}$$

Put $s(n) = \sum_{i=-h_0}^{t(n)-h_0-1} |\delta_i|$. Substituting (6.16) into (6.12) shows that $s(n) \le r(n,m)$ when $1 \le m < n - t(n)$. Thus, we get

$$\sum_{1 \le m < n \le N, m < n-t(n)} \frac{1}{2^{r(n,m)}} \le \sum_{n=1}^{N} \frac{n}{2^{s(n)}}.$$

Let r be a positive integer with $\delta_r \neq 0$ and $\delta_{r+1} = \delta_{r+2} = \cdots = \delta_{3r} = 0$. Then,

$$b^r h\sqrt{b} = b^r \sum_{i=-h_0}^{r} \delta_i b^{-i} + \tau_r, \tag{6.17}$$

where $|\tau_r| \le b^{-2r}$ and $v_r := \sum_{i=-h_0}^{r} \delta_i b^{r-i}$ is an integer of absolute value at most hb^{r+1}. Squaring (6.17) gives

$$b^{2r+1}h^2 = v_r^2 + 2\tau_r v_r + \tau_r^2,$$

which is impossible if $|2\tau_r v_r + \tau_r^2| < 1$, thus, if r is sufficiently large. In particular, every interval $[r+1, 3r]$ with r sufficiently large contains at least an integer i such that $|\delta_i| \ge 1$. Consequently, there exists a positive real number $c < 1$ such that $s(n) \ge 2c(\log 3n) - 1$, for $n \ge 1$, hence,

$$\sum_{n=1}^{N} \frac{n}{2^{s(n)}} \le 2 \sum_{n=1}^{N} (3n)^{1-c} \le 6\, N^{2-c},$$

for $N \ge 1$, and the sum (6.15) converges. This shows that (6.11) holds. We conclude by applying Lemma 1.8 and Theorem 1.2. □

Proof of Theorem 6.5. Assume that $(\log \alpha)/(\log \beta)$ is irrational, and let $(p_k/q_k)_{k\ge1}$ be the sequence of its convergents; see Appendix D. Let p be an integer such that $\beta^p > 3$. Define the measure μ as in the preliminaries with $\lambda = \beta$ and

$$\mathcal{A} = \bigcup_{k\ge1} \{p(q_k+1), p(q_k+2), \ldots, p(\min\{2q_k, q_{k+1}\} - 1)\}.$$

We check that (6.7) is satisfied. Observe that the upper density of $\mathcal{A}$ is at least equal to $1/(4p)$. We thus get from the preliminaries that $\mu(\mathcal{N}(\beta)) = 0$.

Assume that $\mu(\mathcal{N}(\alpha)) < 1$. Then, by Lemma 1.8 and Theorem 1.2, there exists a non-zero integer h such that

$$\sum_{N=1}^{+\infty} \frac{1}{N^3} \sum_{1 \le m < n \le N} \hat{\mu}(h(\alpha^n - \alpha^m)) = +\infty. \tag{6.18}$$

Let m and n be distinct positive integers. It follows from the preliminaries that the inequality

$$\left| h(\alpha^n - \alpha^m) - \sum_{i \in \mathcal{A}, 1 \le i \le R(n,m)} \varepsilon_i \beta^i \right| < 1 \tag{6.19}$$

has at most one solution $(\varepsilon_i)_{i \ge 1}$ such that $\varepsilon_i \in \{0, \pm 1\}$ for $i \ge 1$. If it has no solution, then set $r(n,m) = +\infty$; otherwise, let $r(n,m)$ be the number of non-zero elements of the sequence $(\varepsilon_i)_{i \ge 1}$. The above notation means that $\varepsilon_{R(n,m)} = \pm 1$.

As above, set $t(n) = \max\{1, \lfloor n/(\log 3n)^2 \rfloor\}$ for $n \ge 1$. By (6.18) and the analogues of (6.13) and (6.14), we get

$$\sum_{N=1}^{+\infty} \frac{1}{N^3} \sum_{n=1}^{N} \sum_{m=1}^{n-t(n)} \frac{1}{2^{r(n,m)}} = +\infty$$

and

$$\sum_{n=1}^{+\infty} \frac{1}{n^2} \sum_{m=1}^{n-t(n)} \frac{1}{2^{r(n,m)}} = +\infty.$$

Consequently, there are arbitrarily large integers S such that

$$\sum_{2^S \le n < 2^{S+1}} \; \sum_{1 \le m \le n-t(n)} \frac{1}{2^{r(n,m)}} \ge \frac{2^{2S}}{S^2}.$$

Throughout the rest of this proof, we assume at several places that S is sufficiently large, without mentioning it explicitly. Since $\sum_{m=1}^{n} 1/2 \le 2^S$ for $n < 2^{S+1}$, we deduce that, for these integers S, there is a set $G(S)$, of cardinality at least $2^S/S^3$, of integers in $\{2^S, \ldots, 2^{S+1}-1\}$ such that

$$\sum_{1 \le m \le n-t(n)} \frac{1}{2^{r(n,m)}} \ge \frac{2^S}{S^3},$$

for n in $G(S)$. For each n in $G(S)$, let m be an integer satisfying $1 \le m \le n - t(n)$ and $r(n,m) \le 5 \log S$. We nominate one such $m = m(n)$ for each n in $G(S)$. Fix n in $G(S)$ and suppose that (6.19) holds for the pair $(n, m(n)) = (n,m)$. Set $R(n) = R(n, m(n))$. Then, somewhere in the sequence

$$\varepsilon_{R(n)}, \varepsilon_{R(n)-1}, \ldots, \varepsilon_{R(n)-v(S)},$$

where $v(S) = \lceil 6S^2 \log S \rceil$, there is a block of S^2 consecutive zeros. Let us write $V(n)$ for the largest i such that ε_i is followed by a block of at least S^2 zeros. Since $R(n) - V(n)$ is less than $v(S)$ and $r(n,m)$ does not exceed $5 \log S$, the number of possible choices of

$$\varepsilon_{R(n)}, \varepsilon_{R(n)-1}, \ldots, \varepsilon_{V(n)}$$

does not exceed $(2v(S))^{5 \log S}$, hence is less than $2^{S/2}/S^3$ when S is large enough.

It follows that, for some

$$w = \sum_{k=0}^{M} \varepsilon'_k \beta^{-pk},$$

with $\varepsilon'_0, \varepsilon'_1, \ldots, \varepsilon'_M$ in $\{0, \pm 1\}$, there is a subset $G'(S)$ of $G(S)$ of cardinality at least $2^{S/2}$ such that, for every n in $G'(S)$, we have

$$h\alpha^n (1 + \tau_n) = w\beta^{R(n)} (1 + \sigma_n), \tag{6.20}$$

where

$$\begin{aligned} |\tau_n| &\le \alpha^{-t(n)} \le \mathrm{e}^{-Cn(\log n)^{-2}}, \\ |\sigma_n| &\le \mathrm{e}^{-C(\log n)^2}, \end{aligned} \tag{6.21}$$

for some positive constant C and $R(n)$ in $\mathcal{A}$. Let $n_0 < n_1 < \ldots < n_J$ be these integers n. For $j = 1, \ldots, J$, setting $v_j = n_j - n_0$ and $u_j = R(n_j) - R(n_0)$ and taking logarithms of (6.20) with $n = n_0$ and $n = n_j$ (this allows us to eliminate h and w in (6.20)), we deduce from (6.21) that

$$\gamma := \frac{\log \alpha}{\log \beta} = \frac{u_j}{v_j} + \delta_j, \tag{6.22}$$

where

$$v_j \le n_J \le 2^S \qquad \text{and} \qquad |\delta_j| \le \exp\{-C' S^2\},$$

for some positive constant C'. In particular, we deduce that

$$R(n_j) \ge u_j \ge \gamma v_j / 2. \tag{6.23}$$

Since, by (6.20), the quotient $R(\ell)/\ell$ is bounded independently of the positive integer ℓ, it follows from (6.23) that

$$|\delta_j| \le \exp\{-C' S^2\} < \frac{1}{R(n_j)^3} < \frac{1}{2v_j^2},$$

for $j = 1, \ldots, J$. By (6.22) and Theorem D.7, the rational number u_j/v_j must be a convergent to γ. Thus, there exist positive integers d and r such that $u_j = dp_r$ and $v_j = dq_r$. Moreover, by Theorem D.1 and (6.23), we get

$$\frac{1}{R(n_j)^3} > \left|\gamma - \frac{p_r}{q_r}\right| > \frac{1}{q_r(q_r + q_{r+1})} > \frac{\gamma}{4R(n_j)q_{r+1}},$$

so that $R(n_j) < q_{r+1}$. By the definition of $\mathcal{A}$, this shows that $R(n_j)$ is less than $2pq_r$ and $u_j \le 2pq_r$. It follows from (6.23) that $v_j \le 4pq_r/\gamma$, thus the integer d can take no more than $\lceil 4p/\gamma \rceil$ different values. Since $(q_k)_{k\ge1}$ grows at least exponentially (Theorem D.5) and $q_r \le 2^S$, there are at most S^2 such integers j. This contradicts the fact that $G'(S)$ has $2^{S/2}$ elements. We have shown that $\mu(\mathcal{N}(\alpha)) = 1$. This proves that $\mathcal{N}(\alpha)$ is not contained in $\mathcal{N}(\beta)$ when the quotient $(\log \alpha)/(\log \beta)$ is irrational, as asserted. □

6.3 On the expansions of a real number to two different bases

This section is concerned with the following general question, which was investigated in [155]:

Let r and s be multiplicatively independent positive integers. Are there irrational real numbers whose r-ary expansion and s-ary expansion are, in some sense, both 'simple'?

First, we have to explain what is meant by 'simple'. Actually, there are several possible points of view.

Let $b \ge 2$ be an integer and ξ a real number whose b-ary expansion is given by

$$\xi = \lfloor \xi \rfloor + \sum_{k\ge1} \frac{a_k}{b^k},$$

where a_k is in $\{0, 1, \ldots, b-1\}$ for $k \ge 1$ and an infinity of the a_k are not equal to $b - 1$.

A first point of view is to use the block complexity function $n \mapsto p(n, \xi, b)$ defined in Section 4.4. An irrational real number ξ could then be considered as 'simple' to base b if this function increases very slowly.

A second point of view, addressed in [53], consists in counting the number of non-zero digits of ξ among its n first b-ary digits by setting

$$\mathcal{NZ}(n, \xi, b) = \mathrm{Card}\{k : 1 \le k \le n, a_k \ne 0\}.$$

An irrational real number ξ could then be considered as 'simple' to base b if the function $n \mapsto \mathcal{NZ}(n, \xi, b)$ increases very slowly. This means that ξ has only few non-zero digits.

A third point of view, addressed in [147], consists in estimating the asymptotic behaviour of the number of digit changes in the b-ary expansion of ξ. Following [147], we define the function $\mathcal{DC}$, 'number of digit changes', by

$$\mathcal{DC}(n, \xi, b) = \mathrm{Card}\{k : 1 \le k \le n, a_k \ne a_{k+1}\},$$

for any positive integer n. For every integer $b \ge 2$, the b-ary expansion of any rational number p/q is ultimately periodic, thus there exist integers ℓ_0 and C such that $\mathcal{DC}(n, p/q, b^\ell) \le C$ for $\ell \ge \ell_0$. The growth of the functions $n \mapsto \mathcal{DC}(n, \xi, b^\ell)$ can be used to measure the complexity of the real number ξ. In this respect, the 'simplest' numbers are the rational numbers and an irrational real number ξ could then be considered as 'simple' to base b if the function $n \mapsto \mathcal{NZ}(n, \xi, b)$ increases very slowly. This means that ξ has only few digit changes in its b-ary expansion.

Since, for $n \ge 1$, we have

$$\mathcal{DC}(n, \xi, b) \le 2\mathcal{NZ}(n, \xi, b) + 1, \tag{6.24}$$

a lower bound for $\mathcal{DC}(n, \xi, b)$ implies a lower bound for $\mathcal{NZ}(n, \xi, b)$. However, the converse does not hold.

We show that, if, for some integer $r \ge 2$, the r-ary expansion of an irrational real number ξ has very few digit changes, then, if $s \ge 2$ is coprime with r, the s-ary expansion of ξ must have a certain amount of digit changes.

THEOREM 6.8. *Let $r \ge 2$ and $s \ge 2$ be coprime positive integers. Let ξ be an irrational real number. There exist an integer n_0 and a positive real number κ such that*

$$\mathcal{DC}(n, \xi, r) + \mathcal{DC}(n, \xi, s) \ge \kappa \log n, \quad \textit{for } n \ge n_0, \tag{6.25}$$

and

$$\mathcal{NZ}(n, \xi, r) + \mathcal{NZ}(n, \xi, s) \ge \kappa \log n, \quad \textit{for } n \ge n_0.$$

PROOF. We repeatedly use the elementary fact that, if the r-ary expansion of a rational number ζ reads $0 \cdot a_1 a_2 \ldots a_n aaa \ldots$, with $a, a_1, \ldots a_n$ in $\{0, 1, \ldots, r-1\}$ and $a \ne a_n$, then

$$r^n \zeta = a_1 r^{n-1} + \cdots + a_{n-1} r + a_n + \frac{a}{r-1}$$

and there exists an integer d such that $\zeta = d/(r^n(r-1))$. The latter rational number may not be written under its lowest form. To see this, just observe that

$$\zeta = \frac{(a - a_n) + r(a_n - a_{n-1}) + \cdots + r^{n-1}(a_2 - a_1) + a_1 r^n}{r^n(r-1)},$$

and $a - a_n$ has no reason to be coprime with r. Note however that, since a is not equal to a_n, there exists a prime number p and a positive integer v such that p^v divides r, but p^v does not divide $a_n - a$. This shows that if $\zeta = A/B$ in its reduced form, then p^n divides B.

Set $\lambda = (\log r)/(\log s)$ and $c = 2 + \lceil 2/\lambda \rceil$. Let N be a large positive integer such that $\mathcal{DC}(N, \xi, r) = \mathcal{DC}(2N + c, \xi, r)$. This implies that the $(N+1)$th, $(N+2)$th, ... until the $(2N+c+1)$th digits in the r-ary expansion of ξ are all the same. Let n be the smallest positive integer such that the $(n+1)$th, $(n+2)$th, ... until the $(2N+c+1)$th digits in the r-ary expansion of ξ are all the same. We have $n \le N$ and there exists an integer f such that

$$\left| \xi - \frac{f}{r^n(r-1)} \right| \le \frac{1}{r^{2N+c+1}}.$$

Let h be the integer defined by the inequalities

$$h + 1 \le \lambda(2N - n + c - 1) < h + 2. \tag{6.26}$$

If the integer g satisfies

$$f s^h (s-1) = g r^n (r-1), \tag{6.27}$$

then our choice of n and the above discussion imply that p^n divides $s^h(s-1)$ for a prime divisor p of r. Since r and s are coprime, p does not divide s and we get that $n \le 2 \log s$. This shows that if N is sufficiently large, in terms of r, s and ξ, then no integer g satisfies (6.27). Observe that

$$2r^n s^h (r-1)(s-1) \le 2r^{n+1} r^{2N-n+c-1} \le r^{2N+c+1}.$$

Consequently, the triangle inequality gives

$$\begin{aligned}
\left| \xi - \frac{g}{s^h(s-1)} \right| &\ge \left| \frac{f}{r^n(r-1)} - \frac{g}{s^h(s-1)} \right| - \left| \xi - \frac{f}{r^n(r-1)} \right| \\
&\ge \frac{1}{r^n s^h (r-1)(s-1)} - \frac{1}{r^{2N+c+1}} \\
&\ge \frac{1}{2r^n s^h (r-1)(s-1)} \ge \frac{1}{s^{\lambda(n+1)+h+2}},
\end{aligned}$$

for any integer g, when N (hence, h) is large enough. This implies that, in the s-ary expansion of ξ, the $(h+1)$th, $(h+2)$th, ... until the $(\lambda(n+1) + h + 2)$th digits cannot be all the same, whence

$$\mathcal{DC}(\lambda(n+1)+h+1,\xi,s) \geq \mathcal{DC}(h,\xi,s)+1.$$

It then follows from (6.26) that

$$\begin{aligned}\mathcal{DC}(2\lambda N+\lambda c,\xi,s) &\geq \mathcal{DC}(\lambda N+(c-1)\lambda-2,\xi,s)+1\\ &\geq \mathcal{DC}(\lambda N,\xi,s)+1,\end{aligned}$$

since $n \leq N$ and $(c-1)\lambda - 2 > 0$, by our choice of c.

Set $u_1 = 1$ and $u_{n+1} = 2u_n + c$ for $n \geq 1$. A rapid calculation shows that $u_n \leq (c+1)2^n$ for $n \geq 1$. We thus have proved that

$$\begin{aligned}&\mathcal{DC}(u_{n+1},\xi,r)-\mathcal{DC}(u_n,\xi,r)\\ &\quad+\mathcal{DC}(\lambda u_{n+1},\xi,s)-\mathcal{DC}(\lambda u_n,\xi,s)\geq 1,\end{aligned}$$

for every sufficiently large integer n. Setting $\lambda' = \max\{1,\lambda\}$, we get

$$\mathcal{DC}(\lambda'(c+1)2^n,\xi,r)+\mathcal{DC}(\lambda'(c+1)2^n,\xi,s) \geq n,$$

for every sufficiently large integer n. In view of (6.24), this proves the theorem. □

Further results are given in [155]; see also Exercise 6.2.

For coprime integers $r \geq 2$ and $s \geq 2$, it remains an open problem to find a non-trivial lower bound for $p(n,\xi,r)+p(n,\xi,s)$, valid for every real irrational number ξ and every sufficiently large integer n; see Chapter 10 for related open questions.

6.4 On the representation of an integer in two different bases

Stewart [664] established effectively that an integer cannot have few non-zero digits simultaneously in base 2 and in base 3, a result first obtained by Senge and Strauss [643] but with an ineffective proof; see also [60, 106, 159, 314, 515].

For integers $b \geq 2$ and $n \geq 1$, let $\mathcal{DC}(n,b)$ denote the number of times that a digit different from the previous one is read in the b-ary representation of n. The next theorem, extracted from [159], extends Stewart's result and is established with the same arguments. We omit its proof, which uses Baker's theory of linear forms in the logarithms of algebraic numbers.

THEOREM 6.9. *Let r and s be multiplicatively independent integers. There exists an effectively computable integer c, depending only on r and s, such that, for every integer $n \geq 20$, we have*

$$\mathcal{DC}(n,r) + \mathcal{DC}(n,s) \geq \frac{\log\log n}{\log\log\log n + c} - 1.$$

In particular, Theorem 6.9 implies that n cannot have simultaneously few non-zero digits in its r-ary and in its s-ary representations, when r and s are multipicatively independent integers.

6.5 Exercises

EXERCISE 6.1. Let $\beta > 1$ be a real number. Prove that a real number is normal to base β^2 if it is normal to base $-\beta$. Is the converse true?

EXERCISE 6.2 (*cf.* [155]). Let r and s be positive integers with $\gcd(r, s) \geq 2$. Construct explicitly irrational real numbers ξ for which (6.25) does not hold, whatever the values of $\kappa > 0$ and n_0.

EXERCISE 6.3. Let r and s be positive integers having a common prime divisor. Let $n \geq 1$ be an integer. Give a lower bound for the number of non-zero digits of r^n written in base s.

EXERCISE 6.4 (*cf.* [538]). Let m be a positive integer such that 2^m can be written as a sum of distinct powers of 3. Let k be a positive integer. Show that 2^m can only take 2^{k-1} distinct values modulo 3^k. Deduce that there are only 2^{k-1} residue classes $r_1, \ldots, r_{2^{k-1}}$ modulo $2 \cdot 3^{k-1}$ in which m can lie. For $X \geq 2$, with a suitable choice of k, prove that there are at most $1.62X^{(\log 2)/(\log 3)}$ integers m less than X such that 2^m can be written as a sum of distinct powers of 3.

6.6 Notes

▷ The set of real numbers which are not simply normal to one base but normal to every multiplicatively independent base has Hausdorff dimension one [531–533].

▷ El-Zanati and Transue [278] gave an elementary proof of the following nice result. Let r and s be two multiplicatively independent positive integers. Fix a finite collection G of finite blocks of digits on $\{0, 1, \ldots, s-1\}$ and let $C_s(G)$ be the set of all real numbers whose s-ary expansion does not contain any block from G. If $C_s(G)$ contains a Cantor set, then $C_s(G)$ contains a number whose r-ary expansion contains every finite block of digits on $\{0, 1, \ldots, r-1\}$. This proves that richness to base r does not imply richness to base s, a statement which also follows from Schmidt's proof of Theorem 6.1; see also [342].

▷ Let $r \geq 2$ and $s \geq 2$ be integers. Let ξ be in $(0,1)$ and $\xi_n^{(r)}$ be the rational number obtained by truncating the r-ary expansion of ξ after the nth digit. The problem to determine the largest integer $m_n^{(r,s)}(\xi)$ for which the first $m_n^{(r,s)}(\xi)$ digits of the s-ary expansions of ξ and $\xi_n^{(r)}$ coincide has been studied from the metrical point of view in [124, 211].

▷ Colebrook and Kemperman [201] refined Theorem 6.1 by showing that, given an integer $r \geq 2$ and a real number ξ, there exists a real number η satisfying $V(\eta, r) = V(\xi, r)$ (see the notes at the end of Chapter 4 for the definition) and which is normal to base s, for every integer s such that r and s are multiplicatively independent. This has been further refined by Pushkin [591, 592].

▷ The first use of Riesz product techniques for normality questions appeared in papers by Brown, Moran and Pearce [136, 137]. Subsequent results can be found in [139, 140, 520–522]. It is shown in [136] that, for any integer $b \geq 2$, every real number can be expressed as the sum of four numbers none of which is normal to base b but all of which are normal to all bases multiplicatively independent of b. This has been slightly refined in [134], where it is shown that, in addition, all the four numbers can be taken to be non-simply normal to base b. Further decomposition theorems are given in [137, 138].

▷ An alternative proof of Schmidt's Theorem 6.3 using Riesz product techniques was given in [137]. It extends to non-integer bases. Among other results, it is shown in [137, 138] that if $\mathcal{R}$ and $\mathcal{S}$ are multiplicatively independent sets (this means that r and s are multiplicatively independent for every r in $\mathcal{R}$ and s in $\mathcal{S}$) of algebraic numbers greater than 1, then there are uncountably many real numbers which are normal to every base from $\mathcal{R}$ but not normal to every base from $\mathcal{S}$.

▷ Pollington [573] strengthened Theorem 6.3 by showing that, under the assumptions of that theorem, the set $\mathcal{N}(\mathcal{R}, \mathcal{S})$ of real numbers which are normal to every base from $\mathcal{R}$ but not normal to every base from $\mathcal{S}$ has Hausdorff dimension one; see also [533]. More generally [575], if $(\lambda_n)_{n \geq 1}$ is any real sequence, the set of real numbers ξ such that every translate $x + \lambda_n$, with $n \geq 1$, is normal to every base from $\mathcal{R}$ but not normal to every base from $\mathcal{S}$ has Hausdorff dimension one. This implies that every real number can be expressed as a sum of two numbers from $\mathcal{N}(\mathcal{R}, \mathcal{S})$; see also [138].

▷ Improving an earlier result of Bertrand-Mathis [83], Brown, Moran and Pollington [140] established that, if $\alpha > 1$ is a real number and $r \geq 2$ is an integer, then $r\mathcal{N}(\alpha) = \mathcal{N}(\alpha)$ holds if, and only if, there exists a positive integer k such that α^k or $\alpha^k + \alpha^{-k}$ or $\alpha^k - \alpha^{-k}$ is an integer.

▷ Volkmann [720] proved that for every multiplicatively independent integer r and s, there exist real numbers which are normal to base r and are such that the digits of their expansion to base s have prescribed frequencies; see also [100–102, 588, 590, 721].

▷ The first articles concerned with normality with respect to matrices include [196, 578, 585]. Let $d \geq 2$ be an integer and R, S be rational $d \times d$ matrices with eigenvalues which are algebraic integers other than 0 and roots of unity. A vector $\underline{x}$ is called normal to base R if $(R^n \underline{x})_{n\geq 1}$ is uniformly distributed modulo one in $\mathbb{R}^d$. When R and S commute, Schmidt [629] showed that if the eigenvalues of R are outside the unit circle and $R^m \neq S^n$ for every positive integer m, n, then there are elements of $\mathbb{R}^d$ which are normal to base S but not to base R. For $d \geq 2$, Sigmund [655] gave an elegant proof when R and S are automorphisms. Brown and Moran [133, 135] removed the extra condition on the eigenvalues of R. Schmidt [629] conjectured that the set of vectors normal to base R and the set of vectors normal to base S are equal if, and only if, there exist positive integers m and n such that $R^m = S^n$. This was proved in [141] in the case of two-dimensional matrices; see also [87].

▷ Let R and S be number-theoretical transformations defined on $[0, 1]$ (or on the d-dimensional unit cube). Schweiger [640] showed that, if there exist positive integers m, n such that $R^m = S^n$, then normality to base S (we omit the precise definition) implies normality to base R. He conjectured that the opposite conclusion holds when $R^m = S^n$ has no solution in positive integers m, n. Kraaikamp and Nakada [407] gave two counterexamples to Schweiger's conjecture.

▷ Discrepancy results for the set of real numbers which are not normal to a fixed base $s > 3$, but normal with respect to an arbitrary set of integer bases $r > s$, multiplicatively independent of s, are given in [518].

▷ Feldman and Smorodinsky [302] extended the result of Schmidt [627] to other measures than the Cantor measure. An amusing illustration of their theorem is the following. Flip a coin (perhaps unfair) repeatedly, and write 0 for heads, 1 for tails. This gives the binary expression of a number ξ in $[0, 1]$. Then, unless the coin is almost surely heads or almost surely tails, ξ is with probability one normal to base ten. Kamae [360] made use of specific singular measures; see also [597].

▷ Inspired by a paper of Wagner [722], Kano and Shiokawa [369, 371] proved, for coprime integers b, c with $b, c \geq 2$, the existence of rings of real numbers all of whose non-zero elements are normal to base b but not normal to base bc; see also [357].

▷ Volkmann [717] established that there exist real numbers ξ such that $V_b(\xi) = H_b$ (see the notes at the end of Chapter 4 for the definition) for every integer $b \geq 2$.

▷ Let $b_1, \ldots, b_d$ be integers greater than or equal to 2. Korobov [398] introduced the notion of real numbers $\xi_1, \ldots, \xi_d$ being jointly (or conjunctly) normal to the bases $b_1, \ldots, b_d$; see [402, 431, 434, 583, 661].

▷ Erdős [280] asked how often the ternary expansion of 2^n omits the digit 2. It is likely that there are only finitely many such integers n but the problem is still beyond reach. For $\lambda > 0$ and $X > 1$, let $N_\lambda(X)$ be the number of integers n in $[1, X]$ such that the ternary expansion of the integer part $\lfloor \lambda 2^n \rfloor$ omits the digit 2. Lagarias [418] proved that $N_\lambda(X) \leq 25X^{0.9725}$ for all sufficiently large X. In the case $\lambda = 1$, Narkiewicz [538] obtained the better bound given in Exercise 6.4; see also [378].

7

Diophantine approximation and digital properties

Let ξ be a real number given by its expansion to an integer base $b \geq 2$, that is,

$$\xi = \lfloor \xi \rfloor + \sum_{k \geq 1} \frac{a_k}{b^k} = \lfloor \xi \rfloor + 0 \cdot a_1 a_2 \ldots, \tag{7.1}$$

where the digits a_k are in $\{0, 1, \ldots, b-1\}$ for $k \geq 1$ and infinitely many a_k are different from $b-1$. Looking at (7.1) gives some information on the irrationality exponent of ξ (see Definition E.1). Indeed, a very naïve way to produce good rational approximations to ξ is to search for integers r and s such that $a_{r+1} = \cdots = a_{r+s} = 0$ and to observe that ξ is then close to the rational number $\lfloor \xi \rfloor + \sum_{k=1}^{r} a_k / b^k$. We can elaborate this argument and look merely for repetitions of finite blocks of digits rather than only long strings of 0. Recall that, if $(d_k)_{k \geq 1}$ is a periodic sequence taking its values in $\{0, 1, \ldots, b-1\}$ and such that there are integers r and s with $s \geq 1$ and $d_{r+js+h} = d_{r+h}$ for $j \geq 1$ and $h = 1, \ldots, s$, then there exists an integer p such that

$$\sum_{k \geq 1} \frac{d_k}{b^k} = \frac{p}{b^r (b^s - 1)}. \tag{7.2}$$

Note that the latter fraction may not be written under its lowest form.

Going back to our original discussion, assuming that there are integers r, s, ℓ such that $a_{r+js+h} = a_{r+h}$ for $h = 1, \ldots, s$ and $j = 1, \ldots, \ell$, then $\{\xi\}$ is quite close to the rational number $\xi_{r,s}$ in $[0, 1]$ whose b-ary expansion starts with $a_1 \ldots a_r$ and continues with infinitely many repeated copies of the finite word $a_{r+1} \ldots a_{r+s}$. Since $\{\xi\}$ and $\xi_{r,s}$ have their first $r + (\ell + 1)s$ digits in common, we get

$$|\{\xi\} - \xi_{r,s}| \leq \frac{1}{b^{r+(\ell+1)s}}.$$

The height (see Definition E.4) $H(\xi_{r,s})$ of $\xi_{r,s}$ being at most equal to $b^r(b^s-1)$, we deduce that

$$|\{\xi\} - \xi_{r,s}| \le \frac{1}{H(\xi_{r,s})^{(r+(\ell+1)s)/(r+s)}}. \tag{7.3}$$

If there exist a real number $\mu > 1$ and triples (r, ℓ, s) satisfying (7.3) with $r + s$ arbitrarily large and $(r + (\ell + 1)s)/(r + s) \ge \mu$, then the irrationality exponent of ξ is at least equal to μ. In view of Theorem E.2, this gives non-trivial information only when $\mu > 2$.

These considerations motivate the introduction in Section 7.1 of new exponents of Diophantine approximation which express what can be said on rational approximation to a real number by simply looking at its b-ary expansion. As can be expected, for almost all real numbers these exponents are strictly less than the irrationality exponent. This is discussed in Sections 7.1 and 7.2. Section 7.3 is devoted to Schmidt's (α, β)-games. The next sections are concerned with Diophantine approximation on the middle third Cantor set. Section 7.7 gathers several results on normal numbers with prescribed Diophantine properties. We end this chapter with classical results on the Hausdorff dimension of sets of real numbers with missing digits (or whose digits occur with a prescribed frequency).

7.1 Exponents of Diophantine approximation

Let ξ be a real number given by its expansion in an integer base $b \ge 2$ as in (7.1). If there are positive integers r and s such that $||b^r\xi|| < b^{-s}$, then $a_{r+1}, \ldots, a_{r+s}$ are all equal and their common value is either 0 or $b-1$. This and the discussion below (7.2) motivate the introduction of new exponents of Diophantine approximation [50].

Definition 7.1. Let ξ be an irrational real number. Let $b \ge 2$ be an integer. We denote by $v_b(\xi)$ the supremum of the real numbers v for which the inequality

$$||b^n\xi|| < (b^n)^{-v}$$

has infinitely many solutions in positive integers n. We denote by $v'_b(\xi)$ the supremum of the real numbers v for which the inequality

$$||b^r(b^s-1)\xi|| < (b^{r+s})^{-v} \tag{7.4}$$

has infinitely many solutions in positive integers r and s.

The value $v'_b(\xi)$ coincides with the Diophantine exponent of the infinite word $\boldsymbol{a} = a_1 a_2 \ldots$; see Definition A.2.

Let ξ be an irrational real number and $b \geq 2$ be an integer. Since we can choose $s = 1$ in (7.4), we see that

$$v'_b(\xi) \geq v_b(\xi) \geq 0$$

and

$$\mu(\xi) \geq \max\{v'_b(\xi) + 1, 2\} \geq \max\{v_b(\xi) + 1, 2\}. \tag{7.5}$$

Furthermore, we check (see Exercise 7.1) that

$$v_b(\xi) = v_{b^t}(\xi) \quad \text{and} \quad v'_b(\xi) = v'_{b^t}(\xi), \quad \text{for } t \geq 1. \tag{7.6}$$

The exponent v_b (resp. v'_b) measures how a given real number is approximable by rational numbers whose denominators are powers of b (resp. have the form $b^r(b^s - 1)$ for some positive integers r and s). These rational numbers are not supposed to be written under their lowest form. In particular, if b_1 and b_2 are integers at least equal to 2 and such that the set of prime factors of b_1 is a subset of the set of prime factors of b_2, then $v_{b_2}(\xi)$ cannot be too small when $v_{b_1}(\xi)$ is large. This observation is made more precise in the following lemma.

LEMMA 7.2. *Let b_1 and b_2 be integers at least equal to 2 and such that every prime factor of b_1 divides b_2. Write*

$$b_1 = p_1^{e_1} \ldots p_\ell^{e_\ell}, \qquad b_2 = p_1^{f_1} \ldots p_\ell^{f_\ell} b'_2,$$

where $p_1, \ldots, p_\ell$ are distinct prime numbers, e_i, f_i are positive integers and b'_2 is coprime to $p_1 \ldots p_\ell$. Define

$$\rho(b_1, b_2) := \min_{1 \leq i \leq \ell} \frac{f_i}{e_i}. \tag{7.7}$$

Then, for any real irrational number ξ, we have

$$v_{b_2}(\xi) + 1 \geq \rho(b_1, b_2) \frac{\log b_1}{\log b_2} (v_{b_1}(\xi) + 1).$$

PROOF. Let m_1 and m_2 be positive integers such that $b_1^{m_1}$ divides $b_2^{m_2}$. Set $v = v_{b_1}(\xi)$. Let $\varepsilon = 0$ if $v = 0$ and let ε be in $(0, v)$ otherwise. Let n be a positive integer such that

$$||b_1^n \xi|| < (b_1^n)^{-v+\varepsilon}.$$

Setting $n = (k-1)m_1 + r$ with $0 \leq r < m_1$ and $k \geq 1$, we have

$$||b_2^{m_2 k} \xi|| < (b_1)^{m_1(1+v)-r} (b_2^{m_2}/b_1^{m_1})^k (b_1^{m_1 k})^{-v+\varepsilon},$$

which implies that

$$v_{b_2}(\xi) + 1 \geq \frac{m_1 \log b_1}{m_2 \log b_2} (v_{b_1}(\xi) + 1),$$

since n (hence, k) can be taken arbitrarily large. Hence, we have

$$v_{b_2}(\xi) + 1 \geq \Big(\sup_{m_1, m_2 : b_1^{m_1} | b_2^{m_2}} \frac{m_1}{m_2} \Big) \frac{\log b_1}{\log b_2} (v_{b_1}(\xi) + 1),$$

where, as indicated, the supremum ranges over all positive integers m_1, m_2 such that $b_1^{m_1}$ divides $b_2^{m_2}$. Since $b_1^{m_1}$ divides $b_2^{m_2}$ if, and only if, $m_1 e_i \leq m_2 f_i$ for $i = 1, \ldots, \ell$, we deduce that

$$\sup_{m_1, m_2 : b_1^{m_1} | b_2^{m_2}} \frac{m_1}{m_2} = \min_{1 \leq i \leq \ell} \frac{f_i}{e_i}.$$

This proves the lemma. □

Lemma C.1 implies that almost every real number ξ satisfies $v_b(\xi) = v'_b(\xi) = 0$ for every integer $b \geq 2$. Furthermore, for any given $b \geq 2$, any positive real number v, and any bounded sequence $(d_j)_{j \geq 1}$ of positive integers, we have

$$v_b\Big(\sum_{j \geq 1} \frac{d_j}{b^{\lfloor (v+1)^j \rfloor}} \Big) = v. \tag{7.8}$$

To construct real numbers ξ with a prescribed value for $v'_b(\xi)$ is much more difficult, since we know only very little on the divisors of integers of the form $b^s - 1$. Indeed, the good rational approximations of the form (7.2) may not be written under their lowest form, thus we only have an upper bound for their heights, and, consequently, only a lower bound for $v'_b(\xi)$. To establish that this lower bound is actually the exact value seems to be very difficult.

In view of Lemma 7.2, we propose the following problem. We denote by $\mathcal{B}$ the set consisting of all positive integers which are not perfect powers, thus $\mathcal{B} = \{2, 3, 5, 6, 7, 10, \ldots\}$, and we set $\mathcal{B}_1 = \{1\} \cup \mathcal{B}$. It is convenient to define the function v_1 by

$$v_1(\xi) = \mu(\xi) - 1,$$

for every irrational number ξ.

Problem 7.3. *Let $(v_b)_{b \in \mathcal{B}_1}$ and $(v'_b)_{b \in \mathcal{B}}$ be sequences of real numbers or $+\infty$ satisfying*

$$v_1 \geq 1, \qquad 0 \leq v_b \leq v'_b \leq v_1, \quad \textit{for every } b \in \mathcal{B},$$

and

$$v_{b_2} + 1 \geq \rho(b_1, b_2) \frac{\log b_1}{\log b_2} (v_{b_1} + 1),$$

for every $b_1, b_2 \in \mathcal{B}$ such that every prime factor of b_1 divides b_2. Prove that there exist real numbers ξ such that

$$v_1(\xi) = v_1, \quad v_b(\xi) = v_b \quad \textit{and} \quad v_b'(\xi) = v_b', \quad \textit{for every } b \in \mathcal{B}.$$

Problem 7.3 can be compared with the *Main Problem* studied in [146], where, besides rational approximation, the quality of approximation by algebraic numbers of bounded degree measured by the exponents u_n and w_n^* (see Definition E.12) is also considered.

We end this section with a connection between the exponents v_b and normality. We use an argument already seen in the proof of Theorem 5.7.

PROPOSITION 7.4. *Let ξ be an irrational real number and $b \geq 2$ an integer. If $v_b(\xi)$ is positive, then ξ is not simply normal to base b.*

PROOF. If $v_b(\xi)$ is positive, then there exist an integer $m \geq 4$, a digit d equal to 0 or $b-1$, and arbitrarily large integers N such that, recalling that

$$A_b(d, n, \xi) := \mathrm{Card}\{k : 1 \leq k \leq n, a_k = d\}, \quad \text{for } n \geq 1,$$

we have

$$A_b(d, (m+1)N, \xi) - A_b(d, mN, \xi) = N,$$

thus,

$$\frac{A_b(d, (m+1)N, \xi)}{(m+1)N} - \frac{m}{m+1} \frac{A_b(d, mN, \xi)}{mN} = \frac{1}{m+1}.$$

If $(A_b(d, n, \xi)/n)_{n \geq 1}$ tends to a limit ℓ, then ℓ must satisfy

$$\ell - \frac{m}{m+1} \ell = \frac{1}{m+1},$$

thus $\ell = 1$. Consequently, the digit d does not occur with frequency $1/b$ in the b-ary expansion of ξ. This shows that ξ is not simply normal to base b. □

Note that there exist real numbers ξ satisfying $v_b'(\xi) > 0$ that are normal to base b; see Exercise 7.2.

7.2 Prescribing simultaneously the values of all the exponents v_b

The purpose of this section is to establish a general result, proved by Amou and Bugeaud [50], which solves partially Problem 7.3.

THEOREM 7.5. *Let $(v_b)_{b \in \mathcal{B}_1}$ be a sequence of elements of $\mathbb{R} \cup \{+\infty\}$ satisfying*

$$\frac{1+\sqrt{5}}{2} \le v_b \le v_1, \quad \text{for every } b \in \mathcal{B}, \tag{7.9}$$

and

$$v_{b_2} + 1 \ge \rho(b_1, b_2) \frac{\log b_1}{\log b_2} (v_{b_1} + 1), \tag{7.10}$$

for every $b_1, b_2 \in \mathcal{B}$ such that every prime factor of b_1 divides b_2. There exist uncountably many real numbers ξ such that

$$v_b(\xi) = v_b, \quad \text{for every } b \in \mathcal{B}_1.$$

The lower bound $(1+\sqrt{5})/2$ in (7.9) is typical of proofs whose main argument is the triangle inequality; see [425] or [146, Section 7.7].

To prove Theorem 7.5, we construct inductively the continued fraction expansion of a suitable real number ξ, in such a way that we know all the rational numbers p/q for which $|\xi - p/q|$ is comparable to or smaller than $q^{-(3+\sqrt{5})/2}$. We begin with an auxiliary lemma.

LEMMA 7.6. *Let $b \ge 2$ be an integer and $\nu \ge (3+\sqrt{5})/2$ be a real number. Let m be a positive integer. Let P/Q and R/S be reduced fractions with positive denominators such that*

$$S^\nu \le Q < bS^\nu \tag{7.11}$$

and

$$\frac{1}{Q} \le \left| \frac{P}{Q} - \frac{R}{S} \right| \le \frac{m}{Q}. \tag{7.12}$$

Then, for any reduced fraction A/B with $S \le B < Q$, we have

$$\left| \frac{P}{Q} - \frac{A}{B} \right| > \frac{1}{b(2m)^{1.7} B^\nu}. \tag{7.13}$$

PROOF. Assume first that $S \le B \le S^{\nu-1}/(2m)$. Since this with (7.11) gives $BS \le S^\nu/(2m) \le Q/(2m)$, using

$$\left| \frac{P}{Q} - \frac{A}{B} \right| \ge \left| \frac{A}{B} - \frac{R}{S} \right| - \left| \frac{P}{Q} - \frac{R}{S} \right|$$

together with (7.12), we obtain

$$\left|\frac{P}{Q}-\frac{A}{B}\right| \geq \frac{1}{BS}-\frac{m}{Q} \geq \frac{1}{2BS} \geq \frac{1}{2B^2}.$$

Assume next that $S^{\nu-1}/(2m) < B < Q$. Since this with (7.11) gives $Q < b(2mB)^{\nu/(\nu-1)}$, we obtain

$$\left|\frac{P}{Q}-\frac{A}{B}\right| \geq \frac{1}{BQ} > \frac{1}{bB(2mB)^{\nu/(\nu-1)}},$$

which implies (7.13), since $\nu/(\nu-1) < 1.7$ and $1+\nu/(\nu-1) \leq \nu$. This proves the lemma. □

Proof of Theorem 7.5. Let $(b_j)_{j\geq 1}$ be a sequence of integers from $\mathcal{B}_1$ such that $b_1 = 1$, $b_j \leq j$ for $j \geq 1$, and, for every $b \in \mathcal{B}_1$, there are infinitely many j satisfying $b_j = b$. Let R_0/S_0 be a reduced fraction with $S_0 > 100$. Assume that we have already constructed reduced fractions P_i/Q_i and R_i/S_i for $i = 1, \ldots, j-1$ with $j \geq 1$. (Note that we do not have P_0/Q_0.) Then we construct inductively reduced fractions P_j/Q_j and R_j/S_j as follows.

We first take $Q_j = b_j^{n_j}$ with $n_j \geq 1$ (resp. Q_j prime) if $b_j \geq 2$ (resp. if $b_j = 1$) such that the triple $(S, Q, b) = (S_{j-1}, Q_j, b_j)$ (resp. $(S, Q, b) = (S_{j-1}, Q_j, 2)$) satisfies (7.11) with $\nu = (3+\sqrt{5})/2$, and next take an integer P_j such that P_j/Q_j is reduced and the pair $(R/S, P/Q) = (R_{j-1}/S_{j-1}, P_j/Q_j)$ satisfies (7.12) with $m = m(b) := 2b+2$. Denoting the continued fraction expansion of P_j/Q_j by $[a_0; a_1, \ldots, a_k]$ with $a_k \geq 2$, we define a reduced fraction R_j/S_j by

$$\frac{R_j}{S_j} := [a_0; a_1, \ldots, a_k, \lfloor Q_j^{v-1} \rfloor],$$

where $v = v_{b_j}$ if $v_{b_j} < +\infty$ and $v = j$ if $v_{b_j} = +\infty$.

By construction, we can define a real number ξ by

$$\xi := \lim_{j\to+\infty} \frac{P_j}{Q_j},$$

whose continued fraction expansion has P_j/Q_j and R_j/S_j ($j \geq 1$) among its convergents. We claim that this ξ satisfies the conditions given in the theorem. To this end we consider rational approximations to ξ.

Let j be a positive integer and set $v = v_{b_j}$. Under the above notation, since

$$[a_0; a_1, \ldots, a_k, \lfloor Q_j^{v-1} \rfloor - 1]$$

is the only possible convergent to ξ between P_j/Q_j and R_j/S_j, we have

$$\frac{1}{Q_j(S_j+Q_j)} < \left|\xi - \frac{P_j}{Q_j}\right| < \frac{1}{Q_j(S_j-Q_j)},$$

which implies

$$\frac{1}{2Q_j^{v+1}} < \left|\xi - \frac{P_j}{Q_j}\right| < \frac{2}{Q_j^{v+1}}. \tag{7.14}$$

For the same reason, for any reduced fraction A/B with $Q_j < B < S_j$ $(j \in \mathbb{Z}_{\geq 1})$, we have

$$\left|\xi - \frac{A}{B}\right| \geq \frac{1}{2B^2}.$$

Let A/B be a reduced fraction with $S_{j-1} \leq B < Q_j$ $(j \geq 1)$. Since

$$\left|\frac{P_j}{Q_j} - \frac{A}{B}\right| \geq \frac{1}{BQ_j}, \qquad \left|\xi - \frac{P_j}{Q_j}\right| \leq \frac{2}{Q_j^{v+1}},$$

and since $BQ_j < Q_j^{v+1}/4$, we have

$$\left|\xi - \frac{A}{B}\right| \geq \left|\frac{P_j}{Q_j} - \frac{A}{B}\right| - \left|\xi - \frac{P_j}{Q_j}\right| > \frac{1}{2}\left|\frac{P_j}{Q_j} - \frac{A}{B}\right|.$$

We then infer from Lemma 7.6 that

$$\left|\xi - \frac{A}{B}\right| > \frac{1}{2b(2m(b))^{1.7}B^{(3+\sqrt{5})/2}},$$

where $b = b_j$ and $m(b) = 2b+2$. Consequently, on recalling that $b_j \leq j$ for $j \geq 1$, these estimates prove our claim.

In each step of the inductive procedure, there are at least two choices of P_j/Q_j having the same denominator Q_j; one is less than R_{j-1}/S_{j-1}, and the other is greater than R_{j-1}/S_{j-1}. Hence, we have an infinite directed binary tree of reduced fractions whose infinite paths correspond to real numbers ξ, which are different from each other and have the conditions given in the assertion. This ensures the uncountability of the desired numbers, and completes the proof of the theorem.

For the sake of clarity, we add a few words on (7.10). Let b and b' be distinct elements of $\mathcal{B}$ such that each prime divisor of b' also divides b. Then, for any $Q'_j = (b')^{n'}$ with some positive integer n', we have to estimate from below

$$\left|\xi - \frac{P'_j}{Q'_j}\right| = \left|\xi - \frac{P_j}{Q_j}\right|$$

with $Q_j = b^n$ and $P_j = P'_j(Q_j/Q'_j)$ provided that Q'_j divides Q_j. Since, by (7.14),

$$\left|\xi - \frac{P'_j}{Q'_j}\right| > \frac{1}{2(Q'_j)^{v_{b'}+1}},$$

we have

$$\left|\xi - \frac{P_j}{Q_j}\right| > \frac{1}{2}(Q_j)^{-(v_{b'}+1)(n' \log b')/(n \log b)}.$$

On noting that $n'/n \le \rho(b', b)$ with $\rho(\cdot, \cdot)$ defined in (7.7), we get

$$\left|\xi - \frac{P_j}{Q_j}\right| > \frac{1}{2}(Q_j)^{-\rho(b',\, b)(v_{b'}+1)(\log b')/(\log b)}.$$

In view of the assumption (7.10), this shows that $v_b(\xi) = v_b$. To see this, one should add that, if j is such that $b_j = 1$, then, by assumption, Q_j is a prime number and there are at most finitely many such j for which each prime divisor of Q_j also divides b. □

To conclude this section, we briefly explain why the approach followed in the proof of Theorem 7.5 cannot be applied to construct real numbers ξ with prescribed values for $v'_b(\xi)$, where b is in $\mathcal{B}$. The point is that we do not control all the rational approximations that give the value of $v'_b(\xi)$. Indeed, with the above notation, let us consider the fractions P_j/Q_j for the indices j with $b_j = 1$. Let b be in $\mathcal{B}$. Then, there exist integers T_j, r_j and s_j such that

$$\frac{P_j}{Q_j} = \frac{T_j}{b^{r_j}(b^{s_j} - 1)},$$

where the latter fraction may not be written in reduced form. It may happen that $b^{r_j}(b^{s_j} - 1)$ is not much greater than Q_j. If this is the case for infinitely many j, we may even get that $v'_b(\xi) = v_1(\xi)$. Since there are no ways to control r_j and s_j, we cannot get the exact value of $v'_b(\xi)$.

Additional explicit examples of real numbers with prescribed exponents v_b and v'_b are given in [50].

7.3 Badly approximable numbers to integer bases

Similarly to the classical notion of badly approximable real numbers (see Definition D.8), we introduce the notion of badly approximable numbers to a given integer base.

DEFINITION 7.7. Let $b \geq 2$ be an integer. A real number ξ is *b-badly approximable* if

$$\inf_{n \geq 0} \|\xi b^n\| > 0.$$

Clearly, for an integer $b \geq 2$, a real number is b-badly approximable if, and only if, the blocks of the digit 0 and the blocks of the digit $b-1$ occurring in its b-ary expansion have bounded length.

Although one can easily construct real numbers which are b-badly approximable for some integer $b \geq 2$, it does not seem obvious to confirm the existence of numbers that are b-badly approximable for every integer $b \geq 2$. This was shown in 1966 by Schmidt [631] as one of the many applications of his (α, β)-game. We slightly modify the exposition from [635], taking into account the works of Akhunzhanov [28, 29].

Let $\mathcal{S}$ be a set of real numbers, called the target set. Let α, β and ρ be given positive numbers with $\alpha < 1$ and $\beta < 1$. Consider the following game played by players Black and White. First, Black chooses a closed interval B_1 of length $\lambda(B_1) = |B_1| = 2\rho$ on the real line. Next, White picks a closed interval W_1 contained in B_1 and of length $|W_1| = \alpha |B_1|$. Then, Black chooses a closed interval B_2 contained in W_1 and of length $|B_2| = \beta |W_1|$, etc. In this way, a nested sequence of closed intervals

$$B_1 \supset W_1 \supset B_2 \supset W_2 \supset \dots$$

is constructed, with lengths

$$|B_k| = 2\rho(\alpha\beta)^{k-1} \quad \text{and} \quad |W_k| = 2\rho\alpha(\alpha\beta)^{k-1} \quad (k \geq 1).$$

Clearly, the intersection

$$\bigcap_{k \geq 1} B_k = \bigcap_{k \geq 1} W_k$$

consists of a single point. If this point lies in $\mathcal{S}$, then we say that White wins the game. Furthermore, if White is able to win the game no matter how Black plays, then $\mathcal{S}$ is called an (α, β, ρ)-winning set. A set $\mathcal{S}$ which is (α, β, ρ)-winning for every positive ρ is called an (α, β)-winning set. A set $\mathcal{S}$ which is (α, β)-winning for every β in $(0, 1)$ is called an α-winning set. A set is termed winning if it is α-winning for some positive real number α.

Any (α, β, ρ)-winning set is uncountable. Moreover, Schmidt [631] proved that a winning set has full Hausdorff dimension and that a countable intersection of winning sets is a winning set.

Throughout this section, we make the additional assumption that α is less than 1/3 since this allows some simplification. Our aim is to establish the following statement, essentially proved by Akhunzhanov [28, 29].

THEOREM 7.8. *There are uncountably many real numbers ξ such that*

$$\left|\xi - \frac{p}{q}\right| > \frac{1}{2^{15}q^2}, \quad \text{for all integers } p, q \text{ with } q \geq 1, \tag{7.15}$$

and

$$||\xi b^n|| > b^{-1100b(\log 3b)}, \quad \text{for all integers } b \geq 2 \text{ and } n \geq 1. \tag{7.16}$$

Consequently, there are uncountably many badly approximable real numbers that are b-badly approximable for every integer $b \geq 2$.

We keep the notation from Section 4.4. In order to deduce from Theorem 7.8 the existence of real numbers having maximal entropy to base b for no integer $b \geq 2$, we first claim that, for a real number ξ, an integer $b \geq 2$, and positive integers n and t, we always have

$$p(tn, \xi, b) \leq tb^t p(n, \xi, b^t). \tag{7.17}$$

To see this, observe that any block of tn consecutive digits of ξ to base b is composed of t_1 digits of ξ to base b followed by $n-1$ digits of ξ to base b^t and by $t-t_1$ digits of ξ to base b, for some integer t_1 with $0 \leq t_1 \leq t-1$. If the real number ξ satisfies (7.16), then, putting $t = \lceil 1100b(\log 3b) \rceil$, it has (at least) one missing digit in its expansion to base b^t, yielding that

$$p(n, \xi, b^t) \leq (b^t - 1)^n, \quad \text{for } n \geq 1.$$

We deduce from (7.17) that

$$p(tn, \xi, b) \leq tb^t (b^t - 1)^n,$$

and, by taking the logarithm, dividing by tn and letting n tend to infinity, it then follows that

$$E(\xi, b) \leq \frac{\log(b^t - 1)}{t} \leq \log b - \frac{1}{tb^t}.$$

We have established the following consequence of Theorem 7.8.

COROLLARY 7.9. *There exist uncountably many badly approximable real numbers ξ such that*

$$E(\xi, b) \leq \log b - \frac{1}{b^{1110b(\log 3b)}}, \quad \text{for every } b \geq 2.$$

With some additional effort, it is possible to show that the set of badly approximable real numbers which are b-badly approximable for every integer $b \geq 2$ has Hausdorff dimension one. The same result holds for the set of real numbers having maximal entropy to base b for no integer $b \geq 2$; see [29].

We split the proof of Theorem 7.8 into five auxiliary lemmas. Throughout this section, for given positive real numbers α, β, ρ and for every positive integer k, we set

$$\rho_k = \rho(\alpha\beta)^{k-1} \quad \text{and} \quad \rho'_k = \rho\alpha(\alpha\beta)^{k-1} = \alpha\rho_k.$$

Furthermore, $B(x, \rho)$ denotes the open interval centred at x and of radius ρ.

LEMMA 7.10. *Let α be such that $0 < \alpha < 1/3$. Let $\rho > 0$. Let $B(x, \rho)$ be an open real interval and I be a closed subinterval of $B(x, \rho)$ of length at most $2\alpha\rho$. Then, there exists x' in $B(x, \rho)$ such that $B(x', \alpha\rho) \subset B(x, \rho)$ and $B(x', \alpha\rho)$ has empty intersection with I.*

PROOF. Our assumption on α implies that $\alpha < 1 - 2\alpha$. Consequently, if the centre of I is less than or equal to x, then I is included in the interval $(x - \rho, x + (1 - 2\alpha)\rho)$ and $x' := x + \rho - \alpha\rho$ has the required property. Likewise, if the centre of I is greater than x, then we take $x' := x - \rho + \alpha\rho$. □

LEMMA 7.11. *Let α, β be such that $0 < \alpha < 1/3$, $0 < \beta < 1$. Suppose that Black begins his play with an interval of length 2ρ. Set $c = \min\{\alpha\rho, \alpha^3\beta^2/2\}$. Then, the set of real numbers ξ such that $|\xi - p/q| > cq^{-2}$ for all integers p and $q \geq 1$ is (α, β, ρ)-winning.*

PROOF. Set $R = (\alpha\beta)^{-1/2}$. If $2\rho \leq (\alpha\beta)^2$, then put $k_0 = 1$, otherwise, let k_0 be the integer defined by the inequalities $(\alpha\beta)^2 < 2\rho_{k_0} \leq \alpha\beta$. White chooses $W_1, \ldots, W_{k_0-1}$ arbitrarily. Fix an arbitrary non-negative integer k. We describe White's strategy for choosing W_{k_0+k} such that

$$\left|\xi - \frac{p}{q}\right| > \frac{c}{q^2} \qquad \text{for all } \xi \in W_{k_0+k}, R^k \leq q < R^{k+1}, \gcd(p, q) = 1.$$

Note that, for any distinct $p_1/q_1, p_2/q_2$ with $R^k \leq q_1, q_2 < R^{k+1}$, we have

$$\left|\frac{p_1}{q_1} - \frac{p_2}{q_2}\right| = \left|\frac{p_1 q_2 - p_2 q_1}{q_1 q_2}\right| > \frac{1}{R^{2k+2}}.$$

Since $2\rho_{k_0} \leq \alpha\beta$, we get

$$2\rho_{k_0+k} = 2\rho(\alpha\beta)^{k_0+k-1} = 2\rho_{k_0} R^{-2k} \leq R^{-2k-2},$$

thus B_{k_0+k} contains at most one point p/q with $R^k \le q < R^{k+1}$ and White has to worry only over an interval I of length $2c/q^2$, thus, of length at most $2cR^{-2k}$. Our choices of c and k_0 imply that $2cR^{-2k} < 2\alpha\rho_{k_0+k}$, since we have $2c \le \alpha^3\beta^2 < 2\alpha\rho_{k_0}$ if $k_0 \ge 2$ and $2c \le 2\alpha\rho$ if $k_0 = 1$. Consequently, by Lemma 7.10, White can choose an interval W_{k_0+k} included in B_{k_0+k}, of length $2\rho'_{k_0+k}$, and having empty intersection with I. Thus, for every $\xi \in W_{k_0+k}$ and for every integer p, q with $R^k \le q < R^{k+1}$, we have $|\xi - p/q| > c/q^2$. Since k is arbitrary, we deduce that, for every $\xi \in \cap_{k\ge 0} W_{k_0+k}$, and every integer p, q with $q \ge 1$, we have $|\xi - p/q| > c/q^2$. This proves the lemma. □

LEMMA 7.12. *Let α, β be such that $0 < \alpha < 1/3$, $0 < \beta < 1$. Let $\rho \ge \alpha\beta/4$. Let b be an integer satisfying*

$$b \ge 4(\alpha^2\beta)^{-1}. \tag{7.18}$$

Then, the set of real numbers having no digit 0 and no digit $b-1$ in the b-ary expansion of their fractional part is (α, β, ρ)-winning.

PROOF. Let Black begin with the ball B_1 of length 2ρ. Since $\rho \ge \alpha\beta/4$, there exists a positive integer n_0 such that

$$1/4 > \rho_{n_0} := (\alpha\beta)^{n_0-1}\rho \ge (\alpha\beta)/4. \tag{7.19}$$

For $j \ge 1$, define the positive integer n_j by

$$1/(4b^j) > \rho_{n_j} := (\alpha\beta)^{n_j-1}\rho \ge (\alpha\beta)/(4b^j). \tag{7.20}$$

Since $\alpha\beta > 1/b$, the sequence $(n_j)_{j\ge 0}$ is strictly increasing. We will describe White's strategy to select the balls $W_{n_0}, W_{n_1}, \ldots$

Let $j \ge 1$ be an integer. The real numbers whose jth digit is equal to 0 or to $b-1$ are lying in intervals of length $2b^{-j}$, which are distant from each other by at least

$$b^{1-j}(1 - 2/b) \ge b^{1-j}/2 > 2\rho_{n_{j-1}},$$

using (7.19) and (7.20). Hence, White has to worry over at most one interval I included in $B_{n_{j-1}}$ and of length at most $2b^{-j}$. We infer from (7.19) and (7.20) that $b^{1-j} \le 4\rho_{n_{j-1}}(\alpha\beta)^{-1}$, hence

$$b^{-j} \le 4(\alpha\beta b)^{-1}\rho_{n_{j-1}} \le \alpha\rho_{n_{j-1}},$$

by (7.18) and (7.20). By Lemma 7.10, White can select an interval $W_{n_{j-1}}$ included in $B_{n_{j-1}}$ and having empty intersection with I. This describes the winning strategy of White when the target set is the set of real numbers having no digit 0 and no digit $b-1$ in the b-ary expansion of their fractional part. □

LEMMA 7.13. *Let $\cup_{j=1}^{\infty} P_j$ be a partition of the set of positive integers into disjoint arithmetic progressions P_j with first term m_j and common differences d_j. Given real numbers $0 < \alpha < 1/3$, $0 < \beta < 1$ and $\rho > 0$, let $\beta_j = \beta(\alpha\beta)^{d_j-1}$ and $\rho_j = \rho(\alpha\beta)^{m_j-1}$, for $j \ge 1$. If, for every $j \ge 1$, the set $\mathcal{S}_j$ is an $(\alpha, \beta_j, \rho_j)$-winning set, then the intersection $\cap_{j=1}^{\infty} \mathcal{S}_j$ is an (α, β, ρ)-winning set.*

PROOF. Note that, for $n \ge 1$, the radius of $B_{m_j+(n-1)d_j}$ is

$$\rho(\alpha\beta)^{m_j+(n-1)d_j-1} = (\alpha\beta_j)^{n-1}\rho_j.$$

White has an (α, β, ρ)-winning strategy $W_1, W_2, \ldots$ for the target set $\cap_{j=1}^{\infty} \mathcal{S}_j$. Namely, for $j \ge 1$, he has to play according to the $(\alpha, \beta_j, \rho_j)$-winning strategy for the target set $\mathcal{S}_j$ on his $m_j, m_j + d_j, m_j + 2d_j, \ldots$ turns. Then, for every $j \ge 1$, the intersection $\cap_{i=1}^{\infty} W_i$ belongs to $\mathcal{S}_j$. The lemma follows. □

LEMMA 7.14. *The set of even positive integers can be represented as the disjoint union $\cup_{j=2}^{\infty} P_j$ of arithmetic progressions P_j with first terms m_j and common differences d_j such that*

$$m_j \le d_j \le 100j(\log 3j)^2, \quad \textit{for } j \ge 2. \tag{7.21}$$

PROOF. For $j \ge 2$, put $v_j = 100j(\log 3j)^2$ and let d_j be the power of 2 defined by the inequalities

$$1 \le \frac{v_j}{2} < d_j \le v_j.$$

Observe that the sequence $(d_j)_{j\ge1}$ satisfies $\sum_{j\ge1} d_j^{-1} < 1/2$ and is non-decreasing. Let P_2 be the arithmetic progression of even integers starting with $m_2 = 2$ and with common difference d_2. Let $k \ge 2$ be such that the disjoint arithmetic progressions of even integers $P_2, \ldots, P_k$ have already been constructed. Since the density of the union $P_2 \cup \ldots \cup P_k$ is equal to $d_2^{-1} + d_3^{-1} + \cdots + d_k^{-1}$, which is less than $1/2$, there exists an even positive integer m outside this union. Let m_{k+1} be the smallest positive integer with this property. Since m_{k+1} is not congruent to m_j modulo d_j and d_j divides d_{k+1} for $j = 1, \ldots, k$, the arithmetic progression

$$P_{k+1} := \{m_{k+1} + hd_{k+1} : h \ge 0\}$$

is disjoint from the union $P_2 \cup \ldots \cup P_k$. Furthermore, we deduce from the minimality of m_{k+1} that $m_{k+1} \le d_{k+1}$. This proves the lemma. □

PROOF OF THEOREM 7.8. Set $\alpha = 1/4$, $\beta = 1/2$ and $\rho = 1$. Let $(m_j)_{j\geq 2}$, $(d_j)_{j\geq 2}$ and $(P_j)_{j\geq 2}$ be as in Lemma 7.14. Set $\beta_1 = \alpha\beta^2$ and $\rho_1 = \alpha\beta$. By Lemma 7.11 the set $\mathcal{S}_1$ of real numbers ξ such that $|\xi - p/q| > 2^{-15}q^{-2}$ for every p/q with $q \geq 1$ is $(\alpha, \beta_1, \rho_1)$-winning. Put $d_1 = 2$ and $m_1 = 1$. Denote by P_1 the set of odd integers and observe that $P_1 \cup P_2 \cup \ldots$ is a partition of the set of positive integers into arithmetic progressions.

For $b \geq 2$, set $\beta_b = \beta(\alpha\beta)^{d_b-1}$ and $\rho_b = (\alpha\beta)^{m_b-1}$. Since $d_b \geq m_b$, we get that $\rho_b \geq \alpha\beta_b/4$. Let s_b be the positive integer defined by the inequalities

$$\frac{4}{\alpha^2\beta_b} \leq b^{s_b} < \frac{4b}{\alpha^2\beta_b}. \tag{7.22}$$

By Lemma 7.12, the set $\mathcal{S}_b$ of real numbers having no digit 0 and no digit $b^{s_b} - 1$ in their b^{s_b}-ary expansion is $(\alpha, \beta_b, \rho_b)$-winning. Note that any element ξ of this set satisfies $||\xi b^{ns_b}|| \geq b^{-s_b}$ for every $n \geq 0$, thus $||\xi b^n|| \geq b^{-2s_b}$ for every $n \geq 0$. Further, it follows from (7.21) and (7.22) that

$$b^{s_b} < 128b \cdot 8^{100b(\log 3b)^2} \leq b^{550b(\log 3b)}.$$

Lemma 7.13 implies that the set $\cap_{j\geq 1}\mathcal{S}_j$ is $(\alpha, \beta, 1)$-winning and we have shown that every element ξ in this set satisfies (7.15) and (7.16). This proves the theorem. □

We end this section by pointing out that Lemma 7.11 immediately implies that the set of badly approximable numbers is a winning set, a result established in [631].

7.4 Almost no element of the middle third Cantor set is very well approximable

We have already mentioned that a covering argument shows that $v_b(\xi) = v'_b(\xi) = 0$ for almost every real number ξ and all bases $b \geq 2$. Furthermore, Theorem E.3 asserts that almost every real number ξ satisfies $v_1(\xi) = 1$. Weiss [728] (see also [576, 729]) established that the latter result also holds for almost every real number ξ in the middle third Cantor set K, where 'almost all' and 'almost no' now refer to the standard measure μ_K supported on K; see Definition C.7. Note that the measure μ_K was used for establishing Theorem 6.1, whose proof actually gives the following statement.

THEOREM 7.15. *Almost every real number in the middle third Cantor set K is normal to every integer base which is not a power of* 3. *Consequently, the set of real numbers in K which are normal to every integer base which is not a power of* 3 *has the same Hausdorff dimension as K, namely* $(\log 2)/(\log 3)$.

By Proposition 7.4 and Theorem 7.15, we have $v_b(\xi) = 0$ for μ_K-almost every element ξ of K and every integer $b \geq 2$ which is not a power of 3. Following the method of [728], we slightly extend this result.

THEOREM 7.16. *With respect to the standard measure μ_K on the middle third Cantor set, we have*

$$v_1(\xi) = 1, \quad v_b(\xi) = v_b'(\xi) = 0 \quad (b \geq 2),$$

for μ_K-almost all ξ in the middle third Cantor set.

The key tool for the proofs of the results established in the present section is the fact, established in Lemma C.8, that, setting

$$\gamma = (\log 2)/(\log 3) \quad \text{and} \quad C = 4 \cdot 3^{\gamma},$$

the measure μ_K satisfies

$$\mu_K(B(x, \varepsilon\rho)) \leq C\varepsilon^{\gamma}\mu_K(B(x, \rho)), \tag{7.23}$$

for every x in $[0, 1]$ and every ε, ρ with $0 < \varepsilon \leq 1$ and $0 < \rho < 1$.

PROOF. Observe first that for every non-negative integer Q, every q, q_0 in $\{2^Q, \ldots, 2^{Q+1} - 1\}$ and every integer p, p_0 such that $p/q \neq p_0/q_0$, we have

$$\left|\frac{p}{q} - \frac{p_0}{q_0}\right| \geq \frac{1}{qq_0} > 2^{-2(Q+1)} > 2 \cdot 2^{-2(Q+2)}.$$

Consequently, the intervals $B(p/q, 2^{-2(Q+2)})$ and $B(p_0/q_0, 2^{-2(Q+2)})$ are disjoint. This implies that the μ_K-measure of the union of the intervals $B(p/q, 2^{-2(Q+2)})$ over all the rational numbers p/q with p and q coprime, $0 < p/q < 1$ and $2^Q \leq q < 2^{Q+1}$ is bounded by 1, independently of Q.

Let $w > 1$ be a real number. Let Q_0 be a positive integer such that $q^{-1-w} \leq 2^{-2(Q+2)}$ holds for all Q and q with $Q \geq Q_0$ and $2^Q \leq q < 2^{Q+1}$.

Using (7.23) with $\varepsilon = q^{-1-w}2^{2(Q+2)}$, we get

$$\begin{aligned}
&\sum_{0<p/q<1,(p,q)=1,q\geq 2^{Q_0}} \mu_K(B(p/q, q^{-1-w})) \\
&= \sum_{Q\geq Q_0} \sum_{*} \mu_K(B(p/q, q^{-1-w})) \\
&\leq \sum_{Q\geq Q_0} \sum_{*} C\Big(\frac{q^{-1-w}}{2^{-2(Q+2)}}\Big)^{\gamma} \mu_K(B(p/q, 2^{-2(Q+2)})) \qquad (7.24) \\
&\leq \sum_{Q\geq Q_0} C\,2^{\gamma(-Q(1+w)+2Q+4)} \sum_{*} \mu_K(B(p/q, 2^{-2(Q+2)})) \\
&\leq \sum_{Q\geq Q_0} C\,2^{\gamma(4+Q(1-w))},
\end{aligned}$$

where $\sum_*$ means that the summation is taken over the rationals p/q such that $0 < p/q < 1, \gcd(p,q) = 1$ and $2^Q \leq q < 2^{Q+1}$. The final sum in (7.24) is finite since $w > 1$. Consequently, the sum

$$\sum_{0<p/q<1,(p,q)=1,q\geq 1} \mu_K(B(p/q, q^{-1-w}))$$

converges, and we apply Lemma C.1 to establish the claimed result on v_1.

Let $b \geq 2$ be an integer and v be a real number with $0 < v < 1/2$. Let r, s be integers with $r \geq 0$ and $s \geq 1$. For every integer p with $0 < p < b^r(b^s - 1)$, we have

$$B\Big(\frac{p}{b^r(b^s-1)}, \frac{1}{(b^r(b^s-1))^{1+v}}\Big) \subset B\Big(\frac{p}{b^r(b^s-1)}, \frac{3}{b^{(r+s)(1+v)}}\Big).$$

It follows from (7.23) applied with $\varepsilon = 9b^{-(r+s)v}$ and $\rho = 1/(3b^{r+s})$ that

$$\begin{aligned}
&\mu_K\Big(B\Big(\frac{p}{b^r(b^s-1)}, \frac{1}{(b^r(b^s-1))^{1+v}}\Big)\Big) \\
&\qquad\qquad \leq 4\cdot 3^{3\gamma} b^{-(r+s)v\gamma} \mu_K\Big(B\Big(\frac{p}{b^r(b^s-1)}, \frac{1}{3b^{r+s}}\Big)\Big).
\end{aligned}$$

Since, for fixed r, s and for p varying from 0 to $b^r(b^s-1)$, the intervals

$$B\Big(\frac{p}{b^r(b^s-1)}, \frac{1}{3b^{r+s}}\Big)$$

are disjoint, the sum of their μ_K-measures is at most equal to 1. Consequently,

$$\sum_{0<p<b^r(b^s-1)} \mu_K\Big(B\Big(\frac{p}{b^r(b^s-1)}, \frac{1}{(b^r(b^s-1))^{1+v}}\Big)\Big) \leq 4\cdot 3^{3\gamma}\, b^{-(r-s)v\gamma}$$

and, since v is positive, the sum

$$\sum_{r\geq 0}\sum_{s\geq 1}\sum_{0<p<b^r(b^s-1)} \mu_K\Big(B\Big(\frac{p}{b^r(b^s-1)}, \frac{1}{(b^r(b^s-1))^{1+v}}\Big)\Big) \tag{7.25}$$

converges. Restricting the second sum in (7.25) to $s = 1$, Lemma C.1 implies that μ_K-almost every element ξ in K satisfies $v_b((b-1)\xi) = 0$, hence also $v_b(\xi) = 0$. A further application of Lemma C.1, this time to the converging sum (7.25), yields that μ_K-almost every ξ in K satisfies $v_b(\xi) = v'_b(\xi) = 0$. This completes the proof of the theorem. □

7.5 Playing games on the middle third Cantor set

In Section 7.3, the Schmidt game is played on closed intervals contained in $[0,1]$. As was first observed by Fishman [306], it can be played as well on other closed subsets $\mathcal{K}$ of $[0,1]$. We say that a subset $\mathcal{S}$ of $[0,1]$ is (α,β)-winning on $\mathcal{K}$ if $\mathcal{S}\cap\mathcal{K}$ is (α,β)-winning for Schmidt's game played on the metric space $\mathcal{K}$ with the metric induced from $[0,1]$. Playing the game on $\mathcal{K}$ amounts to choosing balls in $[0,1]$ according to the rules of a game played on $[0,1]$, but with the additional constraint that the centres of all the balls have to lie in $\mathcal{K}$.

We are able to refine Theorem 7.8 as follows. We prove that there are elements on the middle third Cantor set K which are b-badly approximable for every base $b\geq 2$. The key property of K needed for the proof is that the standard measure μ_K enjoys a decay property (Lemma C.9) which allows us to play Schmidt's game on it.

THEOREM 7.17. *There exist a positive real number c and uncountably many real numbers ξ in the middle third Cantor set which are badly approximable and, for all integers $b\geq 2$ and $n\geq 1$, satisfy*

$$||\xi b^n|| > b^{-cb(\log b)}.$$

Since the analogue of Theorem 7.17 holds with the middle third Cantor set replaced by any set of numbers with missing digits, we can deduce the following result, pointed out in [155].

THEOREM 7.18. *Let ε be a positive real number and $b_0\geq 2$ be an integer. There exist a positive real number c, depending only on ε and b_0, and uncountably many real numbers ξ such that*

$$E(\xi, b_0) < \varepsilon$$

and

$$E(\xi, b) \leq \log b - \frac{1}{b^{cb(\log b)}}, \qquad \text{for every } b \geq 2.$$

Before establishing (a more general result than) Theorem 7.17, we introduce the notion of an absolutely decaying measure [131, 383].

DEFINITION 7.19. Let μ be a locally finite Borel measure on $\mathbb{R}$. Let C, γ be positive real numbers. We say that μ is (C, γ)-*absolutely decaying* if there exists $\rho_0 > 0$ such that for all $0 < \rho \leq \rho_0$, x in the support of μ, $y \in \mathbb{R}$ and $\varepsilon > 0$, we have

$$\mu\big(B(x, \rho) \cap B(y, \varepsilon\rho)\big) < C\varepsilon^{\gamma}\mu\big(B(x, \rho)\big)\,. \tag{7.26}$$

We say that μ is absolutely decaying if it is (C, γ)-absolutely decaying for some positive C, γ.

Fishman [306] proved that the Schmidt game can be played on sets which are the supports of absolutely decaying measures. Note that (7.26) clearly holds with $C = \gamma = 1$ when μ is the Lebesgue measure. Furthermore, it follows from Lemma C.9 that the standard measure μ_K on the middle third Cantor set (Definition C.7) is $(2^7, \gamma)$-decaying, with $\gamma = (\log 2)/(\log 3)$.

LEMMA 7.20. *Let $\mathcal{K}$ be the support of a (C, γ)-absolutely decaying measure on $\mathbb{R}$, and let α be such that*

$$0 < \alpha \leq \frac{1}{4}\Big(\frac{1}{3C}\Big)^{1/\gamma}. \tag{7.27}$$

Then for every $0 < \rho < \rho_0$, $x \in \mathcal{K}$ and $z \in \mathbb{R}$, there exists $x' \in \mathcal{K}$ with

$$B(x', \alpha\rho) \subset B(x, \rho) \tag{7.28}$$

and

$$B(x', \alpha\rho) \cap B(z, \alpha\rho) = \emptyset. \tag{7.29}$$

PROOF. If z does not belong to $B(x, 2\alpha\rho)$, then it is sufficient to take $x' = x$. Otherwise, let x_1 and x_2 be the endpoints of $B(x, \rho)$ with $x_1 < x_2$. By (7.26) applied successively with $y = x_1$ and $y = x_2$, we have

$$\mu\big(B(x_1, \alpha\rho)\big) < C\alpha^{\gamma}\mu\big(B(x, \rho)\big) < \frac{\mu\big(B(x, \rho)\big)}{3}$$

and

$$\mu\big(B(x_2, \alpha\rho)\big) < C\alpha^{\gamma}\mu\big(B(x, \rho)\big) < \frac{\mu\big(B(x, \rho)\big)}{3}\,.$$

Furthermore, again by (7.26) with $y = x$, we get

$$\mu(B(x, 4\alpha\rho)) < C(4\alpha)^\gamma \mu(B(x,\rho)) \leq \frac{\mu(B(x,\rho))}{3},$$

so there is a point x' in $\mathcal{K} \cap B(x,\rho)$ which does not belong to $B(x, 4\alpha\rho) \cup B(x_1, \alpha\rho) \cup B(x_2, \alpha\rho)$. Hence, $B(x', \alpha\rho)$ satisfies (7.28) and (7.29), since z belongs to $B(x, 2\alpha\rho)$. □

To establish Theorem 7.17, we follow the proof of Theorem 7.8, replacing the use of Lemma 7.10 by that of Lemma 7.20. This shows that, at every step, White can choose a suitable ball whose centre lies in the support of the absolutely decaying measure. A minor further change is needed. Namely, because of (7.27), we cannot select $\alpha = 1/4$ any more. In the case of the middle third Cantor set, we can take $\alpha = 2^{-18}$. We omit the details and refer the reader to [131].

7.6 Elements of the middle third Cantor set with prescribed irrationality exponent

In this section, following [148], we construct real numbers whose ternary expansion and continued fraction expansion are explicitly given, in order to establish that there are points in the middle third Cantor set K with any prescribed irrationality exponent and that there are badly approximable numbers in K.

THEOREM 7.21. *Let μ be a real number with $\mu \geq 2$. The middle third Cantor set K contains uncountably many elements whose irrationality exponent is equal to μ. For any $\lambda \geq 2$, the real number*

$$\xi_{\mu,\lambda} := \sum_{j \geq 1} \frac{2}{3^{\lfloor \lambda \mu^j \rfloor}}$$

is an element of K with

$$\mu(\xi_{\mu,\lambda}) = \mu \quad \textit{and} \quad v_3(\xi_{\mu,\lambda}) = v_3'(\xi_{\mu,\lambda}) = \mu - 1. \tag{7.30}$$

Furthermore, for any $\lambda \geq 2$, the real number $\xi_{2,\lambda}$ is badly approximable.

The key ingredient of the proof is an elementary lemma from the theory of continued fractions, the so-called *Folding Lemma*, quoted as Theorem D.3.

PROOF OF THEOREM 7.21. Let λ and μ be real numbers at least equal to 2. Set $v = \lfloor \lambda\mu \rfloor$. Since $v \geq 4$, the open real interval with endpoints

$[0;1,1,1,2] = 5/8$ and $[0;1,1,1,4] = 9/14$ contains at least one reduced rational number whose denominator is equal to 3^v. Consequently, there exists a positive integer r, with $1 \le r \le 3^v$ and r coprime with 3, such that the continued fraction of $r/3^v$ reads

$$\frac{r}{3^v} = [0;1,1,a_3,\ldots,a_{h-1},a_h],$$

where $h \ge 4$ and $a_h \ge 2$.

For $k \ge 1$, set

$$u_k = \lfloor \lambda \mu^{k+1} \rfloor - 2\lfloor \lambda \mu^k \rfloor$$

and

$$v_k = u_k + 2u_{k-1} + \cdots + 2^{k-1}u_1 + 2^k v = u_k + 2v_{k-1} = \lfloor \lambda \mu^{k+1} \rfloor,$$

with $v_0 = v$. Applying Theorem D.3 to $r/3^v$ with $t = 3^{u_1}$, then to $r/3^v + (-1)^h/3^{u_1+2v}$ with $t = 3^{u_2}$, and so on, we get a sequence $(d_k)_{k\ge1}$ of integers such that the real number

$$\begin{aligned}\xi_{\mathbf{u}} :&= \frac{r}{3^v} + \frac{(-1)^h}{3^{u_1+2v}} - \frac{1}{3^{u_2+2(u_1+2v)}} - \cdots \\ &\qquad - \frac{1}{3^{u_k+2(u_{k-1}+\cdots+2^{k-2}u_1+2^{k-1}v)}} - \cdots \\ &= [0;1,1,a_3,\ldots,a_{h-1},a_h,3^{u_1}-1,1,a_h-1, \\ &\qquad\qquad a_{h-1},\ldots,a_3,2,3^{u_2}-1,1,1,a_3,\ldots],\end{aligned}$$

associated to the sequence $\boldsymbol{u} = (u_k)_{k\ge1}$, satisfies

$$\frac{1}{3^{u_{k+1}+2v_k}} \le \left|\xi_{\mathbf{u}} - \frac{d_k}{3^{v_k}}\right| \le \frac{2}{3^{u_{k+1}+2v_k}}, \quad \text{for } k \ge 1. \tag{7.31}$$

Here, d_k is the nearest integer to $\xi_{\mathbf{u}} 3^{v_k}$, for $k \ge 1$, and is not divisible by 3. We show now that the sequence $(d_k/3^{v_k})_{k\ge1}$ comprises all the best rational approximations to $\xi_{\mathbf{u}}$. Write

$$\xi_{\mathbf{u}} = [0;a_1,a_2,\ldots], \qquad \frac{p_j}{q_j} = [0;a_1,a_2,\ldots,a_j], \quad j \ge 1,$$

for the continued fraction expansion of $\xi_{\mathbf{u}}$ and for its convergents, respectively. It follows from Theorem D.1 that

$$\frac{1}{(a_j+2)q_{j-1}^2} < \left|\xi_{\mathbf{u}} - \frac{p_{j-1}}{q_{j-1}}\right| < \frac{1}{a_j q_{j-1}^2}, \quad \text{for } j \ge 2. \tag{7.32}$$

Let $m \ge 2$ be an integer. By construction, we have

$$a_{2^m(h+1)} = 3^{u_{m+1}} - 1, \;\; p_{2^m(h+1)-1} = d_m, \;\; q_{2^m(h+1)-1} = 3^{v_m}, \tag{7.33}$$

and we deduce from (7.31) that

$$\frac{1}{q_{2^m(h+1)-1}^{2+u_{m+1}/v_m}} \le \left| \xi_{\mathbf{u}} - \frac{p_{2^m(h+1)-1}}{q_{2^m(h+1)-1}} \right| \le \frac{2}{q_{2^m(h+1)-1}^{2+u_{m+1}/v_m}}. \tag{7.34}$$

Let j satisfy $2^{m-1}(h+1) < j < 2^m(h+1)$. Again by construction, the partial quotient a_j is less than or equal to $3^{u_{m-1}} - 1$ and q_{j-1} exceeds $3^{v_{m-1}}$, by (7.33). Then, (7.32) yields

$$\begin{aligned} \left| \xi_{\mathbf{u}} - \frac{p_{j-1}}{q_{j-1}} \right| &> \frac{1}{(a_j+2)q_{j-1}^2} \\ &\ge \frac{1}{(3^{u_{m-1}}+1)q_{j-1}^2} \ge \frac{1}{2q_{j-1}^{2+u_{m-1}/v_{m-1}}}. \end{aligned} \tag{7.35}$$

Since $(u_k)_{k\ge1}$ is non-decreasing, it follows from (7.34) and (7.35) that

$$\mu(\xi_{\mathbf{u}}) = \limsup_{k\to+\infty} \frac{u_{k+1}}{v_k} = \mu. \tag{7.36}$$

The ternary expansion of the real number

$$2\Big(\frac{r}{3^v} + \frac{(-1)^h}{3^{u_1+2v}} - \xi_{\mathbf{u}}\Big) = \sum_{k\ge2} \frac{2}{3^{v_k}} \tag{7.37}$$

contains only the digits 0 and 2. Let $\boldsymbol{u}' = (u'_k)_{k\ge1}$ be a sequence defined from the above sequence $\boldsymbol{u}$ by setting $u'_1 = 1$, $u'_{2k} = u_k$ and $u'_{2k+1} \in \{1,2\}$ for $k \ge 1$. Then, a similar proof yields that the real number $2(r3^{-v} + (-1)^h 3^{-u'_1-2v} - \xi_{\mathbf{u}'})$ has the same irrationality exponent as $\xi_{\mathbf{u}}$ and its ternary expansion contains only the digits 0 and 2. If $\mu = 2$ and $k \ge 0$, then $\lfloor\lambda\mu^{k+1}\rfloor - 2\lfloor\lambda\mu^k\rfloor$ is in $\{0,1\}$. In that case, the real number $\xi_{\mathbf{u}}$ is badly approximable. Thus, we have constructed uncountably many real numbers with the requested property. Furthermore, (7.30) follows from (7.36), (7.37), (7.5) and (7.8). This completes the proof of the theorem. □

We describe below another class of real numbers having the property that both their b-ary expansion (for some integer $b \ge 2$) and their continued fraction expansion are explicitly determined. It was found by Böhmer [110] in 1927 and rediscovered nearly 50 years later by Danilov [213] and, independently, by Adams and Davison [21], who extended a result of Davison [218]; see also Bundschuh [169] and the references given on [46, p. 297].

THEOREM 7.22. *For a positive real irrational number* $\alpha = [0; a_1, a_2, \dots]$ *in* $(0,1)$ *and an integer* $b \geq 2$, *define*

$$\xi_b(\alpha) = (b-1) \sum_{j=1}^{+\infty} \frac{1}{b^{\lfloor j/\alpha \rfloor}}.$$

For $n \geq 1$, *let* p_n/q_n *denote the nth convergent to* α *and set*

$$t_n := \frac{b^{q_n} - b^{q_{n-2}}}{b^{q_{n-1}} - 1},$$

where $q_{-1} = 0$ *and* $q_0 = 1$. *Then, we have*

$$\xi_b(\alpha) = [0; t_1, t_2, t_3, \dots] \tag{7.38}$$

and the irrationality exponent of $\xi_b(\alpha)$ *is given by*

$$\mu(\xi_b(\alpha)) = 1 + \limsup_{n \to +\infty} [a_n; a_{n-1}, \dots, a_1]. \tag{7.39}$$

Taking $b = 3$ in Theorem 7.22, we get elements of the Cantor set with prescribed irrationality exponent. Note that the set of exponents obtained is included in $[(3+\sqrt{5})/2, +\infty)$. We omit the proof of (7.38) and leave the proof of (7.39) as Exercise 7.5. This computation was done in [2].

7.7 Normal and non-normal numbers with prescribed Diophantine properties

In the present section, we gather various results on the existence of (absolutely) normal numbers with specific Diophantine properties.

THEOREM 7.23. *Let* $b \geq 2$ *be an integer and* $\mu \geq 2$ *be a real number. There exist real numbers that are normal to base* b *and whose irrationality exponent is equal to* μ.

We omit the proof of Theorem 7.23, established in [50]. We only mention that it is constructive and rests on Theorem 5.2 and on an extension of Theorem D.3 worked out by Amou [49]. Namely, for an integer $b \geq 3$ and a real number $\mu > 2$, the real number

$$\sum_{j \geq 1} \frac{1}{b^{\lfloor \mu^j \rfloor} 2^j} \tag{7.40}$$

is normal to base b and its irrationality exponent is equal to μ. Likewise, for $\mu > 2$, the real number

$$\sum_{j\geq 1} \frac{1}{2^{\lfloor \mu^j \rfloor} 3^j} \qquad (7.41)$$

is normal to base b and its irrationality exponent is equal to μ. Replacing μ^j by $j2^j$ in (7.40) and (7.41), we get real numbers with irrationality exponent 2, which are, respectively, normal to base b and to base 2.

On [517, p. 203], Montgomery asked for 'a normal number whose continued fraction coefficients are bounded'. No explicit example of such a number has been exhibited yet. It is written in [517] that, when Kaufman's paper [374] appeared, Roger Baker observed that Kaufman's result combined with Lemma 1.8 and Jarník's estimates [353] for the Hausdorff dimension of the set $S_{\{1,2,\ldots,M\}}$ of real numbers with partial quotients at most equal to some integer $M \geq 9$ (this dimension is at least equal to $1-(M(\log 2)/4)^{-1}$ and does not exceed $1-(8M\log M)^{-1}$) implies the existence of badly approximable real numbers which are absolutely normal.

In view of Lemma 1.8 (see Exercise 7.6), in order to prove that a given real set S contains absolutely normal numbers, it is sufficient to construct a probability measure on S whose Fourier transform does not decrease too slowly at infinity. This is precisely what Kaufman [374, 375] did, first when S is the set of badly approximable numbers and second when S is the set of real numbers whose irrationality exponent is equal to or exceeds μ, where $\mu > 2$ is an arbitrary real number. The first statement of the next theorem has been proved in [599].

THEOREM 7.24. *Let $\mathcal{A}$ be a finite set of positive integers. Let $S_{\mathcal{A}}$ be the set of real numbers with partial quotients in $\mathcal{A}$. If the Hausdorff dimension of $S_{\mathcal{A}}$ exceeds* $1/2$, *then there are real numbers in $S_{\mathcal{A}}$ which are absolutely normal. Moreover, these numbers form a set of Hausdorff dimension* $\dim S_{\mathcal{A}}$. *In particular, the set of badly approximable numbers which are absolutely normal has Hausdorff dimension one.*

Under the assumption of the theorem, for every positive real number δ with $1/2 < \delta < \dim S_{\mathcal{A}}$, Queffélec and Ramaré [596, 599], inspired by [374], have constructed a measure $\mu_{\mathcal{A}}$ supported by the set $S_{\mathcal{A}}$ with the following properties. There exist positive real numbers ε, c_1, c_2 such that

$$\mu_{\mathcal{A}}(B) \leq c_1 |B|^{\delta}, \quad \text{for every Borelian real set } B, \qquad (7.42)$$

and the Fourier transform of $\mu_{\mathcal{A}}$ satisfies $|\hat{\mu}_{\mathcal{A}}(x)| \leq c_2(1+|x|)^{-\varepsilon}$ for every real number x. The dimension result follows from (7.42) and Lemma C.5. This has been pointed out to me by Sanju Velani.

A similar strategy implies the following result, proved by Kaufman [375]; see also [738].

THEOREM 7.25. *Let $\mu > 2$ be given. The set of real numbers with irrationality exponent at least equal to μ and which are absolutely normal has the same Hausdorff dimension as the set of real numbers with irrationality exponent at least equal to μ, namely $2/\mu$.*

A suitable adaptation of the results proved in Chapter 9 of [738] shows that one can replace 'at least equal to μ' by 'equal to μ' in Theorem 7.25.

A little more work is required to apply this strategy to the set of Liouville numbers [144].

THEOREM 7.26. *There are uncountably many Liouville numbers which are absolutely normal.*

The proof of Theorem 7.26 makes use of a result of Bluhm [107, 108], very much inspired by [375].

We conclude this section by two results on numbers normal to no base proved by Pollington [569] and Bugeaud [144], respectively.

THEOREM 7.27. *Let $\mu > 2$ be given. The set of real numbers with irrationality exponent at least equal to μ and which are absolutely non-normal has Hausdorff dimension $2/\mu$.*

Like in Theorem 7.25, one can replace 'at least equal to μ' by 'equal to μ' in the above statement.

THEOREM 7.28. *There are uncountably many Liouville numbers which are absolutely non-normal.*

7.8 Hausdorff dimension of sets with missing digits

We state and prove two results on the size of sets of real numbers with missing digits or whose digits have prescribed frequencies.

THEOREM 7.29. *Let $b \geq 2$ be an integer. Let $\mathcal{S}$ be a subset of $\{0, 1, \cdots, b-1\}$ with $s \geq 2$ elements. Then, the Hausdorff dimension of the set $E_{b,\mathcal{S}}$ of real numbers in $[0, 1]$ having only digits from $\mathcal{S}$ in their b-ary expansion is equal to $(\log s)/(\log b)$.*

PROOF. For $k \geq 1$, denote by E_k the union of the so-called basic intervals $[a/b^k, (a+1)/b^k]$, where a runs through the integers from $\{0, 1, \ldots, b^k - 1\}$ whose digits in their representation in base b all belong to $\mathcal{S}$. The covering of $E_{b,\mathcal{S}}$ consisting of the s^k basic intervals of E_k of length b^{-k} gives that

$$\mathcal{H}^{(\log s)/(\log b)}(E_{b,\mathcal{S}}) \le s^k (b^{-k})^{(\log s)/(\log b)} = 1,$$

thus the Hausdorff dimension of $E_{b,\mathcal{S}}$ is at most equal to $(\log s)/(\log b)$.

Note that $E_{b,\mathcal{S}}$ and its closure differ from at most countably many elements, so they have the same Hausdorff dimension. Let $U_1, \ldots, U_m$ be closed, proper subintervals of $[0,1]$, whose union covers $\overline{E}_{b,\mathcal{S}}$ (since $\overline{E}_{b,\mathcal{S}}$ is compact, we can restrict ourselves to finite coverings). For $i = 1, \ldots, m$, let h_i be the integer such that $b^{-h_i-1} \le |U_i| < b^{-h_i}$. Let k be an integer greater than $h_1, \ldots, h_m$. Then U_i intersects at most

$$s^{k-h_i} = s^k b^{-h_i (\log s)/(\log b)} \le s^k b^{(\log s)/(\log b)} |U_i|^{(\log s)/(\log b)}$$

basic intervals of E_k. However, by assumption, the union $U_1 \cup \ldots \cup U_m$ intersects all s^k basic intervals of length b^{-k}. We then get that

$$s^k \le \sum_{i=1}^{m} s^{k+1} |U_i|^{(\log s)/(\log b)},$$

thus

$$\sum_{i=1}^{m} |U_i|^{(\log s)/(\log b)} \ge 1/s.$$

This gives the requested lower bound for the Hausdorff dimension of $E_{b,\mathcal{S}}$. □

Corollary 7.30. *Let $b \ge 2$ be an integer. The set of real numbers which are not normal to base b has Hausdorff dimension one.*

Proof. For $r \ge 1$, the set of real numbers having only the digits $1, 2, \ldots, b^r - 1$ in their expansion to base b^r has Hausdorff dimension $(\log(b^r - 1))/(\log b^r)$. By Theorem 4.4, these numbers are not normal to base b. As r can be taken arbitrarily large, this proves the corollary. □

We end this section with a metric result of Eggleston [272] on sets of real numbers with prescribed frequencies of digits. In the sequel it is understood that $0 \log 0 = 0$.

Theorem 7.31. *Let $b \ge 2$ be an integer. Let $p_0, p_1, \ldots, p_{b-1}$ be real numbers such that $p_0 + p_1 + \cdots + p_{b-1} = 1$ and $0 \le p_i \le 1$ for $i = 0, \ldots, b-1$. Denote by $F(p_0, \ldots, p_{b-1})$ the set of real numbers ξ in $[0,1]$ such that, for $d = 0, \ldots, b-1$, the digit d occurs with frequency p_d in the b-ary expansion of ξ. Then, the Hausdorff dimension of $F(p_0, \ldots, p_{b-1})$ is equal to*

$$-\frac{1}{\log b} \sum_{d=0}^{b-1} p_d \log p_d. \tag{7.43}$$

PROOF. Let k be a positive integer and $d_1, \dots, d_k$ be in $\{0, 1, \dots, b-1\}$. Define the measure μ on $[0, 1]$ by putting

$$\mu([0 \cdot d_1 \dots d_{k-1} d_k, 0 \cdot d_1 \dots d_{k-1}(d_k + 1)]) = p_{d_1} \cdots p_{d_k},$$

where $0 \cdot d_1 \dots d_{k-1} d_k$ means $\sum_{h=1}^{k} d_h / b^h$. It follows from Lemma 1.8 that, for $d = 0, \dots, b-1$, the digit d occurs with frequency p_d in the b-ary expansion of μ-almost all real numbers ξ. This shows that $\mu(F(p_0, \dots, p_{b-1})) = 1$.

Let ξ be in $[0, 1]$ and write $I_k(\xi)$ for the interval of the form $[a/b^k, (a+1)/b^k]$ to which ξ belongs, where a is a rational integer. Note that

$$\log \mu(I_k(\xi)) = n_{0,k} \log p_0 + \cdots + n_{b-1,k} \log p_{b-1},$$

where $n_{d,k}$ denotes the number of digits d among the first k digits in the b-ary expansion of ξ, for $d = 0, \dots, b-1$. Let s be a positive real number. If ξ is in $F(p_0, \dots, p_{b-1})$, then

$$\begin{aligned} \frac{1}{k} \log \frac{\mu(I_k(\xi))}{|I_k(\xi)|^s} &= \frac{n_{0,k}}{k} \log p_0 + \cdots + \frac{n_{b-1,k}}{k} \log p_{b-1} + s \log b \\ &\underset{k \to +\infty}{\longrightarrow} p_0 \log p_0 + \cdots + p_{b-1} \log p_{b-1} + s \log b. \end{aligned}$$

This shows that

$$\lim_{k \to +\infty} \frac{\mu(I_k(\xi))}{|I_k(\xi)|^s} = 0 \quad \text{or } +\infty,$$

according as s is smaller or greater than the value (7.43). We conclude by applying Lemma C.5. □

Observe that if in Theorem 7.31 we choose $p_0 = \ldots = p_{s-1} = 1/s$, for some integer s with $2 \le s \le b$, then the Hausdorff dimension of the set $F(p_0, \dots, p_{s-1}, 0, \dots, 0)$ is equal to that of the set $E_{b,\{0,1,\dots,s-1\}}$, namely to $(\log s)/(\log b)$.

7.9 Exercises

EXERCISE 7.1. Prove (7.6).

EXERCISE 7.2. Let $b \ge 2$ be an integer. By suitably modifying the b-ary expansion of a number normal to base b and with low discrepancy, prove that there exist real numbers ξ which are normal to base b and satisfy $v_b'(\xi) > 0$.

EXERCISE 7.3 (*cf.* [131, 526, 528]). Let $(t_j)_{j\ge1}$ be a lacunary sequence of positive real numbers greater than 1. Prove that the set of real numbers ξ such that $\inf_{j\ge1} ||\xi t_j||$ is positive is a winning set.

EXERCISE 7.4 (*cf.* [132]). Set $\gamma = (\log 2)/(\log 3)$. Prove that for every ξ in the middle third Cantor set K, there exists a rational number p/q in K such that $|\xi - p/q| < 1/(qQ)$ and $q \le 3^{Q^\gamma}$. Deduce that for every irrational number ξ in K there are infinitely many rational numbers p/q in K such that

$$\left|\xi - \frac{p}{q}\right| < \frac{1}{q} \cdot \left(\frac{\log 3}{\log q}\right)^{1/\gamma}.$$

EXERCISE 7.5. Apply Exercise E.1 to prove equality (7.39).

EXERCISE 7.6. Let μ be a probability measure supported on a set S of real numbers and such that there exists a positive constant c with

$$|\hat{\mu}(x)| \le \bigl(\log(3 + |x|)\bigr)^{-c}, \quad \text{for } x \in \mathbb{R}.$$

Prove that μ-almost all elements of S are normal to every integer base $b \ge 2$.

EXERCISE 7.7. Use the properties of the measure $\mu_{\mathcal{A}}$ introduced below Theorem 7.24 to show that the set of badly approximable real numbers ξ for which (2.34) holds for every prime number p has full Hausdorff dimension.

EXERCISE 7.8. Let v be a positive real number and $b \ge 2$ be an integer. Prove that the Hausdorff dimension of the set of real numbers ξ such that $v_b(\xi) \ge v$ is equal to $1/(v+1)$.

EXERCISE 7.9. Let $\mu \ge 2$ be a real number. Prove that the Hausdorff dimension of the set of real numbers ξ such that $\mu(\xi) \ge \mu$ is equal to $2/\mu$.

7.10 Notes

▷ For every integer $b \ge 2$ and every real number $v \ge 1$, Dubickas [258] constructed explicitly a real number ξ such that $v'_b(\xi) = v$.

▷ By means of the metric number theory, and more precisely of the theory of intersective sets [268, 289, 290], it is proved in [50] that, for every $v \ge 0$, the set of real numbers ξ such that $v_1(\xi) = 2v + 1$ and $v_b(\xi) = v$ for every b in $\mathcal{B}$ has Hausdorff dimension $1/(v+1)$. Combined with Proposition 7.4, this implies that the set of absolutely non-normal real numbers has Hausdorff dimension one. The theory of winning sets and that of intersective sets are two powerful tools to estimate the Hausdorff dimension of countable intersections of sets.

▷ Nilsson [541, 542] studied the sets

$$F(c) := \{\xi \in [0,1] : ||\xi 2^n|| \ge c, \text{for every } n \ge 0\},$$

where c is a positive real number. Among other results, he proved that the Hausdorff dimension of $F(c)$ depends continuously on c, is constant on intervals which form a set of full Lebesgue measure, and is self-similar; see also [354, 416].

▷ An alternative proof of the existence of real numbers which are b-badly approximable for every integer $b \ge 2$ was obtained by Färm in his doctoral thesis [294]. It does not depend on the theory of (α, β)-games.

▷ Let $(t_n)_{n\ge1}$ be a lacunary sequence of positive real numbers greater than 1. The set of real numbers ξ such that $\inf_{n\ge1} ||\xi t_n||$ is positive is a winning set [526, 528] (see also [131] and Exercise 7.3). In particular, for every real number $\alpha > 1$, the set of real numbers ξ such that $\inf_{n\ge1} ||\xi\alpha^n||$ is positive is a winning set.

▷ McMullen [456] introduced the notions of strong and absolute winning sets. Other types of games are described in [550, 715].

▷ The main result of [425] implies that the Hausdorff dimension of the set of real numbers ξ in the middle third Cantor set K for which $v_3(\xi) = v$ is equal to $\gamma/(v+1)$, where $\gamma = (\log 2)/(\log 3)$ is the Hausdorff dimension of K; see also [297]. Kristensen [408] (see also [425]) established that, for any given real number $w \ge 1$, the Hausdorff dimension of the set of ξ in K such that $w_1(\xi) \ge w$ is at most equal to $2\gamma/(w+1)$.

▷ It was proved in [383, 384, 409] that the set of badly approximable numbers lying in the middle third Cantor set K has the same Hausdorff dimension as K. However, μ_K-almost no element of K is badly approximable. This follows from a nice result from [276] asserting that the sequence of partial quotients of almost every element of K contains all finite words on $\{1, 2, \ldots\}$.

▷ Kleinbock, Lindenstrauss and Weiss [383] established that μ_K-almost every element ξ in the middle third Cantor set satisfies $w_n(\xi) = w_n^*(\xi) = n$ for every positive integer n; see also Kristensen [408]. Let $n \ge 2$ be an integer and $w > (2n - 1 + \sqrt{4n^2+1})/2$ be a real number. Set $n_j = \lfloor (w+1)^j \rfloor$ for $j \ge 1$ and $\xi_w := 2\sum_{j\ge1} 3^{-n_j}$. Then, the same proof as that of Theorem 7.8 from [146] shows that $w_n(\xi_w) = w_n^*(\xi_w) = w_1(\xi_w) = w$.

▷ It follows from results of Urbański [700, 701] and arguments from Section 7.5 that, for any (finite or infinite) set $\mathcal{N}$ of at least two positive integers, there exist real numbers which are b-badly approximable for every integer $b \geq 2$ and all of whose partial quotients belong to $\mathcal{N}$. Furthermore, the set of these numbers has the same Hausdorff dimension as the set of real numbers all of whose partial quotients belong to $\mathcal{N}$.

▷ The use of the Folding Lemma to describe the continued fraction expansion of some real numbers defined by their b-ary expansion was first discovered by Shallit [644, 645]. As noted in [395], the continued fraction expansions of the Korobov's numbers, defined in Corollary 5.3, can be given explicitly by repeated use of the Folding Lemma.

▷ Real sets of Lebesgue measure zero can be classified by means of Fourier–Stieltjes transforms. We denote by $M_0(\mathbb{T})$ the set of probability measures on the torus $\mathbb{T}$ whose Fourier coefficients tend to zero at infinity. A closed set of real numbers in $\mathbb{T}$ is called an M_0-set if it carries a probability measure from $M_0(\mathbb{T})$. Kahane and Salem [358] asked whether, for any measure μ in $M_0(\mathbb{T})$, it is true that μ-almost all elements of $\mathbb{T}$ are normal to base 2. A negative answer was given by Lyons [453, 454], who showed the existence of measures μ in $M_0(\mathbb{T})$ such that the set of real numbers which are not normal to base 2 has positive μ-measure. He further gave a lower bound for the speed of convergence of $|\hat{\mu}(n)|$ to 0; see also [551, 596].

▷ Let $b \geq 2$ be an integer and D be a finite word on $\{0, 1, \ldots, b-1\}$. Volkmann [712] computed the Hausdorff dimension of the set of real numbers ξ in the b-ary expansion of which the block D does not occur.

▷ Starting with the seminal papers of Eggleston [272, 273], there is a broad literature on Hausdorff dimension of sets of real numbers with special digit properties. Barreira, Saussol and Schmeling [63, 64] used multifractal analysis to calculate the Hausdorff dimension of sets of real numbers defined by linear or even non-linear relations between the frequencies of their digits; see also [65, 66, 546].

▷ In a series of papers, Volkmann [713, 714] computed the Hausdorff dimension of sets of real numbers whose b-ary expansion enjoys special properties. Li and Dekking [441] generalized Eggleston's result [272] to the situation where the relative frequencies of groups of digits in the expansion are prescribed. Further extensions have been given in [442, 547]. In the latter paper, the authors computed the Hausdorff dimension of the set of real numbers in $[0, 1]$ whose decimal expansion enjoys the following property: a certain group of digits occurs with a certain prescribed frequency and another group of digits also occurs with a (possibly different) prescribed frequency. We emphasize that the groups need not be disjoint, unlike in previous works.

▷ Real numbers ξ such that all the finite blocks occur in their b-ary expansion with prescribed frequencies given by a probability measure μ and such that the sequence $(\xi b^n)_{n\geq 1}$ has a low discrepancy with respect to the measure μ have been constructed in [432].

▷ Färm [296] considered the set of real numbers ξ such that, for every integer $b \geq 2$ and every $d = 0, 1, \ldots, b-1$, the frequency of the digit d in the b-ary expansion of ξ does not exist. He proved that this set has Hausdorff dimension one. This solves a question of Olsen, who obtained partial results in [548, 549]. Further metrical results on non-normal numbers are given in [35, 36].

▷ Kotova [404] proved that there exist uncountably many real numbers ξ such that the digit 1 occurs in their ternary expansion with frequency equal to ξ.

▷ Knichal [385] studied from the metric point of view the set of real numbers ξ whose binary expansion $0{\cdot}a_1a_2\ldots$ satisfies $|\mathrm{Card}\{k : 1 \leq k \leq n : a_k = 0\} - n/2| = O(n^{\delta})$ for some real number δ with $0 < \delta < 1/2$.

▷ Barral and Seuret [61] considered sets of real numbers that are approximated at a certain rate by rational numbers which are selected according to the asymptotic frequencies of their digits in some fixed integer base.

▷ Cesaretto and Vallée [183] studied the Hausdorff dimension of real numbers with bounded digit averages.

8

Digital expansion of algebraic numbers

In 1950, at the end of his paper [116] Émile Borel wrote:

En définitive, le problème de savoir si les chiffres d'un nombre tel que $\sqrt{2}$ *satisfont ou non à* toutes *les lois que l'on peut énoncer pour des chiffres choisis au hasard me paraît toujours être un des problèmes les plus importants qui se posent aux mathématiciens.* [Finally, the problem of knowing whether the digits of a number such as $\sqrt{2}$ satisfy *all* laws that can be stated for randomly chosen digits seems to me to be one of the most important of mathematical problems.]

Not very much has been proved since the publication of Émile Borel's note, and his problem is very far from being solved. There have been some numerical experiments, according to which the binary digits (or the decimal digits) of classical constants like e, π, $\log 2$ and $\sqrt{2}$ are more or less randomly distributed; see e.g. [95, 96, 120, 270, 668].

Throughout the present chapter, b denotes an integer at least equal to 2 and ξ is a real number. There exists a unique infinite sequence $\boldsymbol{a} = (a_k)_{k\geq 1}$ of integers in $\{0, 1, \ldots, b-1\}$ such that

$$\xi = \lfloor \xi \rfloor + \sum_{k\geq 1} \frac{a_k}{b^k} = 0 \cdot a_1 a_2 \ldots, \tag{8.1}$$

and infinitely many a_k are different from $b-1$. The sequence $\boldsymbol{a}$ is ultimately periodic if, and only if, ξ is rational. With a slight abuse of notation, we also denote by $\boldsymbol{a}$ the infinite word $a_1 a_2 \ldots$

We present three results showing that, in various senses, the b-ary expansion of an algebraic irrational number is not 'too simple'. Then, we explain how precise information on the combinatorial structure of Sturmian sequences implies some information on the b-ary expansion of certain real numbers, including e and the badly approximable numbers.

In Section 8.6, we investigate an apparently innocent problem posed and solved by Mahler, concerning the existence of multiples of an irrational number containing infinitely many occurrences of a given finite block of digits.

8.1 A transcendence criterion

Let $b \geq 2$ be an integer. Let ξ be a real algebraic irrational number of degree d. It follows from Liouville's Theorem E.5 that there exists an integer m, depending only on ξ, such that, for any positive integer r, we have

$$||b^r \xi|| \geq b^{-m} b^{-(d-1)r}.$$

This implies that the rth digit of the b-ary expansion of ξ cannot be followed by $(d-1)r+m$ digits 0. By contraposition, we have established the following transcendence criterion.

THEOREM 8.1. *Let $b \geq 2$ be an integer and ξ be an irrational number whose b-ary expansion is given by (8.1). Let $(n_j)_{j\geq 1}$ be the increasing sequence of positive integers composed of the indices k for which $a_k \geq 1$. If $(n_{j+1}/n_j)_{j\geq 1}$ is unbounded, then ξ is transcendental.*

The above discussion shows how a Diophantine inequality can be translated into a combinatorial transcendence criterion involving expansions to an integer base. Replacing the use of Liouville's Theorem E.5 by that of Roth's Theorem E.7 (resp., of Ridout's Theorem E.8), one sees that the same conclusion holds under the weaker assumption that

$$\limsup_{j\to+\infty} n_{j+1}/n_j > 2 \quad (\text{resp., } \limsup_{j\to+\infty} n_{j+1}/n_j > 1).$$

Using the exponent of approximation v_b introduced in Section 7.1, we have explained above that Liouville's theorem (resp., Roth's theorem, Ridout's theorem) implies that ξ is transcendental if $v_b(\xi) = +\infty$ (resp., if $v_b(\xi) > 1$, if $v_b(\xi) > 0$).

This has been formulated in [637, Section I.6] and, in a more general form, in [303, Theorem 1.31]. As is shown in the next theorem, a stronger statement follows from a deep extension of the Roth Theorem E.7, namely from the Schmidt Subspace Theorem E.10.

We begin with a transcendence criterion proved in [15]; see also [6].

THEOREM 8.2. *Let $b \geq 2$ be an integer and ξ be an irrational, real number. If $v_b(\xi)$ or $v'_b(\xi)$ is positive, then ξ is transcendental.*

The statement of Theorem 8.2 is redundant since $v'_b(\xi)$ is always at least equal to $v_b(\xi)$.

PROOF. If $v_b(\xi)$ is positive, then there are a positive real number ε and infinitely many positive integers r such that

$$||b^r\xi|| < (b^r)^{-\varepsilon}.$$

It then follows from Ridout's Theorem E.8 that ξ is transcendental.

If $v'_b(\xi)$ is positive, then there are a positive real number ε and infinitely many pairs of positive integers (r,s) such that

$$||b^r(b^s-1)\xi|| < (b^{r+s})^{-\varepsilon}.$$

Thus, there is an infinite sequence of distinct integer triples $(r_j,s_j,p_j)_{j\geq 1}$ such that $(s_j)_{j\geq 1}$ is non-decreasing, $s_1 \geq 1$ and

$$|b^{r_j+s_j}\xi - b^{r_j}\xi - p_j| < (b^{r_j+s_j})^{-\varepsilon}, \tag{8.2}$$

for every $j \geq 1$. If $(s_j)_{j\geq 1}$ is bounded, then there exists a positive integer s such that $v_b(\xi(b^s-1))$ is positive, and we conclude that $\xi(b^s-1)$, hence also ξ, are transcendental. Consequently, we can assume that the sequence $(s_j)_{j\geq 1}$ is strictly increasing.

We argue by contradiction and assume that ξ is algebraic. Consider the three linearly independent linear forms with real algebraic coefficients:

$$\begin{aligned} L_1(X_1,X_2,X_3) =& X_1, \\ L_2(X_1,X_2,X_3) =& X_2, \\ L_3(X_1,X_2,X_3) =& \xi X_1 - \xi X_2 - X_3. \end{aligned}$$

Fix $j \geq 1$ and observe that

$$\prod_{\ell | b} |b^{r_j+s_j}|_\ell \, |b^{r_j}|_\ell \times \prod_{1\leq i\leq 3} |L_i(b^{r_j+s_j}, b^{r_j}, p_j)| < (b^{r_j+s_j})^{-\varepsilon},$$

where the first product is taken over all the prime divisors of b and, for a prime ℓ, the absolute value $|\cdot|_\ell$ is normalized such that $|\ell|_\ell = \ell^{-1}$. By Theorem E.10, all the triples $(b^{r_j+s_j}, b^{r_j}, p_j)$ with $j \geq 1$ lie in finitely many proper rational subspaces of $\mathbb{Q}^3$. Thus, there exist a non-zero integer triple (z_1,z_2,z_3) and an infinite set of positive integers $\mathcal{J}$ such that

$$z_1 b^{r_j+s_j} + z_2 b^{r_j} + z_3 p_j = 0,$$

for every j in $\mathcal{J}$. Since $(s_j)_{j\geq 1}$ is strictly increasing, z_3 must be non-zero and $p_j/b^{r_j+s_j}$ then tends to the rational number $-z_1/z_3$ when j tends to infinity along $\mathcal{J}$. However, (8.2) implies that $p_j/b^{r_j+s_j}$ tends to ξ as j tends to infinity along $\mathcal{J}$. Since ξ is assumed to be irrational, we get a contradiction. We conclude that ξ is transcendental. The proof is complete. □

8.2 Block complexity of algebraic numbers

As already mentioned in Section 4.4, a natural way to measure the *complexity* of a real number ξ whose b-ary expansion is given by (8.1) is to count the number $p(n, \xi, b) := p(n, \boldsymbol{a}, b)$ of distinct blocks of given length n in the infinite word $\boldsymbol{a} = a_1a_2a_3\ldots$

A first step towards a proof of the normality of irrational algebraic numbers would be a good lower bound for $p(n, \xi, b)$ when ξ is irrational algebraic. The first result of this type, proved in 1997 by Ferenczi and Mauduit [305], asserts that, if ξ is algebraic irrational, then the tail of the expansion of ξ to base b cannot be a Sturmian sequence. Since the Sturmian sequences can be viewed as the 'simplest' non-periodic sequences, this shows that the b-ary expansion of every algebraic irrational number cannot be 'too simple'.

Actually, as pointed out by Allouche [38], the approach of [305] combined with the combinatorial Theorem A.8 of Cassaigne [179] yields a slightly stronger result, namely that

$$\lim_{n\to+\infty} \big(p(n, \xi, b) - n\big) = +\infty, \tag{8.3}$$

for any algebraic irrational number ξ.

The estimate (8.3) follows from a good understanding of the combinatorial structure of Sturmian sequences combined with a combinatorial translation of Ridout's Theorem E.8. The transcendence criterion given in Theorem 8.2, established in [7], yields an improvement of (8.3).

THEOREM 8.3. *For any irrational algebraic number ξ and any integer $b \geq 2$, we have*

$$\lim_{n\to+\infty} \frac{p(n, \xi, b)}{n} = +\infty. \tag{8.4}$$

Although (8.4) considerably strengthens (8.3), it is still very far from what is commonly expected, that is, from confirming that $p(n, \xi, b) = b^n$ holds for every positive n when ξ is algebraic irrational.

Besides Theorem 8.2, the key tool for the proof of Theorem 8.3 is the following combinatorial lemma.

LEMMA 8.4. *Let $b \geq 2$ be an integer. Let $\boldsymbol{w} = w_1 w_2 \ldots$ be an infinite word on the alphabet $\{0, 1, \ldots, b-1\}$ which is not ultimately periodic, and set*

$$\xi = \frac{w_1}{b} + \frac{w_2}{b^2} + \cdots$$

If there exists an integer C such that

$$\liminf_{n \to +\infty} \frac{p(n, \boldsymbol{w}, b)}{n} < C,$$

then $v'_b(\xi) \geq 1/(3C+2)$.

PROOF. By assumption, there exists an infinite set $\mathcal{N}$ of positive integers such that

$$p(n, \boldsymbol{w}, b) \leq Cn, \quad \text{for every } n \text{ in } \mathcal{N}. \tag{8.5}$$

Let n be in $\mathcal{N}$. By (8.5) and the *Schubfachprinzip*, there exists (at least) one block B_n of length n having (at least) two occurrences in the prefix of length $(C+1)n$ of $\boldsymbol{w}$. Thus, there are words U_n, U'_n, V_n and V'_n such that $|U_n| < |U'_n|$ and

$$w_1 \ldots w_{(C+1)n} = U_n B_n V_n = U'_n B_n V'_n.$$

If $|U_n B_n| \leq |U'_n|$, then define X_n such that $U_n B_n X_n = U'_n$. Observe that

$$w_1 \ldots w_{(C+1)n} = U_n (B_n X_n)^{1+|B_n|/|B_n X_n|} V'_n \tag{8.6}$$

and

$$\frac{|B_n|}{|B_n X_n|} \geq \frac{|B_n|}{|U_n B_n X_n|} \geq \frac{n}{Cn} \geq \frac{1}{C}. \tag{8.7}$$

Set $W_n := B_n X_n$ and $c_n := |B_n|/|B_n X_n|$.

If $|U'_n| < |U_n B_n|$, then, recalling that $|U_n| < |U'_n|$, we define B'_n by $U'_n = U_n B'_n$. Since $B_n V_n = B'_n B_n V'_n$ and $|B'_n| < |B_n|$, the word B'_n is a strict prefix of B_n and B_n is a rational power of B'_n. Thus, there are a positive integer x_n and a rational number y_n such that $0 \leq y_n < 2$ and

$$B'_n B_n = {B'_n}^{1+|B_n|/|B'_n|} = {B'_n}^{2x_n + y_n} = ({B'_n}^{x_n})^2 {B'_n}^{y_n}.$$

Observe that

$$2x_n |B'_n| + 2|B'_n| \geq |B'_n B_n|,$$

thus

$$n = |B_n| \leq (2x_n + 1)|B'_n| \leq 3x_n |B'_n|.$$

Consequently, $U_n(B_n'^{x_n})^2$ is a prefix of $\boldsymbol{w}$ such that

$$|B_n'^{x_n}| \geq n/3$$

and

$$\frac{|B_n'^{x_n}|}{|U_n B_n'^{x_n}|} \geq \frac{n}{3} \cdot \frac{1}{(C+1)n - |B_n'^{x_n}|} \geq \frac{1}{(3C+2)}. \tag{8.8}$$

Set $W_n := B_n'^{x_n}$ and $c_n := 1$.

It then follows from (8.6), (8.7) and (8.8) that, for every n in $\mathcal{N}$,

$$U_n W_n^{1+c_n} \quad \text{is a prefix of } \boldsymbol{w}$$

and

$$\frac{|W_n{}^{c_n}|}{|U_n W_n|} \geq \frac{1}{(3C+2)}. \tag{8.9}$$

Thus, ξ is very close to the rational number ξ_n whose b-ary expansion is given by $U_n W_n^\infty$, namely, ξ and ξ_n have at least their first $r_n + (1+c_n)s_n$ (note that this number is an integer) digits in common, where r_n and s_n denote the lengths of U_n and W_n, respectively. Observe that there is an integer p_n such that

$$\xi_n := \frac{p_n}{b^{r_n}(b^{s_n} - 1)}.$$

We then deduce from (8.9) that

$$\left|\xi - \frac{p_n}{b^{r_n}(b^{s_n} - 1)}\right| \leq \frac{1}{b^{r_n + (1+c_n)s_n}} \leq \frac{1}{b^{(r_n+s_n)(1+1/(3C+2))}}.$$

This implies that $v_b'(\xi)$ is at least $1/(3C+2)$, which proves the lemma. □

Combining Theorem 8.2 and Lemma 8.4, we prove that the block complexity of an irrational algebraic number cannot be too small.

PROOF OF THEOREM 8.3. Let ξ be an irrational algebraic number and $b \geq 2$ be an integer. Theorem 8.2 implies that $v_b'(\xi) = 0$, which, in turn, yields (8.4) by Lemma 8.4. □

If we were able to prove that an integer $b \geq 2$ and a real number ξ satisfy $v_b'(\xi) = 0$, then we would immediately get (8.4) by Lemma 8.4. However, we have no single example of a 'classical' transcendental number with this property.

The basic strategy for proving Theorem 8.3 can be summarized as follows: to find a Diophantine property $\mathcal{P}$ shared by the real numbers having small block complexity, but not by the algebraic numbers. At present, any Diophantine property shown to be true for algebraic numbers holds for almost all real numbers. Consequently, the property $\mathcal{P}$ is necessarily satisfied only by almost no numbers. Thus, keeping in mind that almost all real numbers are normal to every integer base, the present strategy cannot be applied to show that algebraic irrational numbers are normal to every integer base.

Theorem 8.3 is too weak for confirming a conjecture of Mahler [473] asserting that there are no algebraic irrational numbers in the middle third Cantor set.

8.3 Zeros in the b-ary expansion of algebraic numbers

As in Section 6.3, for an integer $b \geq 2$, an irrational real number ξ whose b-ary expansion is given by (8.1), and a positive integer n, set

$$\mathcal{NZ}(n, \xi, b) := \mathrm{Card}\,\{k : 1 \leq k \leq n, a_k \neq 0\}.$$

The function $n \mapsto \mathcal{NZ}(n, \xi, b)$ counts the number of non-zero digits among the first n digits of the b-ary expansion of ξ.

It follows from Ridout's Theorem E.8 that

$$\lim_{n \to +\infty} \frac{\mathcal{NZ}(n, \xi, b)}{\log n} = +\infty, \tag{8.10}$$

for every irrational algebraic number ξ; see Exercise 8.4. For the base $b = 2$, this was considerably improved by Bailey, Borwein, Crandall and Pomerance [53], using elementary considerations and ideas from additive number theory. A minor modification of their method allows us to get a similar result for expansions to an arbitrary integer base; see Definition E.4 for the definition of the height of an algebraic number.

THEOREM 8.5. *Let $b \geq 2$ be an integer and ξ an irrational real algebraic number of degree d. Denote by A_d the leading coefficient of the minimal polynomial of $1 + \{\xi\}$ over $\mathbb{Z}$ and by H its height. Then, for any integer n exceeding $(20 b^d d^2 H)^{2d}$, we have*

$$\mathcal{NZ}(n, \xi, b) \geq \frac{1}{b-1} \left(\frac{n}{2(d+1)A_d} \right)^{1/d}.$$

A similar statement was independently proved in [18]. The idea behind the proof of Theorem 8.5 is quite simple and was inspired by a new proof by Knight [386] of the transcendence of the real number $\sum_{j \geq 0} 2^{-2^j}$; see Exercise 8.6. If an irrational real number ξ has only few non-zero digits, then its integer powers $\xi^2, \xi^3, \ldots$, and any finite linear combination of them, cannot have too many non-zero digits. In particular, ξ cannot be a root of an integer polynomial of small degree. This is not at all so simple, since we have to take much care of the carries.

Unlike [53, Theorem 7.1], which depends on Roth's Theorem E.7, Theorem 8.5 above is effective. Using in the proof below Ridout's Theorem E.8 instead of Liouville's Theorem E.5 allows one to show that, under the assumption of Theorem 8.5 and for every positive real number ε, we have

$$\mathcal{NZ}(n, \xi, b) \geq \frac{1-\varepsilon}{b-1} \left(\frac{n}{A_d}\right)^{1/d},$$

provided that n is larger than some integer n_0. This method of proof yields no explicit value for n_0, however.

It immediately follows from Theorem 8.5 that, if $N \geq 10^{12}$ is an integer, then, among the first N digits in the binary expansion of $\sqrt{2}$, there are at least $(N/6)^{1/2}$ digits 1.

We introduce below several quantities that will be used for establishing Theorem 8.5. Let ξ be an irrational real number in $(1, 2)$ whose b-ary expansion is given by (8.1) and set $a_0 = \lfloor \xi \rfloor = 1$. For a positive integer ℓ and non-negative integers n and R, set

$$r_\ell(n, \xi, b) := \sum_{k=1}^{(b-1)^\ell} k \, \mathrm{Card}\{(j_1, \ldots, j_\ell) : j_1 + \cdots + j_\ell = n, a_{j_1} \cdots a_{j_\ell} = k\}$$

and

$$T_\ell(R, \xi, b) = \sum_{m \geq 1} \frac{r_\ell(m + R, \xi, b)}{b^m} = \sum_{m \geq R+1} \frac{r_\ell(m, \xi, b)}{b^{m-R}}. \tag{8.11}$$

Observe that

$$\begin{aligned} r_\ell(n, \xi, b) &\leq (b-1)^\ell \sum_{0 \leq j_1, \ldots, j_\ell \leq n; j_1 + \cdots + j_\ell = n} 1 \\ &= (b-1)^\ell \binom{n + \ell - 1}{\ell - 1} \end{aligned} \tag{8.12}$$

and

$$\xi^\ell = \sum_{m \geq 0} \frac{r_\ell(m, \xi, b)}{b^m}, \tag{8.13}$$

but note that, for $\ell \geq 2$, the sequence $(r_\ell(m, \xi, b))_{m \geq 0}$ is not, in general, the b-ary expansion of ξ^ℓ.

LEMMA 8.6. *For ξ, ℓ and R as above, we have*

$$T_\ell(R,\xi,b) < \frac{(R+\ell)^\ell (b-1)^\ell}{(\ell-1)!(R+1)} \le 2(b-1)^\ell (R+\ell)^{\ell-1}.$$

PROOF. For $\ell \ge 1$, setting

$$\begin{aligned} U_\ell(R,b) &= \sum_{m\ge1} \frac{1}{b^m}\binom{R+m+\ell-1}{\ell-1} \\ &= \sum_{m\ge1} \frac{1}{b^m}\binom{R+m+\ell-2}{\ell-2} + \sum_{m\ge1} \frac{1}{b^m}\binom{R+m+\ell-2}{\ell-1} \\ &= \sum_{m\ge1} \frac{1}{b^m}\binom{R+m+\ell-2}{\ell-2} + \frac{1}{b}\binom{R+\ell-1}{\ell-1} \\ &\qquad + \frac{1}{b}\sum_{m\ge1} \frac{1}{b^m}\binom{R+m+\ell-1}{\ell-1}, \end{aligned} \tag{8.14}$$

where we agree that $\binom{h}{-1} = 0$ for every integer h, it follows from (8.11) and (8.12) that $T_\ell(R,\xi,b) \le (b-1)^\ell U_\ell(R,b)$. Using the recurrence relation

$$U_\ell(R,b) = \frac{b}{b-1}U_{\ell-1}(R,b) + \frac{1}{b-1}\binom{R+\ell-1}{\ell-1}$$

deduced from (8.14), we see that

$$U_\ell(R,b) \le 2U_{\ell-1}(R,b) + \binom{R+\ell-1}{\ell-1}.$$

Since $U_1(R,b) \le 1$, an induction on ℓ gives

$$U_\ell(R,b) \le \sum_{j=0}^{\ell-1}\binom{R+\ell}{j},$$

hence,

$$\begin{aligned} \frac{T_\ell(R,\xi,b)}{(b-1)^\ell} &\le U_\ell(R,b) \\ &< \frac{(R+\ell)^{\ell-1}}{(\ell-1)!}\sum_{n\ge0}\left(\frac{\ell-1}{R+\ell}\right)^n = \frac{(R+\ell)^\ell}{(\ell-1)!(R+1)} \le \frac{\ell(R+\ell)^{\ell-1}}{(\ell-1)!}. \end{aligned}$$

This proves the lemma. □

We are now in a position to establish Theorem 8.5.

Proof of Theorem 8.5. Let ξ in $(1,2)$ be an algebraic number of degree $d \geq 2$ and denote by $A_d X^d + \cdots + A_1 X + A_0$ its minimal defining polynomial over $\mathbb{Z}$ with $A_d \geq 1$. Let $R \geq 0$ and $b \geq 2$ be integers. Set

$$T(R, \xi, b) = \sum_{\ell=1}^{d} A_\ell T_\ell(R, \xi, b). \tag{8.15}$$

It follows from (8.11), (8.13) and (8.15) that

$$\begin{aligned} -b^R A_0 &= b^R \sum_{\ell=1}^{d} A_\ell \sum_{m \geq 0} \frac{r_\ell(m, \xi, b)}{b^m} \\ &= \sum_{m=0}^{R} b^{R-m} \sum_{\ell=1}^{d} A_\ell r_\ell(m, \xi, b) + \sum_{\ell=1}^{d} A_\ell \sum_{m \geq R+1} b^{R-m} r_\ell(m, \xi, b) \\ &= \sum_{m=0}^{R} b^{R-m} \sum_{\ell=1}^{d} A_\ell r_\ell(m, \xi, b) + T(R, \xi, b). \end{aligned}$$

Consequently, $T(R, \xi, b)$ is a rational integer and, for $R \geq 1$,

$$bT(R-1, \xi, b) = T(R, \xi, b) + \sum_{\ell=1}^{d} A_\ell r_\ell(R, \xi, b). \tag{8.16}$$

Set

$$c_1 = (b-1)^{-1} \left(2(d+1)A_d\right)^{-1/d} \quad \text{and} \quad c_2 = (b-1)c_1.$$

Let N be a positive integer such that $\mathcal{NZ}(N, \xi, b) \leq c_1 N^{1/d}$. Since

$$\sum_{0 \leq n \leq N} r_{d-1}(n, \xi, b) \leq (b-1)^{d-1} \left(\mathcal{NZ}(N, \xi, b)\right)^{d-1} \leq c_2^{d-1} N^{1-1/d},$$

the number M of integers R in $[0, N-1]$ for which $r_{d-1}(R, \xi, b) \geq 1$ does not exceed $c_2^{d-1} N^{1-1/d}$. Let us denote these integers by $0 = R_1 < R_2 < \ldots < R_M$, and set $R_{M+1} = N$.

Let I be the subset of $\{1, \ldots, M\}$ consisting of the integers i with $R_{i+1} - R_i \geq c_2^{1-d} N^{1/d}/3$. Observe that

$$\sum_{i \in I} (R_{i+1} - R_i) \geq N - c_2^{1-d} N^{1/d} M/3 \geq 2N/3. \tag{8.17}$$

Let i be in I. Since N exceeds $(20 b^d d^3 H)^d$, we get

$$b^{(R_{i+1} - R_i - d \log N)/(d+1)} \geq 3^{d-1} d^2 H b^{d \log N},$$

and it follows from Liouville's Theorem E.5 that ξ cannot have too long blocks of zeros in its b-ary expansion. More precisely, there is an integer

$$j_i \in [(R_{i+1} - R_i - d\log N)/(d+1), R_{i+1} - R_i - 2d\log N]$$

such that $a_{j_i} \geq 1$, hence $r_1(j_i, \xi, b) \geq 1$ and

$$r_d(R_i + j_i, \xi, b) \geq r_{d-1}(R_i, \xi, b) r_1(j_i, \xi, b) \geq 1.$$

We deduce from (8.11) and (8.15) that

$$\begin{aligned} T(R_i + j_i - 1, \xi, b) &\geq A_d \frac{r_d(R_i + j_i, \xi, b)}{b} \\ &\qquad - \sum_{\ell=1}^{d-1} |A_\ell| \sum_{m \geq 1} \frac{r_\ell(R_i + j_i - 1 + m, \xi, b)}{b^m} \\ &\geq \frac{A_d}{b} - \sum_{\ell=1}^{d-1} |A_\ell| \sum_{m \geq R_{i+1} - R_i - j_i + 1} \frac{r_\ell(R_i + j_i - 1 + m, \xi, b)}{b^m} \\ &= \frac{A_d}{b} - \sum_{\ell=1}^{d-1} |A_\ell| \, b^{R_i + j_i - R_{i+1}} \, T_\ell(R_{i+1} - 1, \xi, b) \\ &\geq \frac{1}{b} - \frac{2}{b^{2d\log N}} \sum_{\ell=1}^{d-1} |A_\ell| \, (R_{i+1} + \ell)^{\ell-1} (b-1)^\ell, \end{aligned}$$

by Lemma 8.6. This gives that

$$T(R_i + j_i - 1, \xi, b) \geq 1/(2b),$$

since N exceeds $(20b^d d^2 H)^{2d}$. Note that, for $R = R_i + 1, R_i + 2, \ldots, R_i + j_i - 1$, we have $r_{d-1}(R, \xi, b) = 0$, and, consequently, $r_1(R, \xi, b) = \ldots = r_{d-2}(R, \xi, b) = 0$. For such an integer R, we get from (8.16) that

$$bT(R-1, \xi, b) = T(R, \xi, b) + A_d r_d(R, \xi, b),$$

and it follows by induction that $T(R, \xi, b)$ is positive for $R = R_i, R_i + 1, \ldots, R_i + j_i - 1$, hence for at least $j_i \geq (R_{i+1} - R_i - d\log N)/(d+1)$ values of R in $[R_i, R_{i+1} - 1]$. Recalling that $T(R, \xi, b)$ is an integer and setting $K = \lceil 2d\log N \rceil$, we get

$$R_i + j_i - 1 \leq R_{i+1} - 2d\log N - 1 \leq N - K.$$

Thus, we have proved that

$$\begin{aligned}\sum_{1\le R\le N-K} |T(R,\xi,b)| &\ge \sum_{i\in I} \frac{R_{i+1}-R_i-d\log N}{d+1} \\ &\ge \frac{2N}{3(d+1)} - M\log N,\end{aligned} \tag{8.18}$$

by (8.17). On the other hand, for $\ell = 1,\ldots,d$, we have

$$\begin{aligned}\sum_{1\le R\le N-K} T_\ell(R,\xi,b) &= \sum_{m\ge 1} b^{-m} \sum_{R\le N-K} r_\ell(R+m,\xi,b) \\ &\le \sum_{m=1}^{K} b^{-m} \sum_{0\le R\le N} r_\ell(R,\xi,b) \\ &\quad + b^{-K} \sum_{m\ge K+1} b^{K-m} \sum_{R\le N-K} r_\ell(R+m,\xi,b) \\ &< \sum_{0\le R\le N} r_\ell(R,\xi,b) + b^{-K} \sum_{K\le R\le N} T_\ell(R,\xi,b),\end{aligned}$$

thus, by Lemma 8.6 and since N exceeds $(20b^d d^2 H)^{2d}$,

$$\begin{aligned}\sum_{1\le R\le N-K} T_\ell(R,\xi,b) &\le (b-1)^\ell \big(\mathcal{N}\mathcal{Z}(N,\xi,b)\big)^\ell \\ &\qquad + N^{-2d(\log b)}(b-1)^\ell (N+d)^d \\ &\le (b-1)^\ell \left(\big(\mathcal{N}\mathcal{Z}(N,\xi,b)\big)^\ell + 1\right).\end{aligned}$$

Consequently, recalling that $T_\ell(R,\xi,b)\ge 0$ for $R\ge 1$, we obtain

$$\begin{aligned}\sum_{1\le R\le N-K} |T(R,\xi,b)| &\le \sum_{\ell=1}^{d} |A_\ell|\,(b-1)^\ell \big(\big(\mathcal{N}\mathcal{Z}(N,\xi,b)\big)^\ell + 1\big) \\ &\le A_d(b-1)^d \big(\mathcal{N}\mathcal{Z}(N,\xi,b)\big)^d \\ &\qquad + b^d dH\big(\mathcal{N}\mathcal{Z}(N,\xi,b)\big)^{d-1} + dHb^d \\ &\le \frac{N}{2(d+1)} + 2b^d dH N^{1-1/d}.\end{aligned}$$

This gives a contradiction to (8.18), since N exceeds $(20b^d d^2 H)^{2d}$. □

8.4 Number of digit changes in the b-ary expansion of algebraic numbers

As in Section 6.3, for an integer $b \geq 2$, an irrational real number ξ whose b-ary expansion is given by (8.1), and a positive integer n, we set

$$\mathcal{DC}(n, \xi, b) := \mathrm{Card}\,\{k : 1 \leq k \leq n, a_k \neq a_{k+1}\},$$

which counts the number of digits followed by a different digit, among the first n digits in the b-ary expansion of ξ. Using this notion for measuring the complexity of a real number, Theorem 8.7 shows that algebraic irrational numbers are 'not too simple'.

THEOREM 8.7. *Let $b \geq 2$ be an integer. For every irrational, real algebraic number ξ, there exists an effectively computable constant $n_0(\xi, b)$, depending only on ξ and b, such that*

$$\mathcal{DC}(n, \xi, b) \geq (\log n)^{5/4}, \tag{8.19}$$

for every integer $n \geq n_0(\xi, b)$.

A weaker result than (8.19), namely that

$$\lim_{n \to +\infty} \frac{\mathcal{DC}(n, \xi, b)}{\log n} = +\infty, \tag{8.20}$$

follows quite easily from Ridout's Theorem E.8. Here, we use the quantitative version of Ridout's theorem given in Theorem E.9 to improve (8.20). We point out that the lower bound in (8.19) does not depend on b. It can be slightly refined by using the same trick as in the proof of Theorem 9.9; see [163].

PROOF. Let the b-ary expansion of ξ be given by (8.1). Assume without loss of generality that $a_1 = b - 1$, that is, $(b-1)/b < \xi < 1$. Define the increasing sequence of positive integers $(n_j)_{j \geq 1}$ by $a_1 = \cdots = a_{n_1}$, $a_{n_1} \neq a_{n_1+1}$ and $a_{n_j+1} = \cdots = a_{n_{j+1}}$, $a_{n_{j+1}} \neq a_{n_{j+1}+1}$ for $j \geq 1$. Observe that

$$\mathcal{DC}(n, \xi, b) = \max\{j : n_j \leq n\}$$

for $n \geq n_1$, and that $n_j \geq j$ for $j \geq 1$. For $j \geq 1$, set

$$\xi_j := \sum_{k=1}^{n_j} \frac{a_k}{b^k} + \sum_{k=n_j+1}^{+\infty} \frac{a_{n_j+1}}{b^k} = \sum_{k=1}^{n_j} \frac{a_k}{b^k} + \frac{a_{n_j+1}}{b^{n_j}(b-1)},$$

and observe that

$$\xi_j = \frac{P_j(b)}{b^{n_j}(b-1)},$$

where $P_j(X) := a_1 X^{n_j} + \cdots + (a_{n_j} - a_{n_j-1})X + (a_{n_j+1} - a_{n_j})$ is an integer polynomial of degree n_j whose constant coefficient is not divisible by b. Consequently, b does not divide $P_j(b)$. We have

$$|\xi - \xi_j| < \frac{1}{b^{n_{j+1}}},$$

and this can be rewritten as

$$\left|(b-1)\xi - \frac{P_j(b)}{b^{n_j}}\right| < \frac{b-1}{b^{n_{j+1}}}. \tag{8.21}$$

By Liouville's Theorem E.5, denoting by H the height of $(b-1)\xi$ and by d its degree, we have

$$\left|(b-1)\xi - \frac{P_j(b)}{b^{n_j}}\right| \geq \frac{1}{d^2 H b^{d-1} b^{n_j d}}.$$

So, if

$$n_j \geq U := 1 + 2H,$$

then

$$n_{j+1} \leq 2dn_j. \tag{8.22}$$

Let $0 < \varepsilon \leq 1$ and let j_1 denote the smallest j such that $n_j \geq \max\{U, 5/\varepsilon\}$. Let j be an integer such that $j \geq j_1$ and $n_{j+1}/n_j \geq 1+2\varepsilon$. We have

$$b^{n_j} > \max\{2H, 2^{4/\varepsilon}\}.$$

Furthermore, it follows from (8.21) and the inequality $\varepsilon n_j \geq 5$ that

$$\left|(b-1)\xi - \frac{P_j(b)}{b^{n_j}}\right| < \frac{b-1}{(b^{n_j})^{1+2\varepsilon}} \leq \frac{1}{(b^{n_j})^{1+\varepsilon}}. \tag{8.23}$$

Since b does not divide $P_j(b)$, Theorem E.9 applied to (8.23) implies that the number T of such integers j with $j \geq j_1$ satisfies

$$T \leq 10^{10} \log(6d)(1+\varepsilon^{-3}) \log\big((1+\varepsilon^{-1})\log(6d)\big).$$

Let J be an integer satisfying

$$J > \max\{n_{j_1}^3, (4d)^6\}$$

and j_2 be the largest integer with

$$n_{j_2} \leq 6dJ^{1/3}.$$

Then $j_1 \leq j_2 < J$ and, since $n_{j_2} \geq U$, we have

$$n_{j_2} \geq \frac{n_{j_2+1}}{2d} \geq 3J^{1/3},$$

by (8.22). Observe that

$$\frac{n_J}{n_{j_2}} = \frac{n_J}{n_{J-1}} \times \frac{n_{J-1}}{n_{J-2}} \times \cdots \times \frac{n_{j_2+1}}{n_{j_2}}$$
$$\leq (1+2\varepsilon)^J (2d)^T.$$

By taking logarithms and choosing $\varepsilon := J^{-2/9}$, we get

$$2 \log n_J \leq 10^{12} J^{7/9} (\log 6d)^2 \leq J^{4/5},$$

when J is large enough. This implies the theorem. □

8.5 On the b-ary expansion of e and some other transcendental numbers

The combinatorial structure of Sturmian sequences leads us to introduce the following exponents of approximation, which measure the initial repetitions occurring in the expansions of a real number to an integer base.

DEFINITION 8.8. Let ξ be an irrational real number. Let b be an integer with $b \geq 2$. We denote by $v_b''(\xi)$ the supremum of the real numbers v for which the inequality

$$||(b^r - 1)\xi|| < (b^r)^{-v}$$

has infinitely many solutions in positive integers r.

The exponent v_b'' measures whether the b-ary expansion of ξ possesses large initial repetitions. It corresponds to the notion of *initial critical exponent* from combinatorics on words, introduced by Berthé, Holton and Zamboni [79].

Clearly, for every $b \geq 2$ and every irrational real number ξ, we have

$$\mu(\xi) \geq 1 + v_b''(\xi). \tag{8.24}$$

We establish the following extension on the results of Ferenczi and Mauduit [305] and Allouche [38] on the b-ary expansion of algebraic numbers (recall that the irrationality exponent of these numbers is equal to 2, by Roth's Theorem E.7).

THEOREM 8.9. *For any real number ξ with irrationality exponent equal to 2 and for any integer $b \geq 2$, we have*

$$\lim_{n \to +\infty} \big(p(n, \xi, b) - n\big) = +\infty.$$

Besides algebraic irrationals, the set of 'classical' real numbers whose irrationality exponent is 2 includes e, $e^{p/q}$, $\tan(1/q)$, for non-zero integers p, q with $q \geq 1$.

Theorem 8.9, which was first published in [1], is an almost immediate consequence of a deep result of Berthé, Holton and Zamboni [79] on the combinatorial structure of Sturmian sequences.

PROOF. Let $b \geq 2$ be an integer. Let ξ be a real number for which there is an integer k such that $p(n, \xi, b) \leq n + k$ for every positive integer n. It follows from Theorem A.8 that there are a non-negative integer s, a Sturmian sequence $\boldsymbol{s}$ on $\{0, 1\}$, and two finite words W_0, W_1 on $\{0, 1, \ldots, b-1\}$ such that the b-ary expansion of (the fractional part of) $b^s\xi$ is the infinite word obtained from $\boldsymbol{s}$ by replacing 0 (resp. 1) by W_0 (resp. W_1). If the slope of $\boldsymbol{s}$ is a badly approximable number, then it follows from Theorem A.6 that $v_b''(b^s\xi)$ exceeds 1, consequently $v_b''(\xi) > 1$. If the slope of $\boldsymbol{s}$ is not a badly approximable number, then, by Theorem A.7, the real number ξ is a Liouville number. Consequently, in any case, we have $\mu(\xi) > 2$, by (8.24). This completes the proof of the theorem. □

It follows from Theorem 8.9 that a real irrational number cannot have simultaneously a 'simple' continued fraction expansion and a 'simple' expansion in some integer base.

COROLLARY 8.10. *For any badly approximable number ξ and any integer $b \geq 2$, we have*

$$\lim_{n \to +\infty} (p(n, \xi, b) - n) = +\infty.$$

Corollary 8.10 is near to best possible, since there exist badly approximable numbers of low complexity. For instance, for $b \geq 2$, the number $\zeta_b = \sum_{j \geq 1} b^{-2^j}$ is badly approximable (by adapting the proof of Theorem 7.21) and it has sublinear complexity, namely $p(n, \zeta_b, b) \leq (2 + \log 3)n + 4$ for $n \geq 1$, as follows from the proof of [320, Lemma 2.4].

8.6 On the digits of the multiples of an irrational number

Let $b \geq 2$ be an integer. Weyl's Theorem 1.3 asserts that, for any irrational number ξ, the sequence $(m\xi)_{m \geq 1}$ is uniformly distributed modulo one. This implies that for any finite block D of digits from $\{0, 1, \ldots, b-1\}$, there exist arbitrarily large integers m such that D occurs at least once in the b-ary expansion of $m\xi$. This does not, however, provide any information regarding the number of occurrences of D in the b-ary expansion of $m\xi$. The question whether there is a positive integer m such that D occurs infinitely often in the b-ary expansion of $m\xi$ was addressed by Mahler [469]. He showed that there is at least one such m below an explicit bound depending on b and on the length of D, but independent of ξ; see Exercise 8.8. His result was subsequently improved by Szüsz and Volkmann [681] and by Berend and Boshernitzan [73].

THEOREM 8.11. *Let $b \geq 2$ and $n \geq 1$ be integers. Let D be a block of digits of length n on $\{0, 1, \ldots, b-1\}$. For any irrational number ξ, there exists an integer m, depending on ξ, b, n, such that*

$$1 \leq m \leq b^{n+1} + b^n - 1 \tag{8.25}$$

and the block of digits D appears infinitely often in the b-ary expansion of $m\xi$.

Theorem 8.11 slightly improves [73, Theorem 1.1], where the upper bound for m is $2b^{n+1} - 1$.

DEFINITION 8.12. For integers $b \geq 2$ and $n \geq 1$, let $M(b,n)$ denote the smallest integer M such that, for every irrational real number ξ and every block D of length n on $\{0, 1, \ldots, b-1\}$, there exists a positive integer m at most equal to M with the property that D has infinitely many occurrences in the b-ary expansion of $m\xi$.

Theorem 8.11 asserts that $M(b,n) \leq b^{n+1} + b^n - 1$ for every integer $b \geq 2$ and $n \geq 1$. The next result, extracted from [73], shows that Theorem 8.11 is not far from being best possible.

THEOREM 8.13. *For every integer $b \geq 2$ and $n \geq 1$, we have*

$$M(b,n) \geq b^n - 1.$$

If b is not a prime power, then, for every positive real number ε, there exists a positive integer n_0 such that

$$M(b,n) \geq (1-\varepsilon) b^{n+1}, \quad \text{for } n \geq n_0.$$

PROOF. For the first assertion, simply take the number $\xi = \sum_{j \geq 1} b^{-2^j}$ and a block D consisting of n consecutive digits $(b-1)$. Then, the b-ary expansion of $m\xi$ contains at most finitely many occurrences of D for $1 \leq m \leq b^n - 2$.

The second assertion is slightly more difficult to prove. Let p be a prime divisor of b. Let ε be a real number with $0 < \varepsilon < 1/4$. Since b

is not a prime power, $(\log p)/(\log b)$ is irrational and, by Theorem D.1, there are arbitrarily large positive integers s, t such that

$$0 < \frac{\log p}{\log b} - \frac{s}{t} < \frac{1}{t^2}.$$

Consequently, there exist positive integers ℓ and r such that $b^\ell < p^r < (1+\varepsilon)b^\ell$. We first claim that the representation in base b of no positive multiple of p^r contains the block D, consisting of $r-\ell$ consecutive digits $(b-1)$, within its r lowest digits. Indeed, if this were the case, then by multiplying this multiple of p^r by an appropriate power of b, we would then get a number of the form mp^r whose block of r lowest digits starts with the block D. The remainder m' in the Euclidean division of mp^r by b^r would satisfy

$$b^r - b^\ell \leq m' < b^r.$$

As m' and b^r are multiples of p^r, this would contradict the fact that p^r exceeds b^ℓ. Thus, the smallest multiple of p^r containing a block consisting of $n \geq r-\ell$ consecutive digits $(b-1)$ is at least equal to

$$\sum_{i=\ell+1}^{\ell+n} (b-1)b^i = b^{\ell+1}(b^n - 1).$$

Now set

$$\xi = \left(\frac{p}{b}\right)^r \sum_{j\geq 1} b^{-2^j}.$$

For $n \geq r-\ell$, the foregoing discussion implies that the smallest m for which $m\xi$ contains the block consisting of n consecutive digits $(b-1)$ infinitely often is at least $b^{\ell+1}(b^n-1)/p^r$, thus greater than $(1-\varepsilon)b^{n+1}$, if n is large enough. □

Proof of Theorem 8.11. Let $\Lambda_b(\xi)$ denote the set of all limit points in $\mathbb{T}$ of the sequence $(\{\xi b^j\})_{j\geq 1}$. We distinguish between several (somewhat overlapping) cases.

Case I. Assume that $\Lambda_b(\xi)$ contains a rational $r = p/q$ with $\gcd(p, q) = 1$ and $b^n < q \leq b^{n+1}$. Since every interval of length b^{-n} (on the torus $\mathbb{T}$) meets the set $\{0, 1/q, \ldots, (q-1)/q\}$ composed of the integer multiples of r modulo 1, there exists an integer m with $1 \leq m \leq q-1$ such that the expansion of $\{mr\}$ begins with $0 \cdot D$ and differs from $0 \cdot D0^\infty$ and from $0 \cdot D(b-1)^\infty$. Using that the sets $\Lambda_b(m\xi)$ and $\{\{m\lambda\} : \lambda \in \Lambda_b(\xi)\}$ are equal, we deduce that $\{mr\}$ is a limit point of the sequence $(\{\xi b^j m\})_{j\geq 1}$.

Consequently, there are infinitely many occurrences of the block D in the b-ary expansion of $m\xi$.

Case II. Assume that 0 is in $\Lambda_b(\xi)$. Replacing ξ by $-\xi$, we may assume that the b-ary expansion of ξ contains arbitrarily long blocks of 0. Let $(m_j)_{j\geq 1}$ be an increasing sequence of integers such that the b-ary expansion of $\{\xi b^{m_j}\}$ begins with 0^j, but that of $\{\xi b^{m_j-1}\}$ does not begin with 0. For any positive integer d, let $r_d = p_d/q_d$ be a limit point of $(\{\xi b^{m_j-d}\})_{j\geq 1}$ written under its reduced form. For $d \geq 2$, we select r_d in such a way that there exists a in $\{0, 1, \ldots, b-1\}$ such that

$$\frac{p_d}{q_d} = \frac{a}{b} + \frac{1}{b} \cdot \frac{p_{d-1}}{q_{d-1}} = \frac{aq_{d-1} + p_{d-1}}{bq_{d-1}}.$$

Consequently, we have $q_{d-1} < q_d \leq bq_{d-1}$ and there exists a suitable d such that $b^n < q_d \leq b^{n+1}$. Since r_d is in $\Lambda_b(\xi)$, we conclude as in Case I.

Case III. Assume that $\Lambda_b(\xi)$ contains a reduced rational number p/q with $q \leq b^{n+1}$. Then we carry out the construction of Case II with ξ replaced by $q\xi$.

Case IV. Let t be a real number with $1 \leq t \leq 2$. Assume that $\Lambda_b(\xi)$ contains no rational point p/q with $q \leq tb^n$. We first claim that there exist a point β in $\Lambda_b(\xi)$ and a rational $r = p/q$ written in its reduced form such that

$$|\beta - p/q| < (tb^{n+1}q)^{-1}, \quad tb^n < q \leq tb^{n+1}. \tag{8.26}$$

Indeed, starting with any β_0 in $\Lambda_b(\xi)$, we can find a reduced rational p_0/q_0 such that $|\beta_0 - p_0/q_0| < (tb^{n+1}q_0)^{-1}$ and $q_0 \leq tb^{n+1}$. Choose inductively points β_i in $\Lambda_b(\xi)$, $i = 1, 2, \ldots$, with $b\beta_{i+1} = \beta_i$ for each i. Next choose reduced rationals $r_i = p_i/q_i$, $i = 1, 2, \ldots$, with $br_{i+1} = r_i$ and $|\beta_i - r_i| = b^{-i}|\beta_0 - r_0|$ for each i. Observe that $q_i \leq q_{i+1} \leq bq_i$ for each i. If the sequence $(q_i)_{i\geq 1}$ is bounded from above by tb^n, then some reduced rational $r' = p'/q'$ with $q' \leq tb^n$ appears infinitely often in the sequence $(r_i)_{i\geq 1}$, in which case r' is in $\Lambda_b(\xi)$, in contradiction with our assumption. Consequently, there exists i such that the rational p_i/q_i satisfies (8.26). This proves our claim.

Instead of choosing $t = 2$ in (8.26) as in [73], we take $t = 1 + 1/b$. We then observe that every interval of length $1/(b^n + b^{n-1})$ (on the torus $\mathbb{T}$) meets the set $\{0, 1/q, \ldots, (q-1)/q\}$ of integer multiples of r taken modulo 1. Since $|m\beta - mr| < 1/(b^{n+1} + b^n)$ for $m = 1, \ldots, q$, we deduce that every interval of length b^{-n} (on the torus $\mathbb{T}$) meets the set $\{\beta, \{2\beta\}, \ldots, \{q\beta\}\}$. Arguing now as in Case I, we get that there exists an integer m, with $1 \leq m \leq q-1$, such that the b-ary expansion of $m\xi$

contains the block D infinitely often. The upper bound for q given by (8.26) implies our statement. □

8.7 Exercises

Exercise 8.1 (*cf.* [19]). Combine Theorem A.11 and Theorem 8.2 to establish that the binary expansion of an algebraic number contains infinitely many occurrences of 7/3-powers.

Exercise 8.2. The complement of a finite word $W = w_1 \ldots w_n$ on $\{0, 1, \ldots, b-1\}$ is the word $W' = w'_1 \ldots w'_n$ such that $w_i + w'_i = b - 1$ for $i = 1, \ldots, n$. Prove that the b-ary expansion of an irrational algebraic number does not begin with arbitrarily large words of the form WW'.

Exercise 8.3 (*cf.* [4]). Let $\boldsymbol{a} = (a_k)_{k\geq 1}$ and $\boldsymbol{a}' = (a'_k)_{k\geq 1}$ be sequences of elements from an alphabet $\mathcal{A}$, that we identify with the infinite words $a_1 a_2 \ldots$ and $a'_1 a'_2 \ldots$, respectively. We say that the pair $(\boldsymbol{a}, \boldsymbol{a}')$ satisfies Condition $(*)$ if there exist three sequences of finite words $(U_n)_{n\geq 1}$, $(U'_n)_{n\geq 1}$ and $(V_n)_{n\geq 1}$ such that:

(i) for any $n \geq 1$, the word $U_n V_n$ is a prefix of the word $\boldsymbol{a}$ and the word $U'_n V_n$ is a prefix of the word $\boldsymbol{a}'$;
(ii) the sequences $(|U_n|/|V_n|)_{n\geq 1}$ and $(|U'_n|/|V_n|)_{n\geq 1}$ are bounded from above;
(iii) the sequence $(|V_n|)_{n\geq 1}$ is increasing.

If, moreover, we add the condition

(iv) the sequence $(|U_n| - |U'_n|)_{n\geq 1}$ is unbounded,

then, we say that the pair $(\boldsymbol{a}, \boldsymbol{a}')$ satisfies Condition $(**)$.

Let $b \geq 2$ be an integer and take $\mathcal{A} = \{0, 1, \ldots, b-1\}$. Assume that the pair $(\boldsymbol{a}, \boldsymbol{a}')$ satisfies Condition $(*)$ and apply Theorem E.10 with $m = 4$ to prove that at least one of the real numbers

$$\xi := \sum_{k=1}^{+\infty} \frac{a_k}{b^k}, \quad \xi' := \sum_{k=1}^{+\infty} \frac{a'_k}{b^k}$$

is transcendental, or the b-ary expansions of ξ and ξ' have the same tail. If, furthermore, the pair $(\boldsymbol{a}, \boldsymbol{a}')$ satisfies Condition $(**)$, then show that at least one of the real numbers ξ, ξ' is transcendental, or ξ and ξ' are both rational and their b-ary expansions have the same tail.

Exercise 8.4. Prove (8.10) and (8.20).

EXERCISE 8.5 (*cf.* [611]). For an integer $b \geq 2$ and a non-negative integer x, let $\mathcal{NZ}(x,b)$ denote the number of non-zero digits in the b-ary representation of x. Prove that, for all positive integers x and y, we have

$$\mathcal{NZ}(x+y,b) \leq \mathcal{NZ}(x,b) + \mathcal{NZ}(y,b)$$

and

$$\mathcal{NZ}(xy,b) \leq \min\{2, b-1\}\,\mathcal{NZ}(x,b)\,\mathcal{NZ}(y,b).$$

Let ξ and η be positive irrational numbers and n be a sufficiently large integer. If $\xi+\eta$ is irrational, prove that

$$\mathcal{NZ}(n,\xi\eta,b) \leq \mathcal{NZ}(n,\xi,b) + \mathcal{NZ}(n,\eta,b) + 1.$$

If $\xi\eta$ is irrational, prove that

$$\mathcal{NZ}(n,\xi\eta,b) \leq 2 \cdot \mathcal{NZ}(n,\xi,b) \cdot \mathcal{NZ}(n,\eta,b) + \log(\xi+\eta+1) + 1.$$

For every positive integer A, prove that

$$\mathcal{NZ}(n,\xi,b) \cdot \mathcal{NZ}(n,A/\xi,b) \geq \frac{1}{2} \cdot (n - 1 - \log(\xi + A/\xi + 1)).$$

Let ξ be an irrational real algebraic number of degree d whose minimal defining polynomial over the integers $A_d X^d + \cdots + A_1 X + A_0$ satisfies $A_0 < 0$ and $A_1, \ldots, A_d \geq 0$. Establish that

$$\mathcal{NZ}(n,\xi,b) \geq \left(2^{d-1}\mathcal{NZ}(A_d,b)\right)^{-1/d} n^{1/d}(1+o(1)),$$

where 2^{d-1} can be replaced by 1 if $b=2$.

EXERCISE 8.6 (*cf.* [386]). Prove that the real number $\xi := \sum_{j\geq 0} 2^{-2^j}$ is transcendental. For $k \geq 1$, let $a(r,k)$ denote the number of ways that the integer $r \geq 0$ can be written as a sum of exactly k powers of 2, where different orderings of the sum are counted as distinct. Let $d \geq 1$ and $m \geq 1$ be integers, and set $N := (2^d - 1)2^m$. Prove that $a(r,k) = 0$ for every integer r, k with $1 \leq k \leq d$ and $N - (2^{m-1} - 1) \leq r \leq N + (2^m - 1)$, except for $(r,k) = (N,d)$, in which case $a(N,d) = d!$. Prove that, if m is large, then, for $k = 1, \ldots, d$, we have

$$\sum_{r=N+2^m}^{+\infty} a(r,k) 2^{-r} < 2^{-N-1}.$$

Deduce that the binary expansions of $1, \xi, \ldots, \xi^{d-1}$ have a common, arbitrarily long block of 0, while the same block for ξ^d has an isolated non-zero 'island'. Conclude that ξ is transcendental.

EXERCISE 8.7. Prove that $M(3,1) = 2$.

EXERCISE 8.8 (*cf.* [469]). Let ξ be an irrational real number with $0 < \xi < 1$ and $b \geq 2$ be an integer. Let n be a positive integer. For an integer m so large that $b^m\xi \geq 1$, prove that there exist integers $x(m,n)$ and $y(m,n)$ such that

$$|b^m\xi x(m,n) - y(m,n)| < b^{-n}, \quad 1 \leq x(m,n) \leq b^n, \quad y(m,n) \geq 1.$$

Deduce that there exist an infinite set $\mathcal{S}_0$ of positive integers and an integer x_0 such that, for every m in $\mathcal{S}_0$ there is an integer $y(m)$ with

$$|b^m\xi x_0 - y(m)| < b^{-n}, \quad 1 \leq x_0 \leq b^n, \quad y(m) \geq 1.$$

Deduce furthermore that there exist an infinite set $\mathcal{S}_1$ of positive integers and an integer x_1 such that, for every m in $\mathcal{S}_1$ there is an integer $y'(m)$ with

$$|b^m\xi x_1 - y'(m)| < b^{-n}, 1 \leq x_1 \leq b^n - 1, \ b \text{ does not divide } x_1, \ y'(m) \geq 1.$$

Denote by $\boldsymbol{a} := a_1a_2\ldots$ the b-ary expansion of ξ and, for a positive integer X, let $\boldsymbol{a}_X := a_{X,1}a_{X,2}\ldots$ be that of $\{X\xi\}$. Set $\xi^* = 1 - \xi$. Establish that $\boldsymbol{a}_{x_1}$ or $\boldsymbol{a}^*_{x_1}$ contains infinitely many blocks of n consecutive digits 0. Denote by $\boldsymbol{a}^\times_{x_1}$ that one of the numbers $\boldsymbol{a}_{x_1}$ or $\boldsymbol{a}^*_{x_1}$ with the latter property.

Let H be a large integer. Let $h_0(H)$ be the smallest integer greater than H such that $a^\times_{x_1,h} = 0$ for $h = h_0(H), \ldots, h_0(N)+n-1$. Let $h_1(H)$ be the smallest integer greater than $h_0(H)$ such that $a^\times_{x_1,h_1(H)} > 0$. Put

$$t = t(H) = \sum_{h \geq h_1(H)} a^\times_{x_1,h} b^{h_1(H)-h-1}$$

and prove that $1/b < t < 1$. Let $b_0 \ldots b_{n-1}$ be a block of digits from $\{0,1,\ldots,b-1\}$ other than the block $00\ldots00$, and put $B = b_0b^{n-1} + \cdots + b_{n-1}$. Prove that there exists an integer $x_2 = x_2(H)$ such that $B < x_2t < B+1$. Show that $1 \leq x_2 \leq b^{n+1} - 1$ and that the block of n digits $a^\times_{x_1x_2,h_1(H)-n} \ldots a^\times_{x_1x_2,h_1(H)-1}$ is identical with the block $b_0 \ldots b_{n-1}$. Deduce that there exists X with $1 \leq X \leq (b^n - 1)(b^{n+1} - 1)$ such that the b-ary expansion of at least one of the two numbers $X\xi$ or $X\xi^*$ contains infinitely many copies of a given block of length n. Conclude.

8.8 Notes

▷ Let $b \geq 2$ be an integer. Mahler [462] investigated how many times a given block of k digits on $\{0, 1, \ldots, b-1\}$ can repeat after the nth digit in the b-ary expansion of an irrational number which is not a Liouville number.

▷ Theorem 8.2 implies that the real number $\sum_{j \geq 0} 2^{-2^j}$, sometimes called the Fredholm number, is transcendental. This result was first proved by Kempner [377] in 1916.

▷ Mahler [473] suggested explicitly to apply the Schmidt Subspace Theorem E.10 exactly as in the proof of Theorem 8.2 to investigate whether the middle third Cantor set contains irrational algebraic elements or not.

▷ An application of the Schmidt Subspace Theorem E.10 similar to that in Theorem 8.2 was used in 1999 by Troi and Zannier [688] to prove the transcendence of the number $\sum 2^{-m}$, where the summation is over the set of integers m which can be represented as sums of distinct terms of the form $2^\ell + 1$ with ℓ being a positive integer.

▷ Alternative expositions of the proof of Theorem 8.3 are given in [12, 99].

▷ By means of a subtle use of the Quantitative Subspace Theorem, it has been proved in [163] that, for any irrational algebraic number ξ and any integer $b \geq 2$, we have

$$\limsup_{n \to +\infty} \frac{p(n, \xi, b)}{n (\log n)^{0.09}} = +\infty.$$

▷ Since its proof ultimately depends on the Schmidt Subspace Theorem, Theorem 8.3 is ineffective. Let $b \geq 2$ be an integer and ξ be an irrational algebraic number. Using a suitable version of the Quantitative Subspace Theorem, Bugeaud [149] gave explicitly a (very small) positive real number δ, expressed only in terms of the degree and the height of ξ, such that $p(n, \xi, b) \geq (1 + \delta)n$ for every positive integer n.

▷ An automatic number is a real number ξ whose expansion in some integer base $b \geq 2$ can be generated by a finite automaton (see [46] for a precise definition of a finite automaton). By a result of Cobham [198], its complexity function satisfies $p(n, \xi, b) = O(n)$. It then straightforwardly follows from Theorem 8.3 that every irrational automatic number is transcendental, confirming a conjecture of Cobham [197], also formulated in [505]. As a particular case, we get that the Thue–Morse–Mahler numbers

$\sum_{k\geq 0} t_k b^{-k}$ are transcendental, where $t_0 t_1 \ldots$ is the Thue–Morse word (Definition A.9) written on $\{0,1\}$. The latter result was first proved by Mahler [460]. Dekking [225] found an alternative proof, reproduced in [46].

▷ Adamczewski and Cassaigne [17] established that the irrationality exponent of an automatic number is finite; see also [70]. Furthermore, Adamczewski and Bugeaud [13] proved that an automatic number cannot be a U-number. Bugeaud [148] used the construction given in the proof of Theorem 7.21 to show that there are automatic real numbers with any prescribed rational irrationality exponent. Conversely, it may be the case that the irrationality exponent of an automatic real number is always rational. In this direction, improving earlier results from [20], Bugeaud [154] established that the irrationality exponent of every Thue–Morse–Mahler number is equal to 2. Furthermore, it is proved in [165, 620] that the exponents v_b and v'_b take rational values at every real number whose expansion to base b can be generated by a finite automaton.

▷ Let $f : \mathbb{Z}_{\geq 1} \rightarrow \mathbb{Z}_{\geq 1}$ be an increasing function. A pair (ξ, ξ') of real numbers is called f-independent if there exist only finitely many integers n for which the prefixes of length $f(n)$ of their b-ary expansions have a block of length n in common. Let $f : \mathbb{Z}_{\geq 1} \rightarrow \mathbb{Z}_{\geq 1}$ be an increasing function such that $n \mapsto f(n)/n$ is bounded. Let ξ and ξ' be two algebraic numbers in $(0,1)$. It was proved in [4] that either the b-ary expansions of ξ and ξ' have the same tail, or the pair (ξ, ξ') is f-independent. Put another way, this shows that if one slightly perturbs the b-ary expansion of an algebraic number ξ and gets a number ξ', then ξ' is transcendental, except in the trivial case when the tails of the expansions of ξ and ξ' coincide; see Exercise 8.3.

▷ Diophantine properties of real numbers whose binary expansion is a Sturmian word were studied in [13, 79, 595].

▷ Using a result of Thue [685] on the structure of ternary infinite words, Adamczewski and Rampersad [19] established that the ternary expansion of an algebraic number contains infinitely many occurrences of squares or infinitely many occurrences of one of the blocks 010 or 02120.

▷ Let $(n_j)_{j\geq 1}$ be a lacunary sequence of positive integers such that $\liminf_{j\rightarrow +\infty} n_{j+1}/n_j > 1$. Applying the Schmidt Subspace Theorem E.10, Corvaja and Zannier [207] established that the function f defined on the open unit disc by $f(z) = \sum_{j \geq 1} z^{n_j}$ takes a transcendental value at every non-zero algebraic point. See also [16] for a weaker conclusion valid, however, for a larger class of functions. Alternative applications of the Subspace Theorem to transcendence results can be found in [171, 513, 630, 750].

▷ Let $b \geq 2$ be an integer. Let $f(X)$ be a quadratic polynomial with rational coefficients and taking positive integer values at positive integers. Let $(d_j)_{j \geq 1}$ be a bounded sequence of non-zero integers. Inspired by [53], Luca [452] established that the real number $\sum_{j \geq 1} d_j b^{-f(j)}$ is neither rational nor quadratic. Neither his method, nor Theorem 8.2, imply that the real number $\sum_{j \geq 1} b^{-j^2}$ is transcendental, a statement established in [92, 271] and which follows from deep results on the values of theta series at algebraic points.

▷ Kaneko [366, 368] established improved lower bounds for the number of digit changes of a class of algebraic numbers. They are similar to the lower bounds for the number of non-zero digits. Inspired by the method of [53], he [367] obtained algebraic independence results for certain real numbers having few non-zero digits.

▷ Let $P(X) = a_0 + a_1 X + \cdots + a_d X^d$ be a real polynomial of degree $d \geq 1$. Theorem 1.4 asserts that the sequence $(P(n))_{n \geq 1}$ is dense modulo one if at least one coefficient among $a_1, \ldots, a_d$ is irrational. If $a_1, \ldots, a_d$ are all rational numbers, denote by $c(P)$ the least common multiple of the denominators of $a_1, \ldots, a_d$. Berend and Boshernitzan [73] proved that, for every given positive real number ε and every $d \geq 1$, there exists $c_0(d)$ such that for every polynomial $P(X)$ in $\mathbb{Q}(X)$ of degree d with $c(P) > c_0$, the set $\{P(n) : n \geq 1\}$ meets every interval of length ε (on the torus $\mathbb{T}$). Further results related to Theorem 8.11 can be found in [47, 71, 74, 75, 321].

▷ Set $u_1 = 1$ and $u_{n+1} = \lfloor \sqrt{2}(u_n + 1/2) \rfloor$ for $n \geq 1$. Graham and Pollak [328] proved that, for $n \geq 1$, the nth binary digit of $\sqrt{2}$ is equal to $u_{2n+1} - 2u_{2n-1}$; see also [665, 666] and the references quoted therein.

9
Continued fraction expansions and β-expansions

Beside b-ary expansions, there are many classical ways to represent a real number, including by means of its continued fraction expansion, or of its β-expansion. In the first section, we define the notion of normal continued fraction and construct explicitly a real number having a normal continued fraction. In the next section, we present several results on the continued fraction expansion of algebraic numbers. Section 9.3 is devoted to a short survey on β-expansions.

9.1 Normal continued fractions

The sequence of partial quotients of an irrational real number ξ in $(0, 1)$ can be obtained by iterations of the Gauss map $T_{\mathcal{G}}$ defined by $T_{\mathcal{G}}(0) = 0$ and $T_{\mathcal{G}}(x) = \{1/x\}$ for $x \in (0, 1)$. Namely, if $[0; a_1, a_2, \ldots]$ denotes the continued fraction expansion of ξ, then $T_{\mathcal{G}}^n(\xi) = [0; a_{n+1}, a_{n+2}, \ldots]$ and $a_n = \lfloor 1/T_{\mathcal{G}}^{n-1}(\xi) \rfloor$ for $n \geq 1$.

It is understood that ξ is a real number in $(0, 1)$, whose partial quotients $a_1(\xi), a_2(\xi), \ldots$ and convergents $p_1(\xi)/q_1(\xi), p_2(\xi)/q_2(\xi), \ldots$ are written $a_1, a_2, \ldots$ and $p_1/q_1, p_2/q_2, \ldots$ when there should be no confusion.

The map $T_{\mathcal{G}}$ possesses an invariant ergodic probability measure, namely the Gauss measure $\mu_{\mathcal{G}}$, which is absolutely continuous with respect to the Lebesgue measure, with density

$$\mu_{\mathcal{G}}(\mathrm{d}x) = \frac{\mathrm{d}x}{(1 + x) \log 2}.$$

For every function f in $L^1(\mu_{\mathcal{G}})$ and almost every ξ in $(0, 1)$, we have ([212, Theorem 3.5.1]; see also Section C.4)

$$\lim_{n\to+\infty} \frac{1}{n}\sum_{k=0}^{n-1} f(T_{\mathcal{G}}^k\xi) = \frac{1}{\log 2}\int_0^1 \frac{f(x)}{1+x}\,\mathrm{d}x \tag{9.1}$$

and

$$\lim_{n\to+\infty} \frac{\log q_n(\xi)}{n} = \frac{\pi^2}{12\log 2}. \tag{9.2}$$

For subsequent results in the metric theory of continued fractions, we refer the reader to [380] and to [212].

DEFINITION 9.1. We say that $[0; a_1, a_2, \ldots]$ is a *normal continued fraction*, if, for every integer $k \geq 1$, every finite word $\mathbf{d} = d_1 \ldots d_k$ on the alphabet $\mathbb{Z}_{\geq 1}$, we have

$$\begin{aligned}\lim_{N\to+\infty} & \frac{\mathrm{Card}\{j : 0 \leq j \leq N-k, a_{j+1} = d_1, \ldots, a_{j+k} = d_k\}}{N} \\ &= \int_{r/s}^{r'/s'} \mu_{\mathcal{G}}(\,\mathrm{d}x) = \mu_{\mathcal{G}}(\Delta_{\mathbf{d}}),\end{aligned} \tag{9.3}$$

where r/s and r'/s' denote the rational numbers $[0; d_1, \ldots, d_{k-1}, d_k]$ and $[0; d_1, \ldots, d_{k-1}, d_k + 1]$ ordered such that $r/s < r'/s'$, and $\Delta_{\mathbf{d}} = [r/s, r'/s']$.

Let $\mathbf{d} = d_1 \ldots d_k$ be a finite word on the alphabet $\mathbb{Z}_{\geq 1}$ (we warn the reader that, throughout this section, $\mathbf{a}$ and $\mathbf{d}$ always denote *finite* words). It follows from Theorem D.1 that the set $\Delta_{\mathbf{d}}$ of real numbers ξ in $(0, 1)$ whose first k partial quotients are $d_1, \ldots, d_k$ is an interval of length

$$\lambda(\Delta_{\mathbf{d}}) = \frac{1}{q_k(q_k + q_{k-1})}.$$

Applying (9.1) to the function f equal to 1 on $\Delta_{\mathbf{d}}$ and to 0 everywhere else, we get that for almost every $\xi = [0; a_1, a_2, \ldots]$ in $(0, 1)$ the limit defined in (9.3) exists and is equal to $\mu_{\mathcal{G}}(\Delta_{\mathbf{d}})$. Thus, we have established the following statement.

THEOREM 9.2. *Almost every ξ in $(0, 1)$ has a normal continued fraction expansion.*

As was shown in Section 4.2, for any integer $b \geq 2$, the number whose b-ary expansion is the concatenation of the representation in base b of the integers $1, 2, \ldots$ is normal to base b. Adler, Keane and Smorodinsky [23] have given a similar construction for a normal continued fraction.

THEOREM 9.3. *Let $1/2, 1/3, 2/3, 1/4, 2/4, 3/4, \ldots$ be the infinite sequence $(r_i)_{i\geq 1}$ obtained in writing the rational numbers in $(0, 1)$ with* *denominator 2, then with denominator 3, denominator 4, etc., ordered with numerators increasing. Let $x_1 x_2 x_3 \dots$ be the sequence of positive integers constructed by concatenating the partial quotients (we choose the continued fraction expansion which does not end with the digit 1) of this sequence of rational numbers. Then, the real number*

$$\xi_{\mathbf{aks}} = [0; x_1, x_2, \dots] = [0; 2, 3, 1, 2, 4, 2, 1, 3, 5, \dots]$$

is a normal continued fraction.

In the statement of Theorem 9.3 and in its proof, the denominator of a rational number p/m is the integer m, even if p and m are not coprime.

PROOF. We follow the proof of [23]. Set $g = \pi^2/(12 \log 2)$. Let δ be a positive real number with $\delta < 1/3$. Let η be a positive real number. Let $m \geq 3$ be an integer and set

$$n = \lfloor (1 - 2\delta)(\log m)/g \rfloor. \tag{9.4}$$

Let $\mathbf{d} = d_1 \dots d_k$ be a finite word on the alphabet $\mathbb{Z}_{\geq 1}$. Let $\Gamma_{m,\delta,\mathbf{d},\eta}$ be the set of all finite words $\mathbf{a} = a_1 \dots a_n$ on $\mathbb{Z}_{\geq 1}$ satisfying

$$\mathrm{e}^{(1-\delta)gn} \leq K_n(a_1, \dots, a_n) \leq \mathrm{e}^{(1+\delta)gn}$$

and

$$\left| \frac{\mathrm{Card}\{j : 0 \leq j \leq n - k, a_{j+1} a_{j+2} \dots a_{j+k} = \mathbf{d}\}}{n} - \mu_{\mathcal{G}}(\Delta_{\mathbf{d}}) \right| < \eta.$$

Here, $K_n(a_1, \dots, a_n)$ denotes the continuant of $a_1, \dots, a_n$; see Definition D.4. Keeping this notation, we say that the rational $r_i = [0; a_{i,1}, a_{i,2}, \dots, a_{i,n_i}]$ is $(m, \delta, \mathbf{d}, \eta)$-*good* if we have

$$(i) \quad K_{n_i}(a_{i,1}, a_{i,2}, \dots, a_{i,n_i}) \leq m,$$

$$(ii) \quad n_i \geq n \quad \text{and} \quad a_{i,1} a_{i,2} \dots a_{i,n} \in \Gamma_{m,\delta,\mathbf{d},\eta}.$$

□

LEMMA 9.4. *Let $\mathbf{a} = a_1 \dots a_n$ be in $\Gamma_{m,\delta,\mathbf{d},\eta}$ and define $p_{n-1}/q_{n-1} = [0; a_1, \dots, a_{n-1}]$ and $p_n/q_n = [0; a_1, \dots, a_n]$. Then we have*

$$\mathrm{Card}\{i : \ r_i = [0; a_{i,1}, a_{i,2}, \dots, a_{i,n}, \dots, a_{i,n_i}] \ \textit{is } (m, \delta, \mathbf{d}, \eta)\textit{-good}$$
$$\textit{and } a_{i,1} a_{i,2} \dots a_{i,n} = \mathbf{a}\} \geq \frac{m^2}{2q_n(q_n + q_{n-1})} - \frac{4m}{q_n}.$$

PROOF. For positive integers s and t with $s < t$, set

$$\phi(s, t) = (tp_n + sp_{n-1}, tq_n + sq_{n-1}).$$

Since $p_n q_{n-1} - p_{n-1} q_n = \pm 1$, the map ϕ is injective. By Theorem D.1, the continued fraction expansion of the rational

$$\frac{tp_n + sp_{n-1}}{tq_n + sq_{n-1}} = \frac{(t/s)p_n + p_{n-1}}{(t/s)q_n + q_{n-1}}$$

is given by $[0; a_1, a_2, \ldots, a_n, a_{n+1}, \ldots, a_k]$, where $[a_{n+1}; a_{n+2}, \ldots, a_k]$ is the continued fraction expansion of t/s. By construction, this rational is $(m, \delta, \mathbf{d}, \eta)$-good as soon as its denominator is less than or equal to m, that is, as soon as

$$tq_n + sq_{n-1} \leq m. \tag{9.5}$$

Since different pairs (s, t) yield different images $\phi(s, t)$, a lower bound for the cardinality of the set defined in the lemma is the number of pairs (s, t) with $1 \leq s < t$ and satisfying (9.5). In the (s, t)-plane, the inequality (9.5) with the extra condition $0 \leq s < t$ defines a triangle of area $A = m^2/(2q_n(q_n + q_{n-1}))$. This triangle contains at least $A - 4m/q_n$ lattice points with positive coordinates. This proves the lemma. □

By definition, the denominator of every $(m, \delta, \mathbf{d}, \eta)$-good rational number is at most m. The next lemma shows that most of the rational numbers of denominator at most m are $(m, \delta, \mathbf{d}, \eta)$-good rational numbers. Observe that, for $m \geq 2$, there are exactly $m - 1$ rational numbers in $(0, 1)$ of denominator m, hence, there are $i(m) := m(m-1)/2$ rational numbers in $(0, 1)$ of denominator at most equal to m.

LEMMA 9.5. *Let m, δ, η and $\mathbf{d}$ be as above. For every positive real number ε, there exists an integer m_0, depending on δ, η and $\mathbf{d}$, such that*

$$\frac{\mathrm{Card}\{i : r_i \text{ is } (m, \delta, \mathbf{d}, \eta)\text{-good}\}}{\mathrm{Card}\{i : \text{the denominator of } r_i \text{ is at most } m\}} > 1 - \varepsilon \tag{9.6}$$

holds for every m exceeding m_0.

PROOF. Let C_m denote the left-hand side of (9.6). It follows from Lemma 9.4 that

$$\begin{aligned} C_m &\geq \frac{2}{m(m-1)} \sum_{\mathbf{a} \in \Gamma_{m,\delta,\mathbf{d},\eta}} \left(\frac{m^2}{2q_n(q_n + q_{n-1})} - \frac{4m}{q_n} \right) \\ &\geq \frac{m}{m-1} \sum_{\mathbf{a} \in \Gamma_{m,\delta,\mathbf{d},\eta}} \frac{1}{q_n(q_n + q_{n-1})} \cdot \left(1 - \frac{16q_n}{m}\right) \\ &= \frac{m}{m-1} \lambda(\tilde{\Gamma}_{m,\delta,\mathbf{d},\eta}) \cdot \left(1 - \frac{16q_n}{m}\right), \end{aligned}$$

where $\tilde{\Gamma}_{m,\delta,\mathbf{d},\eta}$ denotes the set of real numbers in $(0,1)$ whose first n partial quotients $a_1, \ldots, a_n$ are such that the word $a_1 \ldots a_n$ is in $\Gamma_{m,\delta,\mathbf{d},\eta}$.

We deduce from (9.2) that $\lambda(\Gamma_{m,\delta,\mathbf{d},\eta})$ tends to 1 as m tends to infinity. Furthermore, since the rational is assumed to be $(m, \delta, \mathbf{d}, \eta)$-good, we have

$$\frac{16q_n}{m} \leq \frac{16\mathrm{e}^{(1+\delta)gn}}{m} \leq \frac{16\mathrm{e}^{(1+\delta)(1-2\delta)(\log m)}}{m} \leq 16\mathrm{e}^{2g}m^{-\delta},$$

which tends to zero as m tends to infinity. This proves the lemma. □

LEMMA 9.6. *Let m, δ, η and $\mathbf{d}$ be as above. If the rational number $r_i = [0; a_1, \ldots, a_n, a_{n+1}, \ldots, a_{n_i}]$ is an $(m, \delta, \mathbf{d}, \eta)$-good rational number, then we have*

$$n_i - n < 32g\delta n$$

for every m exceeding $\mathrm{e}^{(3g+1)/\delta}$.

PROOF. Write $s/t = [0; a_{n+1}, \ldots, a_{n_i}]$. Then, the denominator q_{n_i} of r_i satisfies

$$m \geq q_{n_i} = tq_n + sq_{n-1} > tq_n,$$

thus,

$$t < \frac{m}{q_n} \leq \frac{m}{\mathrm{e}^{(1-\delta)g(-1+(1-2\delta)(\log m)/g)}} \leq \mathrm{e}^g m^{3\delta}.$$

We get from (D.5) that

$$n_i - n \leq 1 + 3\log t < 1 + 3g + 9\delta \log m \leq 10\delta \log m,$$

if m exceeds $\mathrm{e}^{(3g+1)/\delta}$. By (9.4), this proves the lemma. □

PROOF OF THEOREM 9.3 (CONTINUED). Let $\xi_{\mathbf{aks}} = [0; x_1, x_2, \ldots]$ be the number defined in the theorem. For a positive integer j and a finite word $\mathbf{d} = d_1 \ldots d_k$ on the alphabet $\mathbb{Z}_{\geq 1}$, set

$$\rho_j(\mathbf{d}, \xi_{\mathbf{aks}}) = \mathrm{Card}\{i : 0 \leq i \leq j, x_{i+1} \ldots x_{i+k} = \mathbf{d}\}.$$

For a positive integer m, recall that $i(m) = m(m-1)/2$ is the index of the last rational with denominator m and let $j(m)$ be the index corresponding to the partial quotient $a_{i(m),n_{i(m)}}$ in $\xi_{\mathbf{aks}}$. This means that $x_{j(m)} = a_{i(m),n_{i(m)}}$.

Observe that there are m rationals with denominator $m+1$ and each of them has an expansion of length at most equal to $3\log(m+1)$. This implies that

$$\frac{j(m+1) - j(m)}{j(m)} \leq \frac{3m\log(m+1)}{i(m)},$$

which tends to zero as m tends to infinity. Consequently, to establish the theorem, it is sufficient to show that

$$\lim_{m\to+\infty} \frac{\rho_{j(m)}(\mathbf{d}, \xi_{\mathbf{aks}})}{j(m)} = \mu_G(\Delta_{\mathbf{d}}). \tag{9.7}$$

We estimate $\rho_{j(m)}(\mathbf{d}, \xi_{\mathbf{aks}})$ and, to this end, the number of occurrences of $\mathbf{d}$ in $x_1 x_2 \ldots x_{j(m)+k}$.

Let $\varepsilon, \delta, \eta$ be positive real numbers. By Lemma 9.5, there exists an integer m_0, depending only on δ, η and $\mathbf{d}$ such that, for every m exceeding m_0, at least $(1-\varepsilon)i(m)$ rational numbers are $(m, \delta, \mathbf{d}, \eta)$-good. Then, we have

$$j(m) \geq (1-\varepsilon)ni(m), \tag{9.8}$$

while it follows from Lemma 9.6 that

$$j(m) \leq n(1 + 32g\delta)i(m) + 3\varepsilon i(m)\log m.$$

We know that, by definition, every $(m, \delta, \mathbf{d}, \eta)$-good rational contains at least $n(\mu_G(\Delta_{\mathbf{d}}) - \eta)$ occurrences of $\mathbf{d}$. Consequently,

$$\frac{\rho_{j(m)}(\mathbf{d}, \xi_{\mathbf{aks}})}{j(m)} \geq \frac{ni(m)(1-\varepsilon)(\mu_G(\Delta_{\mathbf{d}}) - \eta)}{n(1+32g\delta)i(m) + 3\varepsilon i(m)\log m},$$

which implies that

$$\liminf_{m\to+\infty} \frac{\rho_{j(m)}(\mathbf{d}, \xi_{\mathbf{aks}})}{j(m)} \geq (\mu_G(\Delta_{\mathbf{d}}) - \eta)\left(\frac{1-\varepsilon}{1+32g(\delta+\varepsilon)}\right). \tag{9.9}$$

On the other hand, the finite word composed of the first n partial quotients of an $(m, \delta, \mathbf{d}, \eta)$-good rational contains at most $n(\mu_G(\Delta_{\mathbf{d}}) + \eta)$ occurrences of $\mathbf{d}$. The number N of other possible occurrences of $\mathbf{d}$ is bounded by the sum of the lengths of the continued fraction expansions of the rational numbers that are not $(m, \delta, \mathbf{d}, \eta)$-good plus the sum of the lengths of the tails of the $(m, \delta, \mathbf{d}, \eta)$-good rationals. Thus, we have

$$N \leq \varepsilon i(m)(3\log m) + 32g\delta ni(m).$$

Consequently, by (9.8),

$$\frac{\rho_{j(m)}(\mathbf{d}, \xi_{\mathbf{aks}})}{j(m)} \leq \frac{ni(m)(\mu_G(\Delta_{\mathbf{d}}) + \eta) + \varepsilon i(m)(3\log m) + 32g\delta ni(m)}{(1-\varepsilon)ni(m)},$$

and we get that

$$\limsup_{m\to+\infty} \frac{\rho_{j(m)}(\mathbf{d}, \xi_{\mathbf{aks}})}{j(m)} \leq \frac{\mu_G(\Delta_{\mathbf{d}}) + \eta}{1-\varepsilon} + \frac{32g(\delta+\varepsilon)}{1-\varepsilon}.$$

Combined with (9.9), this gives (9.7) since ε, η and δ can be taken arbitrarily small. This completes the proof of Theorem 9.3. □

9.2 On the continued fraction expansion of an algebraic number

By Theorem D.6, a real number ξ has a periodic (resp. finite) continued fraction expansion if, and only if, ξ is a quadratic algebraic number (resp. a rational number). Very little is known about the continued fraction expansion of an algebraic number $\xi = [a_0; a_1, a_2, \ldots]$ of degree at least three. On the one hand, it follows from Liouville's Theorem E.5 that the sequence $(a_n)_{n\ge1}$ cannot increase too rapidly (see (9.10) and Theorem 9.9 below), but, on the other hand, we do not even know whether $(a_n)_{n\ge1}$ is unbounded.

A much easier question was posed by Andrzej Schinzel to Harold Davenport. He asked whether, for any given integer N, there exist algebraic numbers of degree at least three which have infinitely many partial quotients greater than N. A positive answer follows from the next theorem, established by Davenport [214] and valid for every irrational number.

THEOREM 9.7. *Let ξ be an irrational number and let p be a prime number. Then, at least one of the numbers*

$$\xi, \xi + 1/p, \ldots, \xi + (p-1)/p$$

has infinitely many partial quotients greater than $p - 2$.

PROOF. Let $(p_n/q_n)_{n\ge1}$ denote the sequence of convergents to the real number $p\xi$. Since, by (D.2), q_n and q_{n+1} are coprime for $n \ge 1$, there exists an infinite set $\mathcal{N}$ of positive integers such that q_n is relatively prime to p for every n in $\mathcal{N}$.

For n in $\mathcal{N}$, let ℓ_n be the integer defined by

$$p_n \equiv \ell_n q_n \pmod p, \quad 0 \le \ell_n \le p-1.$$

There exist an integer ℓ satisfying $0 \le \ell \le p-1$, an infinite subset $\mathcal{N}_1$ of $\mathcal{N}$, and a sequence of integers $(r_n)_{n\in\mathcal{N}_1}$ such that

$$p_n = \ell q_n + p r_n, \quad \text{for } n \text{ in } \mathcal{N}_1.$$

Consequently, for n in $\mathcal{N}_1$, we get

$$\left|\xi - \frac{\ell}{p} - \frac{r_n}{q_n}\right| = \frac{1}{p}\left|p\xi - \frac{p_n}{q_n}\right| < \frac{1}{pq_n^2},$$

and it follows from inequalities (D.3) that infinitely many partial quotients of the real number $\xi - \ell/p$ are greater than $p-2$. □

Pass [553] established that, for any given integer N, there exist algebraic integers of arbitrarily large degree which have infinitely many partial quotients greater than N; see Exercise 9.3. It follows from Theorem 1.11 of Einsiedler, Fishman and Shapira [276] that for any given integer N, every irrational real number has a multiple having infinitely many partial quotients greater than N. Consequently, there exist algebraic integers of every degree ≥ 2 which have infinitely many partial quotients greater than N. We reproduce [276, Theorem 5.5], which should be compared to Theorem 8.11.

THEOREM 9.8. *For any positive integer N there exists a real number M such that, for any irrational number ξ, there exists an integer m between 1 and M for which $m\xi$ has infinitely many partial quotients greater than N.*

When the algebraic number $\xi := [0; a_1, a_2, \ldots]$ is a quadratic number, the sequence $(a_n)_{n\geq 1}$ is ultimately periodic and the sequence $(q_n)_{n\geq 1}$ of the denominators of its convergents is ultimately a linear recurrence sequence. This implies that $(q_n^{1/n})_{n\geq 1}$ is bounded and, even, converges (use Theorem D.5 or see [283, Section 2.4]). When the degree of ξ is greater than two, one generally believes that $(q_n^{1/n})_{n\geq 1}$ also converges or, at least, remains bounded. It is even likely that $(q_n)_{n\geq 1}$ satisfies (9.2). However, we seem to be very far away from a proof (or a disproof).

The first general upper estimate for the rate of increase of $(q_n)_{n\geq 1}$ when ξ is an algebraic number of degree $d \geq 3$ follows by combining Theorem D.1 with Liouville's Theorem E.5. This easily yields that

$$\log\log q_n \leq (\log d)n, \tag{9.10}$$

for every n large enough. A slight sharpening, namely the estimate $\log\log q_n = o(n)$, can be deduced from Roth's Theorem E.7. Sharper upper bounds follow from the quantitative form of Roth's theorem given in Theorem E.9.

THEOREM 9.9. *Let ξ be an irrational, real algebraic number and let $(p_n/q_n)_{n\geq 1}$ denote the sequence of its convergents. Then, for any $\varepsilon > 0$, there exists a constant c, depending only on ξ and ε, such that*

$$\log\log q_n \leq c\,n^{2/3+\varepsilon}.$$

Theorem 9.9 is the analogue for continued fraction expansions of Theorem 8.7. Slightly sharper versions of it are proved in [8, 150].

PROOF. For $\delta > 0$, let $\mathcal{N}(\xi, \delta)$ denote the cardinality of the set of solutions to the inequality

$$\left|\xi - \frac{p}{q}\right| < \frac{1}{q^{2+\delta}},$$

in integers p, q with $\gcd(p, q) = 1$ and $q > 0$. Let $k \geq 2$ be an integer and $\delta_1, \ldots, \delta_k$ be real numbers satisfying $0 < \delta_1 < \ldots < \delta_k < 1$. Let N be a (sufficiently large) integer and put $\mathcal{S}_0 = \{1, 2, \ldots, N\}$. For $j = 1, \ldots, k$, let $\mathcal{S}_j$ denote the set of positive integers n in $\mathcal{S}_0$ such that $q_{n+1} > q_n^{1+\delta_j}$. Observe that $\mathcal{S}_0 \supset \mathcal{S}_1 \supset \ldots \supset \mathcal{S}_k$. It follows from Theorem D.1 that, for any n in $\mathcal{S}_j$, the convergent p_n/q_n gives a solution to

$$\left|\xi - \frac{p}{q}\right| < \frac{1}{q^{2+\delta_j}}.$$

Consequently, the cardinality of $\mathcal{S}_j$ is at most $\mathcal{N}(\xi, \delta_j)$.

Write

$$\mathcal{S}_0 = (\mathcal{S}_0 \setminus \mathcal{S}_1) \cup (\mathcal{S}_1 \setminus \mathcal{S}_2) \cup \ldots \cup (\mathcal{S}_{k-1} \setminus \mathcal{S}_k) \cup \mathcal{S}_k.$$

Let j be an integer with $1 \leq j \leq k$. The cardinality of $\mathcal{S}_0 \setminus \mathcal{S}_1$ is bounded by N and, if $j \geq 2$, the cardinality of $\mathcal{S}_{j-1} \setminus \mathcal{S}_j$ is at most $\mathcal{N}(\xi, \delta_{j-1})$. Furthermore, for any n in $\mathcal{S}_{j-1} \setminus \mathcal{S}_j$, we get

$$\frac{\log q_{n+1}}{\log q_n} \leq 1 + \delta_j. \tag{9.11}$$

Denoting by d the degree of ξ, we infer from Theorem D.1 and Liouville's Theorem E.5 that there exists an integer n_0 such that

$$\frac{\log q_{n+1}}{\log q_n} \leq d, \tag{9.12}$$

for every integer $n \geq n_0$. Assume that N exceeds n_0. Since $\mathcal{S}_k$ has at most $\mathcal{N}(\xi, \delta_k)$ elements, we dedúce from (9.11) and (9.12) that

$$\begin{aligned}\log q_N &= \frac{\log q_N}{\log q_{N-1}} \times \frac{\log q_{N-1}}{\log q_{N-2}} \times \cdots \times \frac{\log q_{n_0+1}}{\log q_{n_0}} \times \log q_{n_0} \\ &\leq (\log q_{n_0})\,(1+\delta_1)^N \prod_{j=2}^{k} (1+\delta_j)^{\mathcal{N}(\xi,\delta_{j-1})}\, d^{\mathcal{N}(\xi,\delta_k)}.\end{aligned}$$

Taking the logarithm of both sides, we get

$$\begin{aligned}\log\log q_N \leq \log\log q_{n_0} &+ N\log(1+\delta_1) \\ &+ \sum_{j=2}^{k} \mathcal{N}(\xi, \delta_{j-1}) \log(1+\delta_j) + (\log d)\mathcal{N}(\xi, \delta_k).\end{aligned} \tag{9.13}$$

We now select $\delta_1, \ldots, \delta_k$. We may assume that $\varepsilon < 1/3$. Set $\nu = 3/(1-\varepsilon)$ and, for $j = 1, \ldots, k$, set

$$\delta_j = N^{-(\nu^k - \nu^{j-1})/(\nu^{k+1}-1)}.$$

We check that $0 < \delta_1 < \ldots < \delta_k < N^{-1/6}$ and deduce from Theorem E.9 that $\mathcal{N}(\xi, \delta) \le \delta^{-\nu}$ for every δ with $0 < \delta \le \delta_k$, if ε is sufficiently small and N sufficiently large. We then infer from (9.13) that

$$\log\log q_N \le 2k\, N^{(\nu^{k+1}-\nu^k)/(\nu^{k+1}-1)} = 2k\, N^{(\nu-1)/(\nu-\nu^{-k})}. \qquad (9.14)$$

Choosing then k equal to $\lceil \log \varepsilon^{-1} \rceil$, we get from (9.14) that

$$\log\log q_N \le 3(\log \varepsilon^{-1})\, N^{\varepsilon + 2/3},$$

if N is large enough. This proves the theorem. □

Since very little is known on the continued fraction expansion of algebraic numbers of degree greater than or equal to three, there is some interest in proving various transcendence criteria for continued fraction expansions. The first results of this type were established by Maillet [475] and A. Baker [58]. The next theorem is extracted from [42]. Sturmian sequences are defined in Theorem A.5.

THEOREM 9.10. *Let a and b be distinct positive integers. Let $(s_n)_{n\ge1}$ be a Sturmian sequence on $\{a, b\}$. Then, the continued fraction $\xi := [0; s_1, s_2, \ldots]$ is transcendental.*

PROOF. Let $(p_n/q_n)_{n\ge1}$ be the sequence of convergents to ξ. By Theorem A.6, there are arbitrarily large integers m such that $s_{m+i} = s_i$ for $i = 1, \ldots, m$. Let $\alpha_m = [0; s_1, \ldots, s_m, s_1, \ldots]$ be the quadratic real number in $(0,1)$ with purely periodic continued fraction expansion of period $s_1, \ldots, s_m$. Since the first $2m$ partial quotients of ξ and α_m coincide, it follows from Theorem D.1 that $|\xi - \alpha_m| < q_{2m}^{-2}$. A rapid calculation shows that α_m is a root of the integer polynomial $P_m(X) = q_{m-1}X^2 + (q_m - p_{m-1})X - p_m$, thus its height $H(\alpha_m)$ is less than q_m. Since $q_{2m} \ge q_m^2$, by Theorem D.5, we get that

$$|\xi - \alpha_m| \le H(\alpha_m)^{-4},$$

and the transcendence of ξ follows from Corollary E.11. □

The key ingredient for the proof of Theorem 9.10 is a result of Schmidt (actually, a special case of the Schmidt Subspace Theorem) asserting that algebraic numbers of degree greater than two cannot be well approximated by quadratic numbers. This result is also crucial in

M. Queffélec's proof [594] of the transcendence of the continued fraction whose sequence of partial quotients is the Thue–Morse sequence on $\{1, 2\}$; see Exercise 9.4 for a (different) proof.

Further combinatorial transcendence criteria for continued fractions have been established in [3, 5, 8, 9, 14]. Their proofs rest on the Schmidt Subspace Theorem E.10, and not only on Corollary E.11. A much stronger result was subsequently obtained in [158]. Let ξ be an irrational real number and write

$$\xi = \lfloor \xi \rfloor + [0; a_1, a_2, \ldots].$$

Let $\boldsymbol{a}$ denote the infinite word $a_1 a_2 \ldots$ over the alphabet $\mathbb{Z}_{\geq 1}$. A natural way to measure the intrinsic *complexity* of ξ is to count the number $p(n, \xi) := p(n, \boldsymbol{a})$ of distinct blocks of given length n in the word $\boldsymbol{a}$. Exactly as in Definition 4.18, we introduce the notion of entropy of a real number. If ξ is a rational number, we agree that $p(n, \xi) = 1$ for $n \geq 1$.

DEFINITION 9.11. The *entropy* of a real number ξ is the quantity

$$E(\xi) := \lim_{n \to +\infty} \frac{\log p(n, \xi)}{n}.$$

Let ξ be a real algebraic number of degree at least three. A first step towards a proof that ξ has unbounded partial quotients would be to get a good lower bound for $p(n, \xi)$. The first result of this type is Theorem 9.10, which gives that $p(n, \xi) \geq n + 2$ holds for every sufficiently large integer n. This is considerably improved in the next statement.

THEOREM 9.12. *For any algebraic number ξ of degree at least three, we have*

$$\lim_{n \to +\infty} \frac{p(n, \xi)}{n} = +\infty.$$

Theorem 9.12, established in [158], is the exact analogue for continued fraction expansions of Theorem 8.3. Its proof, beyond the scope of the present monograph, rests as well on the Schmidt Subspace Theorem.

Actually, Theorem 9.12 is a consequence of the following combinatorial transcendence criterion and of the analogue for continued fraction expansions of Lemma 8.4.

For an irrational and not quadratic real number $\xi = [0; a_1, a_2, \ldots]$, define the exponent $v'_T(\xi)$ as the Diophantine exponent (Definition A.2) of the infinite word $\boldsymbol{a} = a_1 a_2 \ldots$ Denoting by $(p_n/q_n)_{n \geq 1}$ the sequence of convergents to ξ, the following statement is established in [158].

THEOREM 9.13. *With the above notation, if $(q_n^{1/n})_{n\geq 1}$ is bounded and $v'_T(\xi)$ is positive, then ξ is transcendental.*

This is the analogue for continued fraction expansions of Theorem 8.2. We note that Theorem 9.13 applies to a large class of continued fraction expansions with unbounded partial quotients.

9.3 On β-expansions

Recall that if $b \geq 2$ is an integer and if the b-ary expansion of the irrational number ξ in $(0,1)$ is given by $\xi = \sum_{k\geq 1} a_k/b^k$, then $\{b^n\xi\} = \sum_{k\geq n+1} a_k/b^k$ for $n \geq 0$. Denoting by T_b the map defined on $[0,1]$ by $T_b(x) = \{bx\}$, we observe that $a_n = \lfloor bT_b^{n-1}(\xi)\rfloor$ for $n \geq 1$. This shows that, like for the continued fraction expansion, the b-ary expansion of an irrational number is obtained by iterating a suitable map. This point of view is used to define expansions to a non-integer base.

Throughout this section, β denotes a real number greater than 1 and $\lfloor\beta\rceil$ is equal to $\beta - 1$ if β is an integer and to $\lfloor\beta\rfloor$ otherwise. We may consider writing any positive real number as a series $\sum_{n\geq 1} a_n/\beta^n$, where the a_n are non-negative integers. The notion of β-expansion was introduced by Rényi [608] in 1957. We denote by T_β the transformation defined on $[0,1]$ by $T_\beta(x) = \{\beta x\}$.

DEFINITION 9.14. The expansion of a number x in $[0,1]$ to base β, also called the *β-expansion* of x, is the sequence $(a_n)_{n\geq 1}$ of integers from $\{0,1,\ldots,\lfloor\beta\rceil\}$ such that

$$x = \frac{a_1}{\beta} + \frac{a_2}{\beta^2} + \cdots + \frac{a_n}{\beta^n} + \cdots,$$

and defined by one of the following equivalent properties:

$$\sum_{k>n} \frac{a_k}{\beta^k} < \frac{1}{\beta^n}, \quad \text{for all } n \geq 0;$$

$$a_1 = \lfloor\beta x\rfloor, a_2 = \lfloor\beta\{\beta x\}\rfloor, a_3 = \lfloor\beta\{\beta\{\beta x\}\}\rfloor, \ldots$$

$$a_n = \lfloor\beta T_\beta^{n-1}(x)\rfloor, \quad \text{for all } n \geq 1.$$

We then write

$$d_\beta(x) = a_1 a_2 \ldots a_n \ldots$$

For $x < 1$, the β-expansion coincides with the representation of x computed by the 'greedy algorithm'. If β is an integer, then the digits a_i of x lie in the set $\{0, 1, \ldots, \beta - 1\}$ and $d_\beta(x)$ corresponds, for $x \neq 1$, to the usual β-ary expansion of x.

Rényi [608] proved that the transformation T_β has a unique invariant probability measure ν_β which is absolutely continuous with respect to the Lebesgue measure on $[0, 1]$. This measure is ergodic [608] and it is the unique measure of maximal entropy [347].

We endow the set $\{0, 1, \ldots, \lfloor \beta \rceil\}^{\mathbb{Z}_{\geq 1}}$ with the lexicographic order denoted by $<_{\text{lex}}$, the product topology and the one-sided shift operator σ defined by $\sigma((s_n)_{n \geq 1}) = (s_{n+1})_{n \geq 1}$, for any sequence $(s_n)_{n \geq 1}$ in $\{0, 1, \ldots, \lfloor \beta \rceil\}^{\mathbb{Z}_{\geq 1}}$.

Definition 9.15. The closure of the set of all β-expansions of x in $[0, 1]$ is called the *β-shift* and denoted by S_β.

Parry [552] proved that the shift S_β is fully determined by $d_\beta(1)$.

Theorem 9.16. *If $d_\beta(1) = a_1 \ldots a_M 00 \ldots 0 \ldots$, then $\boldsymbol{s} = (s_n)_{n \geq 1}$ belongs to S_β if, and only if,*

$$\sigma^k(\boldsymbol{s}) <_{\text{lex}} a_1 \ldots a_{M-1}(a_M - 1) a_1 \ldots a_{M-1}(a_M - 1) a_1 \ldots, \quad \text{for } k \geq 1.$$

If $d_\beta(1)$ does not terminate with zeros only, then $\boldsymbol{s} = (s_n)_{n \geq 1}$ belongs to S_β if, and only if,

$$\sigma^k(\boldsymbol{s}) <_{\text{lex}} d_\beta(1), \quad \text{for } k \geq 1.$$

It follows from Theorem 9.16 that S_β is contained in $S_{\beta'}$ if, and only if, $\beta \leq \beta'$. The next definition is motivated by Theorem 9.16.

Definition 9.17. A block $d_1 \ldots d_m$ on $\{0, 1, \ldots, \lfloor \beta \rceil\}$ is *admissible* if

$$\sigma^k(d_1 \ldots d_m) <_{\text{lex}} d_\beta(1), \quad \text{for } k = 0, 1, \ldots, m - 1.$$

Blanchard [103] classified the β-shifts into five disjoint classes C_1 to C_5, as follows:

- β belongs to C_1 if $d_\beta(1)$ terminates with zeros only;
- β belongs to C_2 if $d_\beta(1)$ is eventually periodic but does not terminate with zeros only;
- β belongs to C_3 if $d_\beta(1)$ does not contain arbitrarily large strings of 0's and if $d_\beta(1)$ is not eventually periodic;
- β belongs to C_4 if $d_\beta(1)$ does not contain some admissible blocks, but contains arbitrarily large strings of 0's;
- β belongs to C_5 if $d_\beta(1)$ contains all admissible blocks.

Elements of Class $C_1 \cup C_2$ are called *Parry numbers*. The term *simple Parry numbers* usually denotes elements of Class C_1. Classes C_1 and C_2 are both countable and all their elements are algebraic integers. Furthermore, every Pisot number is a Parry number [83, 626]. The fact that all Salem numbers would be Parry numbers is a particular instance of a conjecture of K. Schmidt [626]. This was proved by Boyd [128] for all Salem numbers of degree 4. However, the same author considered in [129] a heuristic suggesting the existence of Salem numbers of degree 8 that are not Parry numbers.

Schmeling [625] proved that C_3 has Hausdorff dimension one and that C_5 has full Lebesgue measure. Explicit natural examples of transcendental numbers in C_3 were given by Allouche and Cosnard [40], and by Chi and Kwon [186]. Explicit examples of transcendental numbers in C_4 were given in [6].

Let $\pi_\beta : [0,1] \to S_\beta$ denote the map assigning to each x in $[0,1]$ its β-expansion. For a block $D_m = d_1 \ldots d_m$ on $\{0, 1, \ldots, \lfloor \beta \rceil\}$, a real number x in $[0,1]$ with $d_\beta(x) = a_1 a_2 \ldots$ and an integer $N \geq 1$, set

$$A_\beta(D_m, N, x) = \mathrm{Card}\{j : 0 \leq j \leq N - m, a_{j+1} = d_1, \ldots, a_{j+m} = d_m\}.$$

Definition 9.18. A real number x in $[0,1]$ is *β-normal* if, for every finite block D on $\{0, 1, \ldots, \lfloor \beta \rceil\}$, the limit $\lim_{N \to +\infty} A_\beta(D, N, x)/N$ exists and is equal to $\nu_\beta(\pi_\beta^{-1} D)$.

It follows from Rényi's result mentioned above that, for any real number $\beta > 1$, almost every real number x in $[0,1]$ is β-normal. Explicit constructions of β-normal numbers have been given in [84, 352] by means of a Champernowne-type construction.

The next definition was formulated by Schmeling [625].

Definition 9.19. A real number β in $(1, +\infty)$ is *self-normal* if $d_\beta(1)$ is β-normal.

Schmeling [625] established that almost all (with respect to the Lebesgue measure) real numbers β in $(1, +\infty)$ are self-normal. By means of a method inspired by Chapernowne's construction, Bertrand-Mathis [90] constructed uncountably many self-normal numbers.

Unlike for expansions to an integer base, it generally remains open to decide whether an algebraic number has or does not have an eventually periodic β-expansion when the base β is algebraic. Moreover, it seems to be very difficult to describe the β-expansion of an algebraic number when this expansion is neither finite, nor eventually periodic. In particular, very little is known on the β-expansion of 1 for an algebraic number β

greater than 1 which is not in the union of the classes C_1 and C_2. Some partial results are given in [6, 262].

Recall that the set $\mathcal{N}(\beta)$, introduced in Definition 6.4, denotes the set of real numbers ξ such that $(\xi\beta^n)_{n\geq 1}$ is uniformly distributed modulo one.

DEFINITION 9.20. For every real number $\beta > 1$, we denote by $\mathcal{D}(\beta)$ the set of real numbers whose β-expansion is β-normal.

There is no reason for the sets $\mathcal{D}(\beta)$ and $\mathcal{N}(\beta)$ to be equal. Bertrand [81] proved that if β is a Pisot number, then $\mathcal{D}(\beta)$ is contained in $\mathcal{N}(\beta)$. We do not know whether or not both sets coincide.

Bertrand [82, 85, 86] established that, if $\beta > 1$ is such that $(\beta^n)_{n\geq 1}$ is uniformly distributed modulo one, then the set $\mathcal{N}(\beta)$ is strictly contained in $\mathcal{D}(\beta)$.

Schmeling [625] proved that every x in $(0, 1]$ is β-normal for almost all β in $(1, +\infty)$.

We extend as follows the notion of b-badly approximable numbers, introduced in Section 7.3.

DEFINITION 9.21. Let $\beta > 1$ be a real number which is not a rational integer. A real number x in $[0, 1]$ is called *β-badly approximable* if the blocks of the digit 0 occurring in its β-expansion have bounded length.

Using a suitable modification of Schmidt's game, Färm, Persson and Schmeling [299] proved that, for any real number β with $1 < \beta < 2$, the set of β-badly approximable numbers is α-winning for any real number α with $0 < \alpha < 1/64$. This was previously established by Färm [295] when $d_\beta(1)$ is finite.

9.4 Exercises

EXERCISE 9.1. Study the rate of approximation of $\xi_{\mathbf{aks}}$ by rational numbers.

EXERCISE 9.2. Prove that there exist real numbers having a normal continued fraction expansion and an arbitrarily prescribed irrationality exponent.

EXERCISE 9.3 (*cf.* [553]). Let p be a large prime number. For $n \geq 1$, set $g_n(X) = X(X+1)\dots(X+n-1)$. Put $f(X) = pg_p(X) - 1$.

(1) Prove that $f(X)$ is irreducible and has a real root α.

(2) Use Theorem 9.1 to prove that there is c in $\{0, 1, \ldots, p-1\}$ such that $p/(\alpha + c)$ is an algebraic integer of degree p with infinitely many partial quotients at least equal to $p-2$.

EXERCISE 9.4. Let $(a_n)_{n \geq 1}$ be a non-ultimately periodic sequence of positive integers. Assume that there are arbitrarily large integers m such that $a_1 \ldots a_m$ is a square (resp., a palindrome). Use Theorem E.10 (resp., Exercise E.3) to prove that $[0; a_1, a_2, \ldots]$ is transcendental. Deduce that the real number $[0; 1, 2, 2, 1, 2, 1, 1, 2, \ldots]$, whose sequence of partial quotients is given by the Thue–Morse infinite word on $\{1, 2\}$, is transcendental.

9.5 Notes

▷ Numerical values and speculation for continued fraction expansions of certain algebraic numbers can be found in [420, 609]. Computational problems connected with the calculation of the continued fraction of algebraic numbers are discussed in [112, 130]. For statistics on the continued fraction expansion of π and other classical numbers, see [228].

▷ Postnikov and Pyateckiĭ [585] (see also [583, Chapter 27]) gave another construction for a normal continued fraction. Postnikov [583] formulated a sufficient condition for a number to have a normal continued fraction which is analogous to Theorem 4.6.

▷ Kraaikamp and Nakada [406] established that a number which is normal with respect to its regular continued fraction expansion is also normal with respect to its continued fraction expansion from below, and that a number is normal with respect to its regular continued fraction expansion if, and only if, it is normal with respect to its nearest integer continued fraction expansion.

▷ Mendès France [503] established that, under the assumptions of Theorem 9.7, there exists j in $\{0, 1, \ldots, p-1\}$ such that 'many' partial quotients of $\xi + j/p$ exceed $p-2$.

▷ Berend and Boshernitzan [74] proved that there exist uncountably many real numbers ξ with the property that the sequence of partial quotients of each integer multiple $n\xi$ tends to infinity.

▷ It follows from Theorem 9.12 that an infinite continued fraction whose sequence of partial quotients is generated by a finite automaton is either quadratic, or transcendental. Transcendence measures for families of continued fractions including the continued fractions defined in Theorem 9.10 are given in [11, 156], following a general method based on the Quantitative Subspace Theorem and explained in [10].

▷ The continued fractions $[0; 1, 2, 3, 4, \ldots]$ and, much more generally, $[0; a_1, a_2, a_3, \ldots]$, where $(a_n)_{n\geq 1}$ is an arithmetic progression of positive integers, are transcendental; see [637, p. 134].

▷ Let p and q be coprime integers with $p > q \geq 2$. Mahler's result [465] established in Exercise 3.9 implies that, for every $\varepsilon > 0$, the first partial quotient of $\|(p/q)^n\|$ is less than $2^{\varepsilon n}$ when n is sufficiently large. For a rational number r, let $\mathcal{L}(r)$ denote the length of the shortest continued fraction equal to r. In 1973, Mendès France [501] asked whether

$$\sup_{n\geq 1} \mathcal{L}\big((p/q)^n\big) = +\infty$$

holds for all coprime integers p and q with $p > q \geq 2$. Choquet [193, 194] gave an affirmative answer to this question. Independently, Pourchet [587] applied Ridout's Theorem E.8 to obtain a stronger statement. He proved that, for all coprime integers p and q with $p > q \geq 2$, for any positive real number ε, the partial quotients of $\|(p/q)^n\|$ are all less than $2^{\varepsilon n}$ when n is sufficiently large. Consequently, $\mathcal{L}\big((p/q)^n\big)$ tends to infinity with n. Pourchet never published his result. Some details of the proof have been given by van der Poorten [579]; see also [209], [747, Exercise II.6] and [150]. Under the above hypotheses, it is proved in [150] that there are arbitrarily large integers n such that $\mathcal{L}((p/q)^n) \geq (\log n)^{1/5}$. Corvaja and Zannier [209] extended Pourchet's theorem to quotients of power sums.

▷ Mendès France [506] asked whether for every real quadratic irrational ξ and every positive M, there exist integers n such that the length of the period of the continued fraction expansion of ξ^n exceeds M. This question was completely solved by Corvaja and Zannier [208]. Results on the length of the period of the continued fraction for values of the square root of power sums have been given in [166, 639]. Further questions on continued fraction expansions of real numbers in a fixed quadratic field are considered in [27, 329, 455, 736].

▷ Mauduit and Moreira [489, 490] computed the generalized Hausdorff dimensions of sets of real numbers having zero entropy.

▷ For additional results on real numbers with bounded partial quotients, see the survey [646].

▷ Fan, Liao and Ma [292] determined the Hausdorff dimension, in terms of a modified variational principle, of sets of real numbers having prescribed frequencies of partial quotients in their continued fraction expansion.

▷ Let ξ be in $(0,1)$ and ξ_n be the rational number obtained by truncating the decimal expansion of ξ after n digits. The problem to determine the largest integer $k_n(\xi)$ for which the first $k_n(\xi)$ partial quotients of the continued fraction expansions of ξ and ξ_n coincide has been studied by Lochs [448]. He established the surprising result (compare with Theorem D.7) that, for almost all real numbers ξ, the ratio $k_n(\xi)/n$ tends to $6\pi^{-2}(\log 2)(\log 10) = 0.9702\ldots$ when n tends to infinity; see also [124, 212, 285–287, 739, 740] and [62, 440] when the decimal expansion is replaced by the β-expansion.

▷ Smorodinsky and Weiss [658] generalized the Champernowne construction to obtain normal sequences for finite-state ergodic Markov shifts and for subshifts with a unique measure maximizing the topological entropy (e.g., the β-shifts). Analogous results were obtained for Markov shifts by Postnikov [583] and for β-shifts by Ito and Shiokawa [352]. Furthermore, Bertrand-Mathis [84] constructed so-called 'Champernowne sequences' for general coding systems. The result of Copeland and Erdős (case $m = 1$ of Theorem 4.10) has been extended in [91] to bases β which are Pisot numbers; see also [327].

▷ Bertrand-Mathis introduced and studied the notion of geometrically normal numbers [85, 86, 89].

▷ Bertrand-Mathis [88] conjectured that, for real numbers $\beta > 1$, $\beta' > 1$ which are not integers, the Rényi measures ν_β and $\nu_{\beta'}$ coincide if, and only if, $\beta' = \beta + 1$ and there exist integers a, b such that $\beta^2 = a\beta + b$ and $1 \le b < a$. She gave some partial results.

▷ Kwon [415] extended Theorem 3.1 to β-expansions.

▷ Durner [269] extended Colebrook's results [200] to β-expansions.

▷ For $\beta > 1$, set $G(\beta) = \nu_\beta(\pi_\beta^{-1}(0))$. This quantity represents the asymptotic proportion of zeros in the β-expansion of ξ for almost every ξ. Brown and Yin [142] considered analytic properties of the map $\beta \mapsto G(\beta)$ and proved, among other results, that if β is not an integer, then $G(\beta)$ exceeds $1/\beta$.

▷ Let α in $(0,1]$ and $\beta > 1$ be algebraic numbers. The asymptotic behaviour of the function that counts the number of digit changes in the β-expansion of α is studied in [151].

▷ Let $v > 0$ and β_0, β_1 be such that $1 < \beta_0 < \beta_1 < 2$. Denote by $D_v(\beta_0, \beta_1)$ the set of real numbers β in (β_0, β_1) for which there are infinitely many integers n such that all the digits $a_n, a_{n+1}, \ldots$ up to $a_{\lfloor (1+v)n \rfloor}$ in the β-expansion of 1 are equal to 0. Persson and Schmeling [558] proved that the Hausdorff dimension of $D_v(\beta_0, \beta_1)$ equals $1/(1+v)$.

▷ For (α, β) in $[0,1) \times (1, +\infty)$ and x in $[0,1]$ define $T_{\alpha,\beta}$ by $T_{\alpha,\beta} x = \{\beta x + \alpha\}$. Faller and Pfister [291] proved that, for every x, α not both 0, the point x is $T_{\alpha,\beta}$-normal (we omit the definition) for almost every β in $(1, +\infty)$. This extends a result of Schmeling [625] who dealt with the case $x = 1$ and $\alpha = 0$.

▷ Färm and Persson [298] (see also [299] for a slightly weaker result) proved that, for every sequence $(\beta_j)_{j \geq 1}$ of real numbers greater than 1, the set of real numbers x such that $(A_{\beta_j}(D, N, x)/N)_{N \geq 1}$ converges for no $j \geq 1$ and no finite block D on $\{0, 1, \ldots, \lfloor \beta_j \rceil\}$ has Hausdorff dimension one.

▷ Chaotic and topological properties of β-expansions are studied in [439]. Simonsen [656] considered β-shifts from the point of view of computability theory.

▷ Dubickas [255] obtained various results on sequences generated by the map $x \mapsto \lceil \beta x + \gamma \rceil$ for real numbers $\beta > 1$ and γ. In particular, if $x_1 = 1$ and $x_{n+1} = \lceil 3x_n/2 \rceil$ for $n \geq 1$, then, setting $w_n = 0$ if x_n is even and $w_n = 1$ otherwise, the infinite word $\boldsymbol{w} = w_1 w_2 \ldots$ on $\{0,1\}$ satisfies $p(n, \boldsymbol{w}, \{0,1\}) \geq 1.709n$ for every sufficiently large integer n.

▷ An interesting discussion on generalizations of continued fractions can be found in [479]. For $(-\beta)$-expansions, see [443] and the references therein. For normality in the context of Q-Cantor series, see [48, 476] and the references therein. Further representations of real numbers by infinite series are studied in [318] and in [393].

10
Conjectures and open questions

We gather open problems encountered in the preceding chapters with several new ones. Instead of formulating them in terms of questions, we merely prefer to propose statements whose validity remains open. In many cases there is no evidence for, or against, the assertion claimed. We do not recall the partial results obtained towards these problems, since they can be (hopefully) easily found in the present book.

A large list of open problems in the general theory of distribution modulo one has been compiled by Strauch and Nair [677]; see also the monograph [678].

The thematic ordering of the problems essentially follows Chapters 1 to 9.

The first problem was posed by Hardy [333] in 1919.

PROBLEM 10.1. *Are there a transcendental number α and a positive real number ξ such that $||\xi\alpha^n||$ tends to 0 as n tends to infinity?*

Very little is known on the sequence of fractional parts of e.

PROBLEM 10.2. *To prove that $||\mathrm{e}^n||$ does not tend to 0 as n tends to infinity.*

The next problem is usually attributed to Mahler although it does not seem to have been stated explicitly in his papers.

PROBLEM 10.3. *To prove that there exists a positive real number c such that $||\mathrm{e}^n|| > \mathrm{e}^{-cn}$, for every $n \geq 1$.*

Waldschmidt [723] conjectured that a stronger result holds, namely that there exists a positive real number c such that $||\mathrm{e}^n|| > n^{-c}$ for every $n \geq 1$. This is supported by metrical results [391].

The *spectrum* of a sequence $(x_n)_{n\geq 1}$ of real numbers is the set of irrational real numbers θ in $(0, 1)$ such that the sequence $(x_n - n\theta)_{n\geq 1}$

is not uniformly distributed modulo one. The next problem was posed by Mendès France [502].

PROBLEM 10.4. *Let ξ be a non-zero real number and $\alpha > 1$ be a real number. The spectrum of the sequence $(\xi\alpha^n)_{n\geq 1}$ is at most countable.*

The next problem was proposed and discussed by Dubickas as Conjecture 2 in [250].

PROBLEM 10.5. *Let $\mathbb{K}$ be a real number field. Then, for any $\varepsilon > 0$, there exists a lacunary sequence $(t_n)_{n\geq 1}$ of positive numbers in $\mathbb{K}$ such that*

$$\limsup_{n\to+\infty} \{\xi t_n\} \geq 1-\varepsilon, \quad \textit{for any real number } \xi \textit{ not in } \mathbb{K}.$$

Moreover, each subinterval of $[0,1]$ of length ε contains a limit point of the sequence $(\{\xi t_n\})_{n\geq 1}$.

As was shown in Section 2.5, for every irrational number ξ, the sequence $(\xi 2^n 3^m)_{m,n\geq 1}$ is dense modulo one. On the other hand, if the sequence $(m_n)_{n\geq 1}$ of positive integers is lacunary, then there exist uncountably many real numbers ξ such that $(\xi m_n)_{n\geq 1}$ is not dense modulo one.

PROBLEM 10.6. *To find a very rapidly increasing sequence $(m_n)_{n\geq 1}$ of positive integers such that $(\xi m_n)_{n\geq 1}$ is dense modulo one for every irrational number ξ.*

The next problem was formulated in [168].

PROBLEM 10.7. *Let ε be a positive real number. Are there arbitrarily large real numbers α such that α is not a Pisot number and all the fractional parts $\{\alpha^n\}$, $n\geq 1$, are lying in an interval of length ε/α?*

We recall the conjecture of de Mathan and Teulié [483] discussed in Section 2.6.

PROBLEM 10.8. *For every real number ξ and every prime number p, we have $\inf_{q\geq 1} q\cdot \|q\xi\|\cdot |q|_p = 0$.*

We continue with a question of Mahler [468].

PROBLEM 10.9. *There are no real numbers ξ such that $0\leq \{\xi(3/2)^n\} < 1/2$ for every positive integer n.*

We keep focusing on fractional parts of integral powers of $3/2$.

PROBLEM 10.10. *The sequence $((3/2)^n)_{n\geq 1}$ is dense modulo one.*

Since Problem 10.10 seems to be much too difficult, we highlight a weaker question, which was posed by Mendès France [506].

Problem 10.11. *The sequence* $(\{(3/2)^n\})_{n\ge1}$ *has an irrational limit point.*

Problem 10.12. *There does not exist a real number* M *such that, for every* $n \ge 1$, *all the partial quotients of* $\{(3/2)^n\}$ *are less than* M.

We continue with a question related to Waring's problem; see Section 3.7.

Problem 10.13. *To find an upper bound for the number of positive integers* n *such that* $||(3/2)^n|| < (3/4)^n$.

Levin [433] established the existence of normal numbers with small discrepancy, but it is unclear whether his result is the best possible.

Problem 10.14. *Let* $b \ge 2$ *be an integer. To find a real number* ξ *and a positive constant* c *such that* $D_N((\xi b^n)_{n\ge1}) \le c(\log 2N)/N$ *for every* $N \ge 1$.

Problem 10.15. *To find a simple construction of a real number normal both to base* 2 *and to base* 3. *To find a simple construction of an absolutely normal real number.*

Problem 10.16. *To prove the existence of a real number* ξ *such that* $D_N((\xi b^n)_{n\ge1}) \le c(\log 2N)^\kappa/N$ *for every* $N \ge 1$ *and every* $b \ge 2$, *where* c *and* κ *are positive constants depending only on* b *and* ξ.

The next problems, suggested by Rivoal [611], are reproduced in [724].

Problem 10.17. *Let* $b \ge 2$ *be an integer. To give an explicit example of a positive real number* ξ *which is simply normal (resp., normal) to base* b *and for which* $1/\xi$ *is not simply normal (resp., not normal) to base* b.

Problem 10.18. *To give an explicit example of a positive real number* ξ *which is absolutely normal and for which* $1/\xi$ *does not share this property.*

The next question is extracted from [25]. Recall that, for an integer $b \ge 2$, the set of real numbers normal to base b is denoted by $\mathcal{N}(b)$.

Problem 10.19. *Do there exist irrational numbers* γ *such that* $\mathcal{N}(b) + \gamma = \mathcal{N}(b)$ *for every integer* $b \ge 2$?

Only very little is known on the expansions of a given real number to different bases. A general question, which was investigated in [155], can be formulated as follows.

Problem 10.20. *Are there irrational numbers having a 'simple' expansion to base* 2 *and a 'simple' expansion to base* 3?

By 'simple', we mean for instance with low block complexity, or with few digit changes, or with few non-zero digits, etc. Of course, 2 and 3 do not play any particular role in Problem 10.20. From Problem 10.21 until Problem 10.29, we denote by r and s any two multiplicatively independent positive integers.

The next problems make the question in Problem 10.20 a little more precise.

PROBLEM 10.21. *Find an increasing function f such that $p(n, \xi, r) + p(n, \xi, s)$ exceeds $f(n)$ for every irrational number ξ and every sufficiently large (in terms of ξ, r, s) integer n.*

PROBLEM 10.22. *There exists a positive real number ξ such that*

$$\lim_{n \to +\infty} \frac{\mathcal{NZ}(n, \xi, r) + \mathcal{NZ}(n, \xi, s)}{n} = 0.$$

PROBLEM 10.23. *Give an explicit example of a real irrational number which is rich neither to base r, nor to base s.*

PROBLEM 10.24. *There are no irrational numbers ξ such that the sequences $(\{\xi r^n\})_{n \geq 1}$ and $(\{\xi s^n\})_{n \geq 1}$ both have only countably many limit points.*

Problem 10.24 is probably easier to solve than [316, Conjecture 2], reproduced below as Problem 10.25. For an integer $b \geq 2$ and a real number ξ, we denote by $\Lambda_b(\xi)$ the set of limit points of the sequence $(\{\xi b^n\})_{n \geq 1}$.

PROBLEM 10.25. *For every irrational number ξ, the sum of the Hausdorff dimensions of $\Lambda_r(\xi)$ and of $\Lambda_s(\xi)$ is at least equal to 1.*

Problem 10.25 is motivated by the following celebrated conjecture of Furstenberg [315, 477].

PROBLEM 10.26. *The Lebesgue measure is the unique non-atomic ergodic measure on the torus which is invariant both under multiplication by 2 and under multiplication by 3.*

Partial results can be found in [301, 346, 348, 349, 355, 356, 444, 495, 615].

We continue on the same topic with [316, Conjecture 2].

PROBLEM 10.27. *In the representation of r^n in base rs every digit and every combination of digits occur as soon as n is sufficiently large.*

Problem 10.27 should be compared with the next problem, which was raised by Schmidt [627].

Problem 10.28. *For any fixed ε and k, the number of $n \le N$ for which the representation of r^n in base s is not (ε, k)-normal (in the sense of Besicovitch) is $o(N)$ as N tends to infinity.*

We display a further question on representations of integers in different integer bases. A finite Sturmian word is any finite block of an infinite Sturmian word.

Problem 10.29. *Prove that there are only finitely many powers of r whose representation in base s is a finite Sturmian word.*

The next problem was proposed by Colebrook and Kemperman [201] and extends a question of Volkmann [716]. For an integer $b \ge 2$ and a real number ξ, the sets $I(b)$ and $V(\xi, b)$ are defined in Section 4.7.

Problem 10.30. *Let $(s_n)_{n\ge1}$ be a sequence of positive integers such that s_n and s_m are multiplicatively independent for any distinct positive integers m and n. For every $n \ge 1$, let V_n be a non-empty and closed subset of $I(s_n)$. There exists a real number ξ such that $V(\xi, s_n) = V_n$ for every $n \ge 1$.*

Hertling [342] asked whether Schmidt's Theorem 6.3 has an analogue with normality replaced by richness.

Problem 10.31. *Let $\mathcal{R} \cup \mathcal{S}$ be a partition of the set of integers greater than or equal to 2 into two classes such that any two multiplicatively dependent integers fall in the same class. There are real numbers which are rich to every base from $\mathcal{R}$ but not rich to every base from $\mathcal{S}$.*

The next problem is a slightly modified form of an open problem given on [46, p. 403].

Problem 10.32. *Is there an irrational real number whose expansions in two multiplicatively independent bases can both be generated by a finite automaton?*

We now radically change our point of view. Rather than looking at the expansions of one given number to several integer bases, we take an infinite word $\boldsymbol{w}$ on $\{0, 1\}$ which we use to define two real numbers, a first one whose binary expansion is given by $\boldsymbol{w}$ and a second one whose ternary expansion is given by $\boldsymbol{w}$. The next problem appeared at the end of a paper of Mendès France [504]; see also [505]. According to him (see the discussion in [46, p. 403]) it was proposed by Mahler; however, we were unable to find any mention of it in Mahler's works.

Problem 10.33. *For an arbitrary infinite sequence $(\varepsilon_k)_{k\ge1}$ of 0's and 1's, the real numbers $\sum_{k=1}^{+\infty} \varepsilon_k/2^k$ and $\sum_{k=1}^{+\infty} \varepsilon_k/3^k$ are algebraic if, and only if, both are rational.*

The following more general question is discussed in [157].

PROBLEM 10.34. *Let $\mathcal{P}$ be a property valid for almost all real numbers. Let $b \geq 2$ be an integer. Let b_1 and b_2 be distinct integers, at least equal to b. Let $(\varepsilon_k)_{k\geq 1}$ be a sequence taking its values in $\{0, 1, \ldots, b-1\}$, which is not ultimately periodic. Is it true that at least one among the numbers*

$$\xi_1 := \sum_{k\geq 1} \frac{\varepsilon_k}{b_1^k} \quad \text{and} \quad \xi_2 := \sum_{k\geq 1} \frac{\varepsilon_k}{b_2^k} \tag{10.1}$$

satisfies property $\mathcal{P}$?

By arguing as in Section 7.6, we see that the answer to Problem 10.34 is negative when $\mathcal{P}$ is the property 'not being a Liouville number' or 'not being badly approximable'. Theorem 1 from [157] gives explicit examples of integers $b \geq 2$ and $b_1 > b$ with $b_1 \neq b^2$ and sequences $(\varepsilon_k)_{k\geq 1}$ taking their values in $\{0, 1, \ldots, b-1\}$ such that, setting $b_2 = b^2$, the irrationality exponents of ξ_1 and ξ_2 defined by (10.1) are different.

We copy an open problem already discussed in Section 7.1, where the notation used in its statement is introduced.

PROBLEM 10.35. *Let $(v_b)_{b\in\mathcal{B}_1}$ and $(v'_b)_{b\in\mathcal{B}}$ be sequences of real numbers or $+\infty$ satisfying*

$$v_1 \geq 1, \qquad 0 \leq v_b \leq v'_b \leq v_1, \quad \text{for every } b \in \mathcal{B},$$

and

$$v_{b_2} + 1 \geq \rho(b_1, b_2)\frac{\log b_1}{\log b_2}(v_{b_1} + 1),$$

for every $b_1, b_2 \in \mathcal{B}$ such that every prime factor of b_1 divides b_2. Prove that there exist real numbers ξ such that

$$v_1(\xi) = v_1, \quad v_b(\xi) = v_b \quad \text{and} \quad v'_b(\xi) = v'_b, \quad \text{for every } b \in \mathcal{B}.$$

By Theorem 7.8, there exist real numbers which are b-badly approximable for every $b \geq 2$.

PROBLEM 10.36. *There exist a real number ξ and a positive real number c such that $\|b^n\xi\| > b^{-c}$ for every base $b \geq 2$ and every integer $n \geq 0$ (resp., every integer n sufficiently large in terms of b).*

There exist Liouville numbers in the middle third Cantor set K and there are Liouville numbers which are normal to base 2. Furthermore, K contains numbers normal to base 2. But we do not know whether there are real numbers with all these three properties.

PROBLEM 10.37. *Prove that the middle third Cantor set contains Liouville numbers which are normal to base* 2.

The above problem concerns the intersection of three sets, any two of them having non-empty intersection.

The exponents w_n and w_n^* are introduced in Definition E.12.

PROBLEM 10.38. *What can be said on the approximation of points in the middle third Cantor set K by algebraic numbers? Are there points ξ in K such that $w_n(\xi)$ differs from $w_n^*(\xi)$ for some integer $n \geq 2$?*

PROBLEM 10.39. *Are there elements of the middle third Cantor set with a prescribed value for the exponent v_2?*

PROBLEM 10.40. *Let $(n_j)_{j\geq 1}$ be an increasing, lacunary sequence of positive integers. Find the irrationality exponent of the real number $\sum_{j\geq 1} 2^{-n_j}$.*

PROBLEM 10.41. *Determine the Hausdorff dimension of the set of badly approximable numbers with prescribed digit frequencies.*

The next problem was proposed in [19] for $b = 2$.

PROBLEM 10.42. *Let $b \geq 2$ be an integer. Is it true that the b-ary expansion of every irrational algebraic number contains arbitrarily large squares? Is it true that the b-ary expansion of every irrational algebraic number contains arbitrarily large palindromes?*

PROBLEM 10.43. *Let $b \geq 2$ be an integer. A real number whose b-ary expansion begins in arbitrarily large palindromes is either rational or transcendental.*

A *morphic number* is a real number ξ whose expansion in some integer base $b \geq 2$ is a morphic sequence (see [46] for a precise definition). As proved in [275], its complexity function satisfies $p(n, \xi, b) = O(n^2)$. Furthermore, any automatic number is a morphic number.

PROBLEM 10.44. *Irrational morphic numbers are transcendental.*

PROBLEM 10.45. *The real number* e *is not an automatic number.*

The notion of *deterministic sequence* has been defined in [359, 604]. The next problem was posed by Mendès France [504].

PROBLEM 10.46. *Let $b \geq 2$ be an integer. Is the b-ary expansion of an irrational algebraic number deterministic?*

Let $b \geq 2$ be an integer. Beside the block complexity of a real number written in base b, we may define as well its *infinite complexity*. For an infinite word $\boldsymbol{w}$ on the alphabet $\{0, 1, \ldots, b-1\}$ and for any positive integer n, we let $p_\infty(n, \boldsymbol{w}, b)$ denote the number of distinct blocks of n letters occurring infinitely often in $\boldsymbol{w}$. Furthermore, for an irrational number ξ whose b-ary expansion is given by $\xi = \lfloor \xi \rfloor + \sum_{k \geq 1} a_k/b^k$, we set $p_\infty(n, \xi, b) = p_\infty(n, \boldsymbol{a}, b)$ with $\boldsymbol{a} = a_1 a_2 \ldots$ Obviously, for $n \geq 1$, we have $1 \leq p_\infty(n, \xi, b) \leq b^n$, and $p_\infty(n, \xi, b) \geq n + 1$ if ξ is irrational.

PROBLEM 10.47. *Give a non-trivial lower bound for $p_\infty(n, \xi, b)$ when ξ is an algebraic irrational number.*

We cannot exclude that $p_\infty(n, \sqrt{2}, b) = n + 1$ for every $b \geq 2$ and every $n \geq 1$.

For integers $b \geq 2$ and $c \geq 2$, let $(c)_b$ denote the sequence of digits of c in its representation in base b. Mahler [471] proved that the real number $0 \cdot (c)_{10}(c^2)_{10} \ldots$ is irrational. This was subsequently reproved and extended to every base $b \geq 2$ by Bundschuh [170] and Niederreiter [539]; see also [59, 172, 647, 652].

PROBLEM 10.48. *With the above notation, prove that, for arbitrary integers $b \geq 2$ and $c \geq 2$, Mahler's number $0 \cdot (c)_b(c^2)_b \ldots$ is transcendental and normal to base b.*

The question of the normality of 0.248163264... to base ten was already posed by Pillai [561].

The next two problems are extracted from [597].

PROBLEM 10.49. *Construct a real number which is normal to some given integer base and whose continued fraction expansion is normal.*

PROBLEM 10.50. *Let $b \geq 2$ be an integer. There is a real number whose continued fraction expansion is normal and whose b-ary expansion has a low block complexity.*

A real number ξ being given, it is very rare that we can say something at the same time on its continued fraction expansion and on its expansion to some integer base (but see Section 7.6).

PROBLEM 10.51. *Is there a real number ξ whose continued fraction expansion is normal and which is not normal to a given integer base $b \geq 2$? which is absolutely non-normal?*

PROBLEM 10.52. *There are irrational numbers with a 'simple' expansion in base 2 and a 'simple' continued fraction expansion.*

PROBLEM 10.53. *There is an irrational real number ξ such that $E(\xi) <$ $\log 2$ and $E(\xi, b) < \log b$ for some integer $b \geq 2$ (resp., for every integer $b \geq 2$).*

Recall that μ_K denotes the standard measure on the middle third Cantor set.

PROBLEM 10.54. *Are μ_K-almost all elements of the middle third Cantor set normal with respect to the Gauss measure?*

Berend and Boshernitzan [74] formulated Problem 10.55 and gave a positive answer to it when ξ is quadratic; see also [276, Question 5.7].

PROBLEM 10.55. *Let ξ be a badly approximable real number. Prove that, for any finite block D of positive integers, there exist arbitrarily large integers n such that the block D appears infinitely often in the sequence of partial quotients of $n\xi$ viewed as an infinite word.*

The next problem was posed by Mendès France; see [487], where further interesting open problems are proposed.

PROBLEM 10.56. *If ξ has a normal continued fraction expansion, then $q\xi$ has the same property for every non-zero rational number q.*

The following problems are motivated by results presented in Chapter 9.

PROBLEM 10.57. *For every integer $N \geq 2$ and $d \geq 2$, there exist algebraic units of degree d which have infinitely many partial quotients greater than N.*

PROBLEM 10.58. *An irrational real number whose sequence of partial quotients is a morphic sequence is quadratic or transcendental.*

PROBLEM 10.59. *For a real number $\beta > 1$, compare the sets $\mathcal{N}(\beta)$ and $\mathcal{D}(\beta)$.*

PROBLEM 10.60. *For almost all β in the middle third Cantor set, the real number $1 + \beta$ is self-normal.*

We conclude with a problem proposed by Mendès France [498].

PROBLEM 10.61. *Let $\alpha > 2$ be a Pisot number and set*

$$C(\alpha) := \big\{(\alpha - 1) \sum_{k \geq 1} \varepsilon_k \alpha^{-k} : \varepsilon_k \in \{0, 1\}\big\}.$$

For every ξ in $C(\alpha)$ the sequence $(\xi\alpha^n)_{n\geq 1}$ is not uniformly distributed modulo one.

Appendix A
Combinatorics on words

Taking a point of view from combinatorics on words has helped to obtain several important theorems in Diophantine approximation, some of which are discussed in the present book. In this appendix, we gather various results, some are classical, others less well known. Standard textbooks on combinatorics on words include [46, 310, 450].

A.1 Definitions

Let $\mathcal{A}$ be a finite or infinite set. A finite word W on the alphabet $\mathcal{A}$ is either the empty word ϵ, or a finite block of elements from $\mathcal{A}$. The set of all finite words on $\mathcal{A}$ is a monoid for the concatenation. The length of a finite word W on the alphabet $\mathcal{A}$, that is, the number of letters composing W, is denoted by $|W|$. For any positive integer ℓ, we write W^ℓ for the word $W \ldots W$ (ℓ times repeated concatenation of the word W) and W^∞ for the infinite word constructed by concatenation of infinitely many copies of W. More generally, for any positive real number x, we denote by W^x the word $W^{\lfloor x \rfloor} W'$, where W' is the prefix of W of length $\lceil (x - \lfloor x \rfloor)|W| \rceil$. In particular, we can write

$$aabaaaabaaaa = (aabaa)^{12/5} = (aabaaaabaa)^{6/5}.$$

Any word W^x with $x > 1$ and W finite is called an x-power (a *square* when $x = 2$ and a *cube* when $x = 3$). An *overlap* is any finite word of the form $wWwWw$, where $w \in \mathcal{A}$ and $W \in \mathcal{A}^*$.

A finite, non-empty word $W = w_1 \ldots w_n$ is a *palindrome* if $w_j = w_{n-j+1}$ for $j = 1, \ldots, n$.

For a finite or infinite word $\boldsymbol{w}$ on the alphabet $\mathcal{A}$ and for any positive integer n, we denote by $p(n, \boldsymbol{w}, \mathcal{A})$ the number of distinct blocks (or subblocks, or factors) of length n occurring in $\boldsymbol{w}$. Obviously, putting $\text{Card}\mathcal{A} = +\infty$ if $\mathcal{A}$ is infinite, we have

$$1 \leq p(n, \boldsymbol{w}, \mathcal{A}) \leq (\text{Card}\mathcal{A})^n,$$

and both inequalities are sharp. Moreover, the function $n \mapsto p(n, \boldsymbol{w}, \mathcal{A})$ is non-decreasing. It measures the complexity of the word $\boldsymbol{w}$.

DEFINITION A.1. An infinite word $\boldsymbol{w} = w_1 w_2 \ldots$ is *ultimately periodic* if there exist integers $r \geq 0$ and $s \geq 1$ such that

$$w_{n+s} = w_n, \quad \text{for every } n \geq r + 1.$$

The word $w_{r+1} w_{r+2} \ldots w_{r+s}$ is a period of $\boldsymbol{w}$. If r can be chosen equal to 0, then $\boldsymbol{w}$ is *periodic*, otherwise, $w_1 \ldots w_r$ is a preperiod of $\boldsymbol{w}$.

If $b \geq 2$ is an integer and $\mathcal{A}$ is the alphabet $\{0, 1, \ldots, b-1\}$, then we write $p(\cdot, \boldsymbol{w}, b)$ instead of $p(\cdot, \boldsymbol{w}, \mathcal{A})$.

The Diophantine exponent of an infinite word $\boldsymbol{a}$, written over a finite or an infinite alphabet, has been introduced in [6].

Let $\rho \geq 1$ be a real number. We say that $\boldsymbol{a}$ satisfies *Condition* $(*)_\rho$ if there exist two sequences of finite words $(U_n)_{n\geq 1}$, $(V_n)_{n\geq 1}$ and a sequence of positive real numbers $(w_n)_{n\geq 1}$ such that:

(i) for any $n \geq 1$, the word $U_n V_n^{w_n}$ is a prefix of the word $\boldsymbol{a}$;
(ii) for any $n \geq 1$, $|U_n V_n^{w_n}|/|U_n V_n| \geq \rho$;
(iii) the sequence $(|V_n^{w_n}|)_{n\geq 1}$ is strictly increasing.

DEFINITION A.2. The *Diophantine exponent* of the infinite word $\boldsymbol{a}$, denoted by $\text{Dio}(\boldsymbol{a})$, is the supremum of the real numbers ρ for which $\boldsymbol{a}$ satisfies Condition $(*)_\rho$.

It follows from its definition that the Diophantine exponent of every ultimately periodic word is infinite.

A.2 Sturmian words

We begin this appendix with a seminal result from Morse and Hedlund [524, 525].

THEOREM A.3. *Let $\boldsymbol{w}$ be an infinite word over a finite or infinite alphabet $\mathcal{A}$. If $\boldsymbol{w}$ is ultimately periodic, then there exists a positive constant C such that $p(n, \boldsymbol{w}, \mathcal{A}) \leq C$ for every positive integer n. Otherwise, we have $p(n+1, \boldsymbol{w}, \mathcal{A}) \geq p(n, \boldsymbol{w}, \mathcal{A}) + 1$ for every $n \geq 1$, thus,*

$$p(n, \boldsymbol{w}, \mathcal{A}) \geq n + 1.$$

PROOF. Throughout the proof, we write $\boldsymbol{w} = w_1 w_2 \ldots$ and $p(\cdot, \boldsymbol{w})$ instead of $p(\cdot, \boldsymbol{w}, \mathcal{A})$.

Let $\boldsymbol{w}$ be an ultimately periodic infinite word, and assume that it has a preperiod of length r and a period of length s. Fix $h = 1, \ldots, s$ and let n be a positive integer. For every $j \geq 1$, the block of length n starting at w_{r+js+h} is the same as that starting at w_{r+h}. Consequently, there cannot be more than $r+s$ distinct blocks of length n, thus, $p(n, \boldsymbol{w}) \leq r+s$.

Let $\boldsymbol{w}$ be an infinite word for which there exists an integer $n_0 \geq 1$ such that $p(n_0, \boldsymbol{w}) = p(n_0+1, \boldsymbol{w})$. This means that every block of length n_0 extends uniquely to a block of length n_0+1. It implies that $p(n_0, \boldsymbol{w}) = p(n_0+j, \boldsymbol{w})$ holds for every positive integer j. By definition of $p(n_0, \boldsymbol{w})$, two among the words $w_j \ldots w_{n_0+j-1}$, $j = 1, \ldots, p(n_0, \boldsymbol{w})+1$, are the same. Consequently, there are integers k and ℓ with $0 \leq k < \ell \leq p(n_0, \boldsymbol{w})$ and $w_{k+m} = w_{\ell+m}$ for $m = 1, \ldots, n_0$. Since every block of length n_0 extends uniquely to a block of length n_0+1, this gives $w_{k+m} = w_{\ell+m}$ for every positive integer m. This proves that the word $\boldsymbol{w}$ is ultimately periodic.

Consequently, if the infinite word $\boldsymbol{w}$ is not ultimately periodic, then the inequality $p(n+1, \boldsymbol{w}) \geq p(n, \boldsymbol{w})+1$ holds for every positive integer n. Then, one has $p(1, \boldsymbol{w}) \geq 2$ and an immediate induction shows that $p(n, \boldsymbol{w}) \geq n+1$ for every $n \geq 1$. The proof of the theorem is complete. □

COROLLARY A.4. *Let $\boldsymbol{w}$ be an infinite word over a finite alphabet $\mathcal{A}$. Let n be a positive integer. If $\boldsymbol{w}$ is not ultimately periodic, then there exist letters a, a', b, b' in $\mathcal{A}$ and finite words V, W of length n such that $a \neq a'$, $b \neq b'$ and each of the words $aV, a'V, Wb, Wb'$ occurs infinitely often in $\boldsymbol{w}$.*

PROOF. For $m \geq 0$, let $\boldsymbol{w}_m$ be the word $\boldsymbol{w}$ deprived of its first m letters. By Theorem A.3, since $p(n+1, \boldsymbol{w}_m, \mathcal{A}) \geq 1 + p(n, \boldsymbol{w}_m, \mathcal{A})$, there exist finite words V_m, W_m of length n and letters a_m, a'_m, b_m, b'_m such that $a_m \neq a'_m$, $b_m \neq b'_m$ and $a_m V_m, a'_m V_m, W_m b_m, W_m b'_m$ occur in $\boldsymbol{w}_m$. Since the alphabet $\mathcal{A}$ is finite, there are finite words V, W of length n and letters a, a', b, b' in $\mathcal{A}$ such that $a \neq a'$, $b \neq b'$ and $aV, a'V, Wb, Wb'$ occur infinitely often in $\boldsymbol{w}$. □

The next theorem shows the existence of uncountably many infinite words $\boldsymbol{w}$ on $\{0, 1\}$ such that $p(n, \boldsymbol{w}, 2) = n+1$ for every positive integer n.

THEOREM A.5. *Let θ and ρ be real numbers with $0 < \theta < 1$ and θ irrational. For $n \geq 1$, set*

$$s_n := \lfloor (n+1)\theta + \rho \rfloor - \lfloor n\theta + \rho \rfloor, \quad s'_n := \lceil (n+1)\theta + \rho \rceil - \lceil n\theta + \rho \rceil,$$

and define the infinite words

$$\boldsymbol{s}_{\theta,\rho} := s_1 s_2 s_3 \dots, \quad \boldsymbol{s}'_{\theta,\rho} := s'_1 s'_2 s'_3 \dots$$

Then we have

$$p(n, \boldsymbol{s}_{\theta,\rho}, 2) = p(n, \boldsymbol{s}'_{\theta,\rho}, 2) = n + 1, \quad \textit{for } n \geq 1.$$

The infinite words $\boldsymbol{s}_{\theta,\rho}$ *and* $\boldsymbol{s}'_{\theta,\rho}$ *are called the* Sturmian words with slope θ and intercept ρ. *Conversely, for every infinite word* $\boldsymbol{w}$ *on* $\{0,1\}$ *such that* $p(n, \boldsymbol{w}, 2) = n + 1$ *for* $n \geq 1$, *there exist real numbers* θ_w *and* ρ_w *with* $0 < \theta_w < 1$ *and* θ_w *irrational such that* $\boldsymbol{w} = \boldsymbol{s}_{\theta,\rho}$ *or* $\boldsymbol{s}'_{\theta,\rho}$.

For θ and ρ as in Theorem A.5, the words $\boldsymbol{s}_{\theta,\rho}$ and $\boldsymbol{s}'_{\theta,\rho}$ differ only by their first letters. Classical references on Sturmian words include [310, Chapter 6] and [450, Chapter 2].

The combinatorial structure of Sturmian words has been much studied. A first result was proved in [305], and the current state of the art is summarized in the next two theorems.

THEOREM A.6. *Let* θ *be a real irrational number in* $(0, 1)$ *and* $\boldsymbol{s}$ *be a Sturmian word of slope* θ. *There are arbitrarily long finite words* W *such that* $\boldsymbol{s}$ *begins with* W^2. *Moreover, if the sequence of partial quotients of* θ *is bounded, then there exist a positive real number* ε *and arbitrarily long finite words* W *such that* $\boldsymbol{s}$ *begins with* $W^{2+\varepsilon}$.

The first statement of Theorem A.6 was proved in [42]; see also [256]. The last statement is more difficult. It was shown by Berthé, Holton and Zamboni [79], but, since it is not pointed out there, we briefly explain how it follows from the results proved in that paper.

PROOF. Since the sequence $a_1, a_2, \dots$ of partial quotients of θ is bounded, there are positive integers $s \geq 2$ and t such that either

$$a_k = s, \ a_{k+1} = t, \quad \text{for infinitely many } k,$$

or

$$a_k = a_{k+1} = 1, \ a_{k+2} = t, \quad \text{for infinitely many } k.$$

Setting $\varepsilon = (4(s+1)(t+1)+1)^{-1}$, the combination of Propositions 5.1 and 5.2 from [79] asserts then that, whatever the value of the intercept ρ, there are arbitrarily long finite words W such that the Sturmian word $\boldsymbol{s}_{\theta,\rho}$ begins with $W^{2+\varepsilon}$. □

Theorem A.6 does not cover the (easier) case of Sturmian words whose slope has an unbounded sequence of partial quotients, which was treated in [13].

THEOREM A.7. *Let θ in $(0,1)$ be a real number having an unbounded sequence of partial quotients. Let $\boldsymbol{s}$ be a Sturmian word of slope θ. Then, the Diophantine exponent of $\boldsymbol{s}$ is infinite.*

PROOF. This follows from the proof of [13, Proposition 11.1]. □

The next result, established by Cassaigne [179], describes the non-ultimately periodic infinite words with very low complexity.

THEOREM A.8. *Let $b \geq 2$ be an integer and $\boldsymbol{w}$ be an infinite word on $\{0, 1, \ldots, b-1\}$. Assume that there are positive integers k and n_0 such that*

$$p(n, \boldsymbol{w}, b) = n + k, \quad \text{for } n \geq n_0.$$

Then there exist finite words W, W_0, W_1 on $\{0, 1, \ldots, b-1\}$ and a Sturmian word $\boldsymbol{s}$ on $\{0, 1\}$ such that

$$\boldsymbol{w} = W\phi(\boldsymbol{s}),$$

where $\phi(\boldsymbol{s})$ denotes the infinite word obtained by replacing in $\boldsymbol{s}$ every 0 by W_0 and every 1 by W_1.

A word $\boldsymbol{w}$ as in Theorem A.8 is sometimes called a *quasi-Sturmian* word; see [42].

A.3 The Thue–Morse infinite word

The Thue–Morse word was introduced by Thue [685] and Morse [523], independently; see [45, 488] and [46, pp. 208–209] for further references.

DEFINITION A.9. The Thue–Morse word $\boldsymbol{t} = (t_n)_{n \geq 0}$ on the alphabet $\{a, b\}$ is defined for $n \geq 0$ by $t_n = a$ if the sum of digits in the binary representation of n is even and by $t_n = b$ otherwise. Thus, we have

$$\boldsymbol{t} = abbabaabbaababbabaababba\ldots$$

Among the many definitions of $\boldsymbol{t}$, we emphasize that $\boldsymbol{t}$ is the fixed point starting with a of the morphism τ defined by $\tau(a) = ab$ and $\tau(b) = ba$.

THEOREM A.10. *The Thue–Morse word $\boldsymbol{t}$ contains no overlap. In particular, it is not ultimately periodic.*

This is an important property of the infinite word $\boldsymbol{t}$. Although there are uncountably many overlap-free words on $\{a, b\}$, the word $\boldsymbol{t}$ is, roughly speaking, the 'canonical' example.

PROOF. This is Theorem 1.6.1 in [46]. An alternative proof is given in [488]; see Exercise A.2. □

In this section, we are mainly interested in less classical properties of the infinite word $\boldsymbol{t}$ and of infinite words related to it.

We begin with a structure theorem of Karhumäki and Shallit [372] for the infinite words having only short overlaps. Recall that ϵ denotes the empty word.

THEOREM A.11. *Let $\boldsymbol{x}$ be an infinite word on $\{a, b\}$ which does not contain any 7/3-powers. Then, there exist W in $\{\epsilon, a, b, aa, bb\}$ and an infinite word $\boldsymbol{y}$ on $\{a, b\}$ such that $\boldsymbol{x} = W\tau(\boldsymbol{y})$ and $\boldsymbol{y}$ does not contain any 7/3-powers.*

We introduce an infinite word closely related to $\boldsymbol{t}$.

DEFINITION A.12. Let $\boldsymbol{z}$ be the fixed point of the morphism σ defined by $\sigma(1) = 2$ and $\sigma(2) = 211$. Thus, we have

$$\boldsymbol{z} = 2112221121121122211222 11\ldots$$

The infinite word $\boldsymbol{z}$ is not ultimately periodic; see Exercise A.3. Set $A_0 = 2$, $A_1 = 211$, and, for an integer $k \geq 2$, set $A_k = \sigma^k(2)$. The sequence of finite words $(A_k)_{k \geq 0}$ converges to the infinite word $\boldsymbol{z}$. An immediate induction shows that $A_k = A_{k-1}A_{k-2}A_{k-2}$ for $k \geq 2$.

Let us define an ordering $\succ$ for the (finite and infinite) words over the alphabet $\{1, 2\}$. Let $\boldsymbol{v}$ and $\boldsymbol{v}'$ be distinct words such that neither word is a prefix of the other. Permuting $\boldsymbol{v}$ and $\boldsymbol{v}'$ if necessary, there exists a smallest positive integer k such that $\boldsymbol{v}$ and $\boldsymbol{v}'$ have the same prefix of length $k-1$ and their kth letters are 2 and 1, respectively. Then, we put $\boldsymbol{v} \succ \boldsymbol{v}'$ if k is odd and $\boldsymbol{v}' \succ \boldsymbol{v}$ if k is even.

As in the proof of Corollary A.4, for a non-negative integer m and an infinite word $\boldsymbol{v}$, we write $\boldsymbol{v}_m$ for the word obtained from $\boldsymbol{v}$ by deleting its first m letters.

THEOREM A.13. *The word $\boldsymbol{z}$ is the smallest non-periodic infinite word satisfying $\boldsymbol{z} \succ \boldsymbol{z}_m$ for every integer $m \geq 1$.*

PROOF. Let $\mathcal{W}$ be the set of all infinite non-periodic words $\boldsymbol{v}$ over the alphabet $\{1, 2\}$ satisfying $\boldsymbol{v} \succ \boldsymbol{v}_m$ for every positive integer m. We need to show that $\boldsymbol{z}$ is in $\mathcal{W}$ and that, for any $\boldsymbol{v} \in \mathcal{W} \setminus \{\boldsymbol{z}\}$ we have $\boldsymbol{v} \succ \boldsymbol{z}$.

Let $\boldsymbol{v}$ be in $\mathcal{W}$. Since $\boldsymbol{v}$ is non-periodic, it contains infinitely many occurrences of 2, thus, it must begin with a 2. Similarly, it contains infinitely many blocks 21, thus it must begin with 21. If $\boldsymbol{v}$ begins with 212 or with 2111, then $\boldsymbol{v} \succ \boldsymbol{z}$. So, we can assume that $\boldsymbol{v}$ belongs to the subset $\mathcal{W}_1$ of $\mathcal{W}$ composed of the words beginning with 2112. In particular, $\boldsymbol{v}$ contains no occurrence of 212 and of 2111. It must be obtained by concatenation of the blocks

$$A_0 = 2 \quad \text{and} \quad A_1 = 211.$$

Since $\boldsymbol{v}$ is non-periodic, it contains infinitely many occurrences of the block $A_1A_0 = 2112$. If $\boldsymbol{v}$ begins with $A_1A_0A_1$ or $A_1A_0A_0A_0A_0$ or $A_1A_0A_0A_0A_1$, then $\boldsymbol{v} \succ \boldsymbol{z}$, thus $\boldsymbol{v}$ begins with $A_1A_0A_0A_1$ and it belongs to the subset $\mathcal{W}_2$ of $\mathcal{W}$ composed of the words beginning with A_2 and obtained by concatenation of the blocks A_1 and A_2. Since the lengths of A_2 and A_1 have the same parity, as do the lengths of A_1 and A_0, we can repeat the same argument with A_2 and A_1 in place of A_1 and A_0, respectively, and so on. We thus obtain a sequence of sets

$$\cdots \subset \mathcal{W}_3 \subset \mathcal{W}_2 \subset \mathcal{W}_1 \subset \mathcal{W}.$$

Here, $\mathcal{W}_k$ is the subset of $\mathcal{W}$ composed of the words $\boldsymbol{v}$ beginning with A_k and obtained by concatenation of the blocks A_{k-1} and A_k, and such that $\boldsymbol{v} \succ \boldsymbol{z}$. But the intersection of all the sets $\mathcal{W}_k$ is the word beginning with A_k for every $k \geq 1$, that is, the word $\boldsymbol{z}$. We have also established that $\boldsymbol{z}$ has no subword or the form 212, 2111, $A_kA_{k-1}A_k$, $A_kA_{k-1}A_{k-1}A_{k-1}A_{k-1}$, $A_kA_{k-1}A_{k-1}A_{k-1}A_k$ for $k \geq 1$. □

The next result follows from the proof of Theorem A.13.

THEOREM A.14. *Let $\boldsymbol{v}$ be a non-periodic word and $m \geq 2$ be an integer. Then, either A_m occurs infinitely many times in $\boldsymbol{v}$, or there exists a finite word W such that $W \succ \boldsymbol{z}$ and W occurs infinitely many times in $\boldsymbol{v}$.*

PROOF. Assume that A_m appears in $\boldsymbol{v}$ only finitely many times. If $m = 2$, then $\boldsymbol{v}$ contains infinitely many occurrences either of 212 or of 2111. Otherwise, $\boldsymbol{v}$ contains infinitely many occurrences either of 212, or of 2111, or, for some $\ell = 1, \ldots, m-1$, of $A_\ell A_{\ell-1}A_\ell$, or, for some $\ell = 1, \ldots, m-2$, of $A_\ell A_{\ell-1}A_{\ell-1}A_{\ell-1}A_{\ell-1}$, or of $A_\ell A_{\ell-1}A_{\ell-1}A_{\ell-1}A_\ell$. Since these words are all greater than $\boldsymbol{z}$, the theorem is proved. □

LEMMA A.15. *With respect to the ordering $\succ$, the largest word of length four which is a subblock of $\boldsymbol{z}$ is* 2112 *and the smallest word of lenth five which is a subblock of $\boldsymbol{z}$ is* 12112.

PROOF. The four largest words of length four are

$$2121 \succ 2122 \succ 2111 \succ 2112$$

and the four smallest words of length five are

$$12121 \prec 12122 \prec 12111 \prec 12112.$$

The definition of $\boldsymbol{z}$ implies that 111 and 212 are not subblocks of $\boldsymbol{z}$. But 2112 and 12112 are. □

Let $\boldsymbol{u} = u_1u_2\ldots = 12112221121\ldots$ be the word formed by the number of consecutive symbols in the Thue–Morse word $\boldsymbol{t}$. The word $\boldsymbol{u}$ is not ultimately periodic. It is written on the alphabet $\{1, 2\}$, since $\boldsymbol{t}$ does not contain cubes.

THEOREM A.16. *The word $\boldsymbol{z}$ coincides with the infinite word $u_2u_3\ldots$*

PROOF. This statement can be proved using the strategy described in [39]. The details are left to the reader as Exercise A.4. □

A.4 Exercises

EXERCISE A.1. Let $\boldsymbol{v} = v_0v_1v_2\ldots$ and $\boldsymbol{v}' = v_0'v_1'v_2'\ldots$ be distinct infinite words on the alphabet $\mathbb{Z}_{\geq 1}$. Establish that $\boldsymbol{v} \succ \boldsymbol{v}'$ if, and only if, $[v_0; v_1, v_2, \ldots] > [v_0'; v_1', v_2', \ldots]$.

EXERCISE A.2. Prove that every subblock W of length ≥ 4 of $\boldsymbol{t}$ can be factorized in a unique way under the form $\varepsilon_1 W \varepsilon_2 = \tau(W')$, with $|W'| < |W|$. Assume that there exists a subblock of $\boldsymbol{t}$ of the form $w_1\ldots w_\ell w_1\ldots w_\ell w_1$, where $\ell \geq 4$ and $w_1, \ldots, w_\ell$ are in $\{a, b\}$. Prove that the above property of unique facorization implies that ℓ must be even. For ℓ even, show that there exists a word W' of first letter w' and of length $|W'| < \ell$ such that $W'W'w'$ is a subblock of $\boldsymbol{t}$. Conclude.

EXERCISE A.3. Prove that the infinite word $\boldsymbol{z}$ is not ultimately periodic.

EXERCISE A.4. Prove Theorem A.16.

Appendix B
Some elementary lemmata

Several proofs require some classical lemmas from number theory, which we state and establish below.

We use the following notation. For a non-zero integer a and a prime number p, we write $\mathrm{ord}_p(a)$ for the largest integer e such that p^e divides a. Furthermore, if p does not divide a and if h is a positive integer, then $\mathrm{ord}(a, p^h)$ denotes the smallest positive integer ℓ such that a^ℓ is congruent to 1 modulo p^h.

THEOREM B.1. *Let $n \geq 2$ be an integer, and x, y be non-zero relatively prime integers. Then we have*

$$\gcd\left(x - y, \frac{x^n - y^n}{x - y}\right) = \gcd(x - y, n).$$

PROOF. Observe that the binomial theorem gives

$$\begin{aligned}\frac{x^n - y^n}{x - y} &= \frac{((x-y)+y)^n - y^n}{x - y} \\ &= (x-y)^{n-1} + \binom{n}{1} y(x-y)^{n-2} \\ &\quad + \cdots + \binom{n}{n-2} y^{n-2}(x-y) + ny^{n-1} \\ &= a(x-y) + ny^{n-1},\end{aligned}$$

for a suitable positive integer a. Since $x - y$ and y are coprime, this implies the theorem. □

LEMMA B.2. *Let a and b be distinct non-zero integers. Let p be an odd prime number which divides $a - b$ but does not divide b. Then, for every positive integer r, we have*

$$\mathrm{ord}_p(a^{p^r} - b^{p^r}) = r + \mathrm{ord}_p(a - b). \tag{B.1}$$

If 2 divides $a-b$ and b is odd, then, for every positive integer r, we have

$$\mathrm{ord}_2(a^{2^r} - b^{2^r}) \geq r + \mathrm{ord}_2(a-b). \tag{B.2}$$

PROOF. Write $a = b + kp^e$ with $e \geq 1$ and $\gcd(k,p) = 1$. By the binomial theorem, we get that

$$a^p = b^p + \sum_{j=1}^{p-1} \binom{p}{j} b^{p-j}(kp^e)^j + k^p p^{pe}.$$

Since p divides $p(p-1)/2$, we see that

$$\mathrm{ord}_p(a^p - b^p - pb^{p-1}kp^e) \geq \min\{1+2e, 3e\},$$

hence

$$\mathrm{ord}_p(a^p - b^p) = \mathrm{ord}_p(pb^{p-1}kp^e) = 1 + e = 1 + \mathrm{ord}_p(a-b).$$

This proves (B.1) when $r = 1$. An easy induction on r gives then (B.1).

If 2 divides $a-b$ and a is odd, then $\mathrm{ord}_2(a+b) \geq 1$ and

$$\mathrm{ord}_2(a^2 - b^2) = \mathrm{ord}_2(a+b) + \mathrm{ord}_2(a-b) \geq 1 + \mathrm{ord}_2(a-b).$$

This proves (B.2) when $r = 1$. An easy induction on r gives then (B.2). □

COROLLARY B.3. *Let p be an odd prime and b be an integer not divisible by p. For any positive integer e, we have*

$$\mathrm{ord}(b, p^e) = p^{e-h}\mathrm{ord}(b,p),$$

with $h = \min\{e, \mathrm{ord}_p(b^{\mathrm{ord}(b,p)} - 1)\}$.

PROOF. Set $\ell = \mathrm{ord}(b,p)$. Let t be a positive integer and write $t = p^e m$, with $e \geq 0$ and m not divisible by p. By Theorem B.1 applied with $n = m$, $x = b^\ell$ and $y = 1$, we get that p does not divide $(b^{\ell m} - 1)/(b^\ell - 1)$, since p divides $b^\ell - 1$ but does not divide m. This implies that

$$\mathrm{ord}_p(b^{\ell m} - 1) = \mathrm{ord}_p(b^\ell - 1).$$

We then deduce from Lemma B.2 that

$$\mathrm{ord}_p(b^{\ell m p^e} - 1) = e + \mathrm{ord}_p(b^{\ell m} - 1),$$

hence

$$\mathrm{ord}_p(b^{\ell t} - 1) = \mathrm{ord}_p(t) + \mathrm{ord}_p(b^\ell - 1).$$

This proves the corollary. □

Corollary B.4. *Let a be a positive integer. For any positive integer m, the integer $(a+1)^{a^m}-1$ is divisible by a^{m+1}.*

Proof. Let p be a prime divisor of a and set $v = \mathrm{ord}_p(a)$. Let m be a positive integer. By Lemma B.2, we have

$$\mathrm{ord}_p\bigl((a+1)^{p^{vm}}-1\bigr) \geq vm + \mathrm{ord}_p(a) = v(m+1),$$

thus $(a+1)^{p^{vm}}-1$ is divisible by $p^{v(m+1)}$. Since p^{vm} divides a^m, the integer $(a+1)^{a^m}-1$ is divisible by $(a+1)^{p^{vm}}-1$, hence, by $p^{v(m+1)}$. Letting p run through all the prime divisors of a, we get the lemma. □

Appendix C
Measure theory

In this appendix we recall some basic facts from measure theory. We begin with the easy half of the Borel–Cantelli lemma and the definition of the Hausdorff dimension. Then, we define the standard measure on the middle third Cantor set and establish some of its useful properties. We conclude with a few words on ergodic theory.

C.1 The easy half of the Borel–Cantelli lemma

We start with an easy and well-known lemma, often referred to as the (easy half of the) Borel–Cantelli lemma. Since Cantelli pointed out that the total independence of the events was not needed in the proof of Lemma C.1, the next lemma should perhaps be called the Cantelli lemma.

LEMMA C.1. *Let S be a set equipped with a measure μ. Let $(E_j)_{j\geq 1}$ be a sequence of measurable sets in S and set*

$$E_\infty := \{s \in S : s \in E_j \text{ for infinitely many } j \geq 1\}.$$

If the sum $\sum_{j\geq 1} \mu(E_j)$ converges, then we have $\mu(E_\infty) = 0$.

PROOF. By definition, the set E_∞ is the limsup set

$$E_\infty = \bigcap_{N=1}^{+\infty} \bigcup_{j=N}^{+\infty} E_j.$$

Consequently, for every positive integer N, the set E_∞ is included in the infinite union $E_N \cup E_{N+1} \cup \ldots$ and

$$\mu(E_\infty) \leq \sum_{j=N}^{+\infty} \mu(E_j).$$

By a suitable choice of N, the latter sum can be made arbitrarily small since $\sum_{j\geq 1} \mu(E_j)$ converges. This proves that the μ-measure of E_∞ is zero. □

C.2 Hausdorff dimension

Hausdorff's idea [339] consists in measuring a set by covering it by an infinite, countable family of sets of bounded diameter, and then in looking at what happens when the maximal diameter of these covering sets tends to 0. The reader interested in the theory of Hausdorff dimension is directed, for example, to the books of Rogers [613], Falconer [288, 289] and Mattila [484].

If U is a bounded subset of $\mathbb{R}$, we denote by $|U|$ the length of the shortest interval containing U. Let J be a finite or infinite set of indices. If for some positive real number δ the sets E and U_j satisfy $E \subset \bigcup_{j\in J} U_j$ and $0 < |U_j| \leq \delta$ for any j in J, then $\{U_j\}_{j\in J}$ is called a *δ-covering* of E.

Let s be a positive real number. For any positive real number δ, set

$$\mathcal{H}^s_\delta(E) := \inf \sum_{j\in J} |U_j|^s,$$

where the infimum is taken over all the countable δ-coverings $\{U_j\}_{j\in J}$ of E. Clearly, the function $\delta \mapsto \mathcal{H}^s_\delta(E)$ is non-increasing. Consequently,

$$\mathcal{H}^s(E) := \lim_{\delta\to 0} \mathcal{H}^s_\delta(E) = \sup_{\delta\to 0} \mathcal{H}^s_\delta(E)$$

is well-defined and lies in $[0, +\infty]$. If E_1 and E_2 are two real subsets with E_1 included in E_2, we then have $\mathcal{H}^s(E_1) \leq \mathcal{H}^s(E_2)$. Furthermore, $\mathcal{H}^s$ is subadditive and is a regular outer measure for which the Borelian sets are measurable (see [484, 613]), called the s-dimensional Hausdorff measure.

Lemma C.2. *Let r, s, t be real numbers with $0 < r < s < t$. If the set E satisfies $0 \leq \mathcal{H}^s(E) < +\infty$ (resp. $0 < \mathcal{H}^s(E) \leq +\infty$), we then have $\mathcal{H}^t(E) = 0$ (resp. $\mathcal{H}^r(E) = +\infty$).*

Proof. Assume that E satisfies $0 \leq \mathcal{H}^s(E) < +\infty$ and let ε be a positive real number. There exists $\delta > 0$ such that $x^t \leq \varepsilon x^s/(\mathcal{H}^s(E)+1)$ for any x with $0 < x < \delta$. For any positive real number δ' less than δ, there exists a countable δ'-covering $\{U_j\}_{j\geq 1}$ of E such that

$$\sum_{j\geq 1} |U_j|^s \leq \mathcal{H}^s(E) + 1,$$

whence

$$\sum_{j\geq 1} |U_j|^t \leq \varepsilon.$$

This means that $\mathcal{H}^t_{\delta'}(E) \leq \varepsilon$ and yields $\mathcal{H}^t(E) = 0$. Replacing the real numbers s and t by r and s, respectively, we get by contraposition the result asserted in brackets. □

We define $\mathcal{H}^0$ as the counting measure: $\mathcal{H}^0(E)$ is equal to the cardinality of the set E. Furthermore, $\mathcal{H}^1$ coincides with the Lebesgue measure on $\mathbb{R}$. We check that the function $s \mapsto \mathcal{H}^s(E)$ is non-increasing on $\mathbb{R}_{\geq 0}$.

We display an immediate consequence of Lemma C.2.

COROLLARY C.3. *Let E be a real set. If there exists $s \geq 0$ such that $\mathcal{H}^s(E) < +\infty$, then $\mathcal{H}^{s+\varepsilon}(E) = 0$ for any $\varepsilon > 0$. If there exists $s > 0$ such that $\mathcal{H}^s(E) > 0$, then $\mathcal{H}^{s-\varepsilon}(E) = +\infty$ for any ε in $]0, s]$.*

Corollary C.3 shows that there is a critical value of s at which $\mathcal{H}^s(E)$ 'jumps' from $+\infty$ to 0. This value is called the *Hausdorff dimension* of E.

DEFINITION C.4. The *Hausdorff dimension* of a real set E, denoted by $\dim E$, is the unique non-negative real number s_0 such that

$$\mathcal{H}^s(E) = 0 \qquad \text{if } s > s_0$$

and

$$\mathcal{H}^s(E) = +\infty \qquad \text{if } 0 < s < s_0.$$

In other words, with the notation of Definition C.4, we have

$$\dim E = \inf\{s : \mathcal{H}^s(E) = 0\} = \sup\{s : \mathcal{H}^s(E) = +\infty\}.$$

The main properties of Hausdorff dimension for real sets $E, E_1, \ldots$ are (see e.g. [146, p. 93]):

(i) $\dim E \leq 1$;
(ii) if $\lambda(E)$ is positive, then $\dim E = 1$;
(iii) if $E_1 \subset E_2$, then $\dim E_1 \leq \dim E_2$;
(iv) $\dim \cup_{j=1}^{+\infty} E_j = \sup\{\dim E_j : j \geq 1\}$;
(v) the Hausdorff dimension of a finite or countable set of points is zero;
(vi) two sets differing by a countable set of points have the same Hausdorff dimension.

Observe that there exist sets of Hausdorff dimension zero which are uncountable (e.g. the set of Liouville numbers) as well as uncountable sets of Lebesgue measure zero and with Hausdorff dimension one (e.g. the set of badly approximable real numbers).

By the property (iv), to determine the Hausdorff dimension of a real subset E, it is enough to know the Hausdorff dimension of intersections of E with bounded intervals.

The Hausdorff dimension is a useful tool to discriminate between sets of zero Lebesgue measure. We can be even more precise by replacing the functions $x \mapsto x^s$ in the definition of $\mathcal{H}^s$ by general functions $x \mapsto f(x)$ which satisfy $\lim_{x \to 0} f(x) = 0$ and are strictly increasing and continuous on some open interval $(0, t)$ with t positive. This allows us to discriminate between sets having the same Hausdorff dimension.

Different methods, more or less sophisticated, allow one to bound from below and from above the Hausdorff dimension of a real set. It is often possible to bound the Hausdorff dimension of a Cantor-type set from below by using the so-called mass distribution principle (or the easy half of the Frostman lemma [313]). This corresponds to the first part of Lemma C.5.

LEMMA C.5. *Let μ be a probability measure with support in a bounded real set $\mathcal{K}$. If there exist positive real numbers s, κ and δ such that*

$$\mu(J) \le \kappa |J|^s$$

for any interval J with length $|J| \le \delta$, then we have $\dim \mathcal{K} \ge s$. *If there exist positive real numbers s, κ and δ such that*

$$\mu(J) \ge \kappa |J|^s$$

for any interval J with length $|J| \le \delta$, then we have $\dim \mathcal{K} \le s$.

Lemma C.5 is a consequence of [289, Proposition 4.9].

We are often in a position to apply the next theorem to bound from below the Hausdorff dimension of sets arising in Diophantine approximation.

THEOREM C.6. *Let $\mathcal{E}_0$ be a bounded real interval. Assume that, for any positive integer k, there exists a finite family $\mathcal{E}_k$ of disjoint compact intervals in $\mathcal{E}_0$ such that any interval belonging to $\mathcal{E}_k$ is contained in exactly one of the intervals of $\mathcal{E}_{k-1}$ and contains at least two intervals belonging to $\mathcal{E}_{k+1}$. We also suppose that the maximum of the lengths of the intervals in $\mathcal{E}_k$ tends to 0 when k tends to infinity. For $k \ge 0$, we denote by E_k the union of the intervals belonging to the family $\mathcal{E}_k$. Assume further that there exists a positive integer k_0 such that, for any*

$k \ge k_0$, each interval of E_{k-1} contains at least $m_k \ge 2$ intervals of E_k, these being separated by at least ε_k, where $0 < \varepsilon_{k+1} < \varepsilon_k$. We then have

$$\dim \bigcap_{k=1}^{+\infty} E_k \ge \liminf_{k \to +\infty} \frac{\log(m_1 \dots m_{k-1})}{-\log(m_k \varepsilon_k)}.$$

PROOF. See [289, Chapter 4] or [146, Section 5.3]. □

The combination of Lemma C.1 and Theorem C.6 shows that the Hausdorff dimension of the middle third Cantor set equals $(\log 2)/(\log 3)$.

C.3 The standard measure on the middle third Cantor set

In this section, we establish several classical results on the standard measure on the middle third Cantor set. This measure is the restriction of the $(\log 2)/(\log 3)$-dimensional Hausdorff measure to the middle third Cantor set; see [288, Theorem 1.14] or [289, p. 55].

DEFINITION C.7. Set $E_0 = [0, 1]$. For $k \ge 1$, denote by E_k the union of the closed intervals

$$\left[\frac{a_1 3^{k-1} + \dots + a_{k-1} 3 + a_k}{3^k}, \frac{a_1 3^{k-1} + \dots + a_{k-1} 3 + a_k + 1}{3^k}\right],$$

over $a_1, \dots, a_k$ in $\{0, 2\}$. The middle third Cantor set K is the set

$$K := \bigcap_{k \ge 0} E_k.$$

The *standard measure* on K, denoted by μ_K, is such that, for $k \ge 0$, each of the 2^k intervals of length 3^{-k} in E_k carries a mass 2^{-k}.

It follows from Definition C.7 that, for a real number ξ in E_0 which can be expressed as $\xi = a_1 3^{-1} + a_2 3^{-2} + \cdots$, with $a_k \in \{0, 1, 2\}$ for $k \ge 1$, we have

$$\mu_K([0, \xi]) = \sum_{k : a_k \ge 1} 2^{-k}.$$

Denoting by μ_k the measure on $[0, 1]$ giving to each interval of length 3^{-k} in E_k a mass 2^{-k}, Definition C.7 asserts that μ_K is the weak star limit of the sequence of measures $(\mu_k)_{k \ge 1}$. In particular, the Fourier transform of μ_K is the pointwise limit of the Fourier transforms of the μ_k, thus, for any continuous function f on $[0, 1]$, we have

$$\int_0^1 f(\xi)\, \mathrm{d}\mu_K(\xi) = \lim_{k \to +\infty} 2^{-k} \sum_{a_1, \dots, a_k \in \{0, 2\}} f(a_1 3^{-1} + \dots + a_k 3^{-k}).$$

Our first auxiliary lemma points out a decay property of the measure μ_K. For a real number x and a positive real number ρ, we denote by $B(x,\rho)$ the open interval $(x-\rho, x+\rho)$.

LEMMA C.8. *For every x in K and every real number ε and ρ with $0<\rho<1$ and $0<\varepsilon\le 1$, we have*

$$(4\cdot 3^\gamma)^{-1}\varepsilon^\gamma \mu_K(B(x,\rho)) \le \mu_K(B(x,\varepsilon\rho)) \le 4\cdot 3^\gamma \varepsilon^\gamma \mu_K(B(x,\rho)), \quad \text{(C.1)}$$

where $\gamma = (\log 2)/(\log 3)$.

PROOF. Let k be the integer such that $3^{-k-1} \le |B(x,\varepsilon\rho)| < 3^{-k}$. Then, $B(x,\varepsilon\rho)$ intersects exactly one of the intervals of E_k, and, since x is in K, it contains at least one of the intervals of E_{k+2}. Consequently, we have

$$\mu_K(B(x,\varepsilon\rho)) \ge 2^{-k-2} \ge 2^{-2}\cdot 3^{-\gamma k} \ge 2^{-2}(2\varepsilon\rho)^\gamma \quad \text{(C.2)}$$

and

$$\mu_K(B(x,\varepsilon\rho)) \le 2^{-k} \le 3^{-\gamma k} \le (6\varepsilon\rho)^\gamma. \quad \text{(C.3)}$$

Since (C.2) and (C.3) hold for $\varepsilon = 1$, we have

$$2^{-2}(2\rho)^\gamma \le \mu_K(B(x,\rho)) \le (6\rho)^\gamma,$$

which, combined with (C.2) and (C.3), gives (C.1). □

The second auxiliary lemma shows that the measure μ_K is absolutely decaying in the sense of [131]; see Definition 7.19.

LEMMA C.9. *For every x in K, every y in $[0,1]$, and every real number ε and ρ with $0<\varepsilon<1$ and $0<\rho<1/3$, we have*

$$\mu_K(B(x,\rho)\cap B(y,\varepsilon\rho)) \le (4\cdot 3^\gamma)^2(3\varepsilon)^\gamma \mu_K(B(x,\rho)).$$

PROOF. Without any loss of generality, we assume that $\mu_K(B(x,\rho)\cap B(y,\varepsilon\rho))$ is positive and we let y' be in $K\cap B(x,\rho)\cap B(y,\varepsilon\rho)$. Then, we have

$$\begin{aligned}\mu_K(B(x,\rho)\cap B(y,\varepsilon\rho)) &\le \mu_K(B(y',2\varepsilon\rho))\\ &\le 4\cdot 3^\gamma\varepsilon^\gamma\mu_K(B(y',2\rho))\\ &\le 4\cdot 3^\gamma\varepsilon^\gamma\mu_K(B(x,3\rho)) \le (4\cdot 3^\gamma)^2(3\varepsilon)^\gamma\mu_K(B(x,\rho)),\end{aligned}$$

by Lemma C.8, since $B(y',2\rho)$ is contained in $B(x,3\rho)$. □

C.4 Ergodic theory

We very briefly discuss certain basic notions from ergodic theory and state several results without proof. Classical references include the monographs [212, 726].

Let $(X, \mathcal{B}, \mu)$ be a probability space and $T : X \to X$ be a transformation. We say that T is *measurable* if $T^{-1}(B)$ is included in $\mathcal{B}$ for every set B in $\mathcal{B}$. If furthermore $\mu(T^{-1}(B)) = \mu(B)$ holds for every set B in $\mathcal{B}$, then T is *measure-preserving*.

Let $T : (X, \mathcal{B}, \mu) \to (X, \mathcal{B}, \mu)$ be a measure-preserving transformation. We say that T is *ergodic* with respect to μ if, for every set B in $\mathcal{B}$ such that $T^{-1}(B) = B$, we have $\mu(B) \in \{0, 1\}$.

A fundamental result in ergodic theory, namely Birkhoff's (pointwise) ergodic theorem, can be stated as follows. We keep the above notation.

THEOREM C.10. *Let $T : (X, \mathcal{B}, \mu) \to (X, \mathcal{B}, \mu)$ be a measure-preserving transformation. Let f be in $L^1(\mu)$. If T is ergodic with respect to μ, then*

$$\lim_{n \to +\infty} \frac{1}{n} \sum_{i=0}^{n-1} f(T^i(x)) = \int_X f \, \mathrm{d}\mu, \quad \text{for } \mu\text{-almost all } x \in X.$$

Let $b \geq 2$ be an integer. Since the transformation $T_b : [0, 1] \to [0, 1)$ defined by $T_b(x) = \{bx\}$ for x in $[0, 1]$ is ergodic with respect to the Lebesgue measure, it follows from Theorem C.10 that, for almost all real numbers ξ, the sequence $(\xi b^n)_{n \geq 1}$ is uniformly distributed modulo one.

Since the Gauss map $T_{\mathcal{G}}$ defined by $T_{\mathcal{G}}(0) = 0$ and $T_{\mathcal{G}}(x) = \{1/x\}$ for x in $(0, 1)$ is ergodic with respect to the Gauss measure defined in Section 9.1, it follows from Theorem C.10 that almost all real numbers have a normal continued fraction expansion.

Appendix D
Continued fractions

In this appendix we review basic results on continued fractions and state several slightly less-known theorems. We omit most of the proofs and refer the reader to a text of Van der Poorten [580] and to the books of Bugeaud [146], Cassels [181], Dajani and Kraaikamp [212], Hardy and Wright [334], Iosifescu and Kraaikamp [351], Khintchine [380], Perron [557], Schmidt [635] and Schweiger [641], among many others.

Let $x_0, x_1, \ldots$ be real numbers with $x_1, x_2, \ldots$ positive. A *finite continued fraction* denotes any expression of the form

$$[x_0; x_1, x_2 \ldots, x_n] = x_0 + \cfrac{1}{x_1 + \cfrac{1}{x_2 + \cfrac{1}{\cdots + \cfrac{1}{x_n}}}}.$$

We call any expression of the above form or of the form

$$[x_0; x_1, x_2, \ldots] = x_0 + \cfrac{1}{x_1 + \cfrac{1}{x_2 + \cfrac{1}{\cdots}}} = \lim_{n \to +\infty} [x_0; x_1, x_2 \ldots, x_n]$$

a *continued fraction*, provided that the limit exists. Throughout this book, when we refer to a continued fraction expansion, we implicitly mean that x_0 is an integer and $x_1, x_2, \ldots$ are positive integers.

Any rational number r has exactly two different continued fraction expansions. These are $[r]$ and $[r-1; 1]$ if r is an integer and, otherwise, one of them reads $[a_0; a_1, \ldots, a_{n-1}, a_n]$ with $a_n \geq 2$, and the other one is $[a_0; a_1, \ldots, a_{n-1}, a_n - 1, 1]$. Any irrational number has a unique expansion in continued fraction.

Theorem D.1. *Let $\xi = [a_0; a_1, a_2, \ldots]$ be an irrational number. For $k \geq 1$, set $p_k/q_k := [a_0; a_1, a_2, \ldots, a_k]$ with p_k and q_k coprime. Let n be a positive integer. Putting*

$$p_{-1} = 1, \ q_{-1} = 0, \ p_0 = a_0 \ \text{and} \ q_0 = 1,$$

we have

$$p_n = a_n p_{n-1} + p_{n-2}, \qquad q_n = a_n q_{n-1} + q_{n-2}, \tag{D.1}$$

and

$$p_{n-1} q_n - p_n q_{n-1} = (-1)^n. \tag{D.2}$$

Furthermore, setting $\xi_{n+1} = [a_{n+1}; a_{n+2}, a_{n+3}, \ldots]$, we have

$$\xi = [a_0; a_1, \ldots, a_n, \xi_{n+1}] = \frac{p_n \xi_{n+1} + p_{n-1}}{q_n \xi_{n+1} + q_{n-1}},$$

thus

$$q_n \xi - p_n = \frac{(-1)^n}{q_n \xi_{n+1} + q_{n-1}},$$

and

$$\begin{aligned} \frac{1}{(a_{n+1}+2)q_n^2} < \frac{1}{q_n(q_n + q_{n+1})} < \left| \xi - \frac{p_n}{q_n} \right| \\ < \frac{1}{q_n q_{n+1}} < \frac{1}{a_{n+1} q_n^2} < \frac{1}{q_n^2}. \end{aligned} \tag{D.3}$$

Under the assumption of Theorem D.1, the rational number p_k/q_k is called the kth convergent to ξ and the positive integers $a_1, a_2, \ldots$ are its partial quotients. The next result is sometimes termed the *mirror formula*.

Theorem D.2. *Let $n \geq 2$ be an integer and $a_1, \ldots, a_n$ be positive integers. For $k = 1, \ldots, n$, set $p_k/q_k = [0; a_1, \ldots, a_k]$. Then, we have*

$$\frac{q_{n-1}}{q_n} = [0; a_n, a_{n-1}, \ldots, a_1].$$

Proof. We get from (D.1) that

$$\frac{q_n}{q_{n-1}} = a_n + \frac{q_{n-2}}{q_{n-1}},$$

for $n \geq 1$. The theorem then follows by induction. □

The next theorem is a particular case of the Folding Lemma, first proved by Mendès France [501] and later rediscovered by many authors; see [157] for references.

THEOREM D.3. *Let* $a/m = [0; 1, 1, a_3, \ldots, a_{n-1}, a_n]$, *with* $n \geq 4$ *and* $a_n \geq 2$, *be a rational number. Then we have*

$$\frac{ma + (-1)^n}{m^2} = [0; 1, 1, a_3, \ldots, a_{n-1}, a_n + 1, a_n - 1, a_{n-1}, \ldots, a_3, 2]$$

and, for any integer $t \geq 2$,

$$\frac{tma + (-1)^n}{tm^2} = [0; 1, 1, a_3, \ldots, a_{n-1}, a_n, t - 1, 1, a_n - 1, a_{n-1}, \ldots, a_3, 2].$$

PROOF. We treat only the case $t \geq 2$ and leave the remaining case to the reader. For $k = 3, \ldots, n$, set $p_k/q_k = [0; 1, 1, a_3, \ldots, a_k]$. We deduce from Theorem D.2 that

$$\begin{aligned} \beta &:= [0; t-1, 1, a_n - 1, a_{n-1}, \ldots, a_3, 2] \\ &= \cfrac{1}{t - 1 + \cfrac{1}{1 + \cfrac{1}{\frac{q_n}{q_{n-1}} - 1}}} = \frac{1}{t - 1 + \frac{q_n - q_{n-1}}{q_n}} = \frac{q_n}{tq_n - q_{n-1}}. \end{aligned}$$

By Theorem D.1, we have

$$\begin{aligned} [0; a_1, \ldots, a_{n-1}, a_n + \beta] &= \frac{p_{n-1}(a_n + \beta) + p_{n-2}}{q_{n-1}(a_n + \beta) + q_{n-2}} \\ &= \frac{p_n + \beta p_{n-1}}{q_n + \beta q_{n-1}} = \frac{p_n(tq_n - q_{n-1}) + q_n p_{n-1}}{q_n(tq_n - q_{n-1}) + q_n q_{n-1}}. \end{aligned}$$

It then follows from (D.2) and the above equalities that

$$[0; a_1, \ldots, a_{n-1}, a_n, t-1, 1, a_n - 1, a_{n-1}, \ldots, a_3, 2] = \frac{tp_n q_n + (-1)^n}{tq_n^2}.$$

This proves the theorem. □

We continue with a useful result on continuants.

DEFINITION D.4. Let $m \geq 1$ and $a_1, \ldots, a_m$ be positive integers. The *continuant* of $a_1, \ldots, a_m$, usually denoted by $K_m(a_1, \ldots, a_m)$, is the denominator of the rational number $[0; a_1, \ldots, a_m]$.

THEOREM D.5. *For any positive integers* $a_1, \ldots, a_m$ *and any integer* k *with* $1 \leq k \leq m - 1$, *we have*

$$K_m(a_1, \ldots, a_m) = K_m(a_m, \ldots, a_1), \tag{D.4}$$

$$2^{(m-1)/2} \leq K_m(a_1, \ldots, a_m) \leq (1 + \max\{a_1, \ldots, a_m\})^m, \tag{D.5}$$

and

$$K_k(a_1,\dots,a_k)\cdot K_{m-k}(a_{k+1},\dots,a_m) \le K_m(a_1,\dots,a_m) \le 2\,K_k(a_1,\dots,a_k)\cdot K_{m-k}(a_{k+1},\dots,a_m). \tag{D.6}$$

PROOF. The first statement is an immediate consequence of Theorem D.2. For the second statement, letting a be the maximum of $a_1,\dots,a_m$, it follows from the recursion (D.1) that

$$K_m(1,\dots,1) \le K_m(a_1,\dots,a_m) \le K_m(a,\dots,a) \le (a+1)^m.$$

Since $K_1(1)=1$, $K_2(1,1)=2$ and $2^{\ell/2} \le 2^{(\ell-1)/2} + 2^{(\ell-2)/2}$ for $\ell \ge 2$, an immediate induction gives the first inequality of (D.5).

Combining

$$K_m(a_1,\dots,a_m) = a_m K_{m-1}(a_1,\dots,a_{m-1}) + K_{m-2}(a_1,\dots,a_{m-2})$$

with (D.4), we get

$$K_m(a_1,\dots,a_m) = a_1 K_{m-1}(a_2,\dots,a_m) + K_{m-2}(a_3,\dots,a_m),$$

which implies (D.6) for $k=1$. Let k be in $\{1,2,\dots,m-2\}$ such that

$$\begin{aligned} K_m :=& K_m(a_1,\dots,a_m) = K_k(a_1,\dots,a_k)\cdot K_{m-k}(a_{k+1},\dots,a_m) \\ &+ K_{k-1}(a_1,\dots,a_{k-1})\cdot K_{m-k-1}(a_{k+2},\dots,a_m), \end{aligned} \tag{D.7}$$

where we have set $K_0 = 1$. We then have

$$\begin{aligned} K_m &= K_k(a_1,\dots,a_k)\cdot \big(a_{k+1}K_{m-k-1}(a_{k+2},\dots,a_m) \\ &\qquad\qquad + K_{m-k-2}(a_{k+3},\dots,a_m)\big) \\ &\quad + K_{k-1}(a_1,\dots,a_{k-1})\cdot K_{m-k-1}(a_{k+2},\dots,a_m) \\ &= K_{m-k-1}(a_{k+2},\dots,a_m)\cdot \big(a_{k+1}K_k(a_1,\dots,a_k) \\ &\qquad\qquad + K_{k-1}(a_1,\dots,a_{k-1})\big) \\ &\quad + K_k(a_1,\dots,a_k)\cdot K_{m-k-2}(a_{k+3},\dots,a_m), \end{aligned}$$

giving (D.7) for the index $k+1$. This shows that (D.7) and, a fortiori, (D.6) hold for $k=1,\dots,m-1$. □

We conclude this appendix with two classical results and one definition.

THEOREM D.6. *The real irrational number* $\xi = [a_0; a_1, a_2, \dots]$ *has a periodic continued fraction expansion (that is, there exist integers* $r \ge 0$ *and* $s \ge 1$ *such that* $a_{n+s} = a_n$ *for all integers* $n \ge r$*) if, and only if,* ξ *is a quadratic irrationality.*

The 'only if' part of Theorem D.6 is due to Euler [282], and the 'if' part was established by Lagrange [419] in 1770. Theorem D.7 was proved by Legendre [422].

THEOREM D.7. *Let ξ be an irrational real number. If the rational number a/b satisfies $|\xi - a/b| < 1/(2b^2)$, then a/b is a convergent to ξ.*

Theorem D.7 gives a partial converse to the right inequalities of (D.3).

DEFINITION D.8. An irrational real number $\xi := [a_0; a_1, a_2, \ldots]$ is a *badly approximable number* if there exists a positive constant c such that $|\xi - p/q| > c/q^2$ holds for every rational number p/q with $q \geq 1$.

It follows from (D.3) that an irrational real number $\xi := [a_0; a_1, a_2, \ldots]$ is a badly approximable number if, and only if, the sequence $(a_n)_{n \geq 0}$ is bounded.

Appendix E
Diophantine approximation

In this appendix, we survey classical results on approximation to real (algebraic) numbers by rational numbers and, more generally, by algebraic numbers of bounded degree. For additional results (and proofs), the reader is directed to the monographs [111, 466, 635].

E.1 Rational approximation

DEFINITION E.1. The irrationality exponent $\mu(\xi)$ of the real number ξ is the supremum of the real numbers μ for which the inequality

$$0 < \left|\xi - \frac{p}{q}\right| < \frac{1}{q^\mu}$$

has infinitely many solutions in non-zero integers p and q.

We begin with an easy result.

THEOREM E.2. *Let ξ be a real number. We have $\mu(\xi) = 1$ if ξ is rational, and $\mu(\xi) \geq 2$ otherwise.*

PROOF. Since $||q\xi|| < 1$ for every positive integer q, we get that $\mu(\xi) \geq 1$. If ξ is the reduced rational a/b and if p/q differs from ξ, then $|\xi - p/q|$ is at least equal to $1/|bq|$ and $\mu(\xi) \leq 1$. If ξ is irrational, then, by (D.3), there are infinitely many rational numbers p/q such that $|\xi - p/q| < 1/q^2$, which shows that $\mu(\xi) \geq 2$. □

THEOREM E.3. *With respect to the Lebesgue measure, almost all real numbers have an irrationality exponent equal to 2.*

PROOF. Let ε be a positive real number. Since the sum

$$\sum_{q \geq 1} q \frac{2}{q^{2+\varepsilon}}$$

converges, it follows from Theorem C.1 that the set of real numbers ξ in $[0, 1]$ such that

$$\left|\xi - \frac{p}{q}\right| < \frac{1}{q^{2+\varepsilon}}$$

holds for infinitely many rational numbers p/q with $q \geq 1$ has zero Lebesgue measure. This proves that the set of real numbers whose irrationality exponent exceeds 2 has zero Lebesgue measure, as asserted. □

The irrationality exponent of a real number ξ can be read on its continued fraction expansion; see Exercise E.1. When ξ is given by its expansion in an integer base or by some formula, it is in general very difficult to determine the exact value of $\mu(\xi)$. The case of algebraic numbers is of special interest and has a long history.

DEFINITION E.4. The *height* of a polynomial with complex coefficients

$$P(X) = a_d X^d + \cdots + a_1 X + a_0 = a_d(X - \alpha_1) \ldots (X - \alpha_d),$$

denoted by $H(P)$, is the maximum of the moduli of its coefficients. The height of an algebraic number α, denoted by $H(\alpha)$, is the height of its minimal polynomial over $\mathbb{Z}$ (that is, of the integer polynomial of lowest positive degree, with coprime coefficients and positive leading coefficient, which vanishes at α).

We begin with a result of Liouville [445, 446] proved in 1844.

THEOREM E.5. *The irrationality exponent of a real algebraic number does not exceed its degree. More precisely, if ξ is an irrational, real algebraic number of degree d and height at most H, then*

$$\left|\xi - \frac{p}{q}\right| \geq \frac{1}{d^2 H (1 + |\xi|)^{d-1} q^d} \tag{E.1}$$

for all rational numbers p/q with $q \geq 1$.

PROOF. Inequality (E.1) is true when $|\xi - p/q| \geq 1$. Let p/q be a rational number satisfying $|\xi - p/q| < 1$. Denoting by $P(X)$ the minimal defining polynomial of ξ over $\mathbb{Z}$, we have $P(p/q) \neq 0$ and $|q^d P(p/q)| \geq 1$. By Rolle's theorem, there exists a real number t lying between ξ and p/q such that

$$|P(p/q)| = |P(\xi) - P(p/q)| = |\xi - p/q| \times |P'(t)|.$$

Hence, we have $|t - \xi| \leq 1$ and

$$|P'(t)| \leq d^2 H (1 + |\xi|)^{d-1}.$$

This proves the theorem. □

Liouville applied Theorem E.5 to prove the existence of transcendental numbers by constructing explicit examples of such numbers. Indeed, it follows from Theorem E.5 that any real number whose irrationality exponent is infinite is transcendental.

DEFINITION E.6. A *Liouville number* is a real number whose irrationality exponent is infinite.

Thue [684] established in 1909 the first significant improvement on Liouville's Theorem E.5. There was subsequent progress by Siegel, Dyson and Gelfond, until Roth [614] proved in 1955 that, as far as the irrationality exponent is concerned, the irrational, real numbers do behave like almost all real numbers.

THEOREM E.7. *The irrationality exponent of every irrational, real algebraic number is equal to* 2.

Theorem E.7 can be restated as follows. Let ξ be a real algebraic number of degree at least 2. Then, for any positive real number ε, there exists a positive constant $c(\xi, \varepsilon)$ such that

$$\left|\xi - \frac{p}{q}\right| > \frac{c(\xi, \varepsilon)}{q^{2+\varepsilon}}$$

for any rational number p/q with $q \geq 1$.

For a prime number ℓ and a non-zero rational number x, we set $|x|_\ell := \ell^{-u}$, where $u \in \mathbb{Z}$ is the exponent of ℓ in the prime decomposition of x. Furthermore, we set $|0|_\ell = 0$. The next theorem, proved by Ridout [610], extends Theorem E.7.

THEOREM E.8. *Let S be a finite set of prime numbers. Let ξ be a real algebraic number. Let ε be a positive real number. The inequality*

$$\prod_{\ell \in S} |pq|_\ell \cdot \min\left\{1, \left|\xi - \frac{p}{q}\right|\right\} < \frac{1}{q^{2+\varepsilon}} \tag{E.2}$$

has only finitely many solutions in non-zero integers p, q.

Theorems E.7 and E.8 are ineffective, in the sense that their proofs do not allow us to compute explicitly either a suitable value for $c(\xi, \varepsilon)$, or an integer q_0 such that (E.2) has no solution with q greater than q_0. Nevertheless, we are able to bound explicitly the *number* of primitive solutions to an inequality like (E.2). The first result in this direction was proved in 1955 by Davenport and Roth [217]. Here, we quote a consequence of [163, Theorem 5.1].

THEOREM E.9. *Let ξ be a real algebraic number of degree $d \geq 1$ and height at most H. Let ε be a positive real number. The inequality*

$$\left|\xi - \frac{p}{q}\right| < \frac{1}{q^{2+\varepsilon}}$$

has at most

$$10^{10}(1+\varepsilon^{-1})^3 \log(6d) \log((1+\varepsilon^{-1}) \log(6d))$$

solutions in non-zero coprime integers p, q with $q > \max\{2H(\xi), 2^{4/\varepsilon}\}$. Let $b \geq 2$ be an integer. The inequality

$$\left|\xi - \frac{p}{b^n}\right| < \frac{1}{(b^n)^{1+\varepsilon}}$$

has at most

$$10^{10}(1+\varepsilon^{-1})^3 \log(6d) \log((1+\varepsilon^{-1}) \log(6d))$$

solutions in non-zero integers p, n with $b^n > \max\{2H(\xi), 2^{4/\varepsilon}\}$ and p not divisible by b.

E.2 The Schmidt Subspace Theorem

The Schmidt Subspace Theorem [632, 633, 635] is a powerful multidimensional extension of the Roth Theorem, with many outstanding applications [99, 153, 747]. We quote below a version of it (proved by Schlickewei [624]) which is suitable for our purpose, but the reader should keep in mind that there are more general formulations.

THEOREM E.10. *Let $m \geq 2$ be an integer. Let S be a finite set of prime numbers. Let $L_1, \ldots, L_m$ be m linearly independent linear forms with real algebraic coefficients. Let ε be a positive real number. Then, the set of solutions $\underline{x} = (x_1, \ldots, x_m)$ in $\mathbb{Z}^m$ to the inequality*

$$\prod_{\ell \in S} \prod_{i=1}^{m} |x_i|_\ell \cdot \prod_{i=1}^{m} |L_i(\underline{x})| \leq (\max\{|x_1|, \ldots, |x_m|\})^{-\varepsilon} \tag{E.3}$$

lies in the union of finitely many proper subspaces of $\mathbb{Q}^m$.

Note that Theorem E.10 is ineffective in the sense that its proof does not yield an explicit upper bound for the heights of the subspaces containing all the solutions to (E.3). However, Schmidt [636] was able to bound the number of these subspaces; see [284] for a common generalization of Theorems E.9 and E.10, usually called the Quantitative Subspace Theorem.

The next corollary extends Theorem E.7 to the approximation of real algebraic numbers by real algebraic numbers of smaller degree.

COROLLARY E.11. *Let ξ be a real algebraic number of degree $d \geq 2$. Let n be an integer with $1 \leq n \leq d-1$. For any positive real number ε, there exists a positive constant $c(\xi, n, \varepsilon)$ such that*

$$|\xi - \alpha| > c(\xi, n, \varepsilon) H(\alpha)^{-n-1-\varepsilon},$$

for any algebraic number α of degree at most n.

Corollary E.11 shows that, as far as approximation by algebraic numbers of degree bounded by n is concerned, algebraic numbers of degree greater than n do behave like almost all real numbers.

E.3 Approximation by algebraic numbers

The set of real numbers splits into algebraic and transcendental numbers. These two subsets are far from having the same size, the former being countable, while the latter has the power of continuum. Such a crude classification of real numbers is rather unsatisfactory, and one aims to find some way to classify the transcendental numbers. A possibility could be to take their irrationality exponent into account; this would, however, not meet a natural requirement asking that two algebraically dependent real numbers should belong to the same class.

An attempt towards a 'reasonable' classification was made in 1932 by Mahler [461], who proposed to subdivide the set of transcendental real numbers into three classes according to, roughly speaking, their properties of approximation by algebraic numbers. In Mahler's classification any two algebraically dependent transcendental real numbers belong to the same class. It has been widely studied, as well as the closely related Koksma's classification, which was proposed a few years later [390]; see [146, Chapter 3].

DEFINITION E.12. Let ξ be a real number and n a positive integer. Let $w_n(\xi)$ denote the supremum of the real numbers w for which there exist infinitely many integer polynomials $P(X)$ of degree at most n satisfying

$$0 < |P(\xi)| \leq H(P)^{-w}.$$

Let $w_n^*(\xi)$ denote the supremum of the real numbers w^* for which there exist infinitely many real algebraic numbers α of degree at most n satisfying

$$0 < |\xi - \alpha| \leq H(\alpha)^{-w^*-1}. \tag{E.4}$$

To justify the -1 occurring in (E.4), we recall that if $P(X)$ is an integer polynomial of degree $n \geq 1$ and if ξ is a real number such that $P(\xi)$ is non-zero, then $P(X)$ has a root α satisfying $|\xi - \alpha| \leq n\,|P(\xi)|/|P'(\xi)|$. Furthermore, $|P'(\xi)|$ can be as large as $H(P)$ and, when $|P(\xi)|$ is small, $|P'(\xi)|$ has no reason to be small as well (unless $P(X)$ has two or more roots close to ξ). Consequently, $|\xi-\alpha|$ is 'often' less than $H(P)^{-1}|P(\xi)|$.

According to Mahler, we classify the real numbers by means of the exponents w_n.

DEFINITION E.13. Let ξ be a real number and set

$$w(\xi) = \limsup_{n\to+\infty} w_n(\xi)/n.$$

We say that ξ is an

$$\begin{aligned} &A\textit{-number}, \text{ if } w(\xi) = 0;\\ &S\textit{-number}, \text{ if } 0 < w(\xi) < +\infty;\\ &T\textit{-number}, \text{ if } w(\xi) = +\infty \text{ and } w_n(\xi) < +\infty \text{ for any } n \geq 1;\\ &U\textit{-number}, \text{ if } w(\xi) = +\infty \text{ and } w_n(\xi) = +\infty \text{ from some } n \text{ onwards.}\end{aligned}$$

Following Koksma, we define the classes A^*, S^*, T^* and U^* exactly as in Definition E.13, with, however, the functions w_n replaced everywhere by the functions w_n^*.

The next theorem gathers several basic and classical results on the classifications of Mahler and Koksma.

THEOREM E.14. *The classifications of Mahler and of Koksma coincide, in the sense that any S-number (resp. T-number, U-number) is an S^*-number (resp. T^*-number, U^*-number). The A^*-numbers (resp. A-numbers) are exactly the algebraic numbers. If ξ is an algebraic number of degree d, then*

$$w_n^*(\xi) = w_n(\xi) = \min\{n, d-1\}, \quad \textit{for } n \geq 1.$$

Every transcendental real number ξ satisfies

$$w_n(\xi) \geq n, \quad \textit{for } n \geq 1.$$

Almost every real number ξ is an S-number and satisfies

$$w_n^*(\xi) = w_n(\xi) = n, \quad \textit{for } n \geq 1.$$

The set of T-numbers (resp. of U-numbers) is non-empty and has Hausdorff dimension zero. For every $n \geq 2$, there exist real numbers ξ for which $w_n(\xi) \neq w_n^*(\xi)$.

The reader is referred to Chapters 3 and 5 of [146] for the proof of Theorem E.14 and for subsequent results on the functions w_n and w_n^*.

E.4 Exercises

EXERCISE E.1. Let $\xi = [a_0; a_1, a_2, \ldots]$ be an irrational real number with convergents $p_0/q_0, p_1/q_1, \ldots$ Prove that we have

$$\mu(\xi) = 2 + \limsup_{n \to +\infty} \frac{\log a_n}{\log q_{n-1}},$$

if this limsup is finite, and $\mu(\xi) = +\infty$ otherwise.

EXERCISE E.2. Prove that Theorem E.10 implies Theorem E.8.

EXERCISE E.3. Let ξ be an algebraic number of degree at least three. Let $\varepsilon > 0$ be a real number. Apply Theorem E.10 to prove that there are only finitely many triples (p_1, p_2, q) of integers such that $q \geq 1$ and $\max\{|q\xi - p_1|, |q\xi^2 - p_2|\} \leq q^{-1/2-\varepsilon}$.

Appendix F
Recurrence sequences

In this appendix, we recall some basic facts on recurrence sequences.

A homogeneous linear recurrence sequence with constant coefficients (*recurrence sequence* for short) is a sequence $(u_n)_{n\geq 0}$ of complex numbers such that

$$u_{n+k} = v_{k-1}u_{n+k-1} + v_{k-2}u_{n+k-2} + \cdots + v_0 u_n \quad (n \geq 0), \tag{F.1}$$

for some complex numbers $v_0, v_1, \ldots, v_{k-1}$ with $v_0 \neq 0$ and with initial values $u_0, \ldots, u_{k-1}$ not all zero. The positive integer k is called the *order* of the recurrence.

The *companion polynomial* to a recurrence as above is given by

$$G(X) = X^k - v_{k-1}X^{k-1} - \cdots - v_0. \tag{F.2}$$

Let

$$G(X) = \prod_{j=1}^{s} (X - \omega_j)^{a_j} \tag{F.3}$$

be its factorization over $\mathbb{C}$, where the *roots* $\omega_1, \ldots, \omega_s$ are distinct and $a_1, \ldots, a_s$ are positive integers.

A recurrence sequence may satisfy different relations of the form (F.1), however, for every recurrence sequence, there is a unique recurrence of minimal order. In the sequel, the order, the recurrence coefficients, the roots of a recurrence sequence are always all meant with respect to this unique recurrence of minimal order.

The following theorem is fundamental in the theory of recurrence sequences.

THEOREM F.1. *Let $(u_n)_{n\geq 0}$ be a sequence of complex numbers satisfying relation (F.1) with $v_0 \neq 0$. Let $G(X)$, $\omega_1, \ldots, \omega_s$ and $a_1, \ldots, a_s$ be determined by (F.2) and (F.3). Let $\mathbb{K}$ be the field generated over $\mathbb{Q}$ by*

$u_0, \ldots, u_{k-1}, \omega_1, \ldots, \omega_s$. Then there exist uniquely determined polynomials $f_j(X)$ in $\mathbb{K}[X]$ of degree less than a_j $(j = 1, \ldots, s)$ such that

$$u_n = \sum_{j=1}^{s} f_j(n)\omega_j^n \quad (n \geq 0). \tag{F.4}$$

Furthermore, putting $u(z) = \sum_{n\geq 0} u_n z^n$, there exist uniquely determined complex numbers β_{ij} in $\mathbb{K}$ and a polynomial $P(X)$ with coefficients in $\mathbb{Q}(u_0, \ldots, u_{k-1}, v_0, \ldots, v_{k-1})$ such that

$$u(z) = \sum_{j=1}^{s} \sum_{i=1}^{a_j} \frac{\beta_{ij}}{(1-\omega_j z)^i} = \frac{P(z)}{z^k G(1/z)}.$$

Conversely, let $\omega_1, \ldots, \omega_s$ be distinct complex numbers and $a_1, \ldots, a_s$ be positive integers. Set $k = a_1 + \cdots + a_s$ and define $v_0, v_1, \ldots, v_{k-1}$ by (F.3) and (F.2). For $j = 1, \ldots, s$, let $f_j(X)$ be a polynomial of degree less than a_j. Then the sequence $(u_n)_{n\geq 0}$ defined by (F.4) satisfies the recurrence relation (F.1).

PROOF. See e.g. [651, Chapter C]. □

A recurrence is called algebraic (rational, integral) if all the initial values and recurrence coefficients are algebraic (rational, integral, respectively). The resulting sequence is then called an algebraic (rational, integral, respectively) recurrence sequence.

THEOREM F.2. *A recurrence sequence of algebraic (rational, integral) numbers is an algebraic (rational, integral) recurrence sequence.*

PROOF. The statement for an algebraic (rational) recurrence sequence is an easy consequence of Cramer's rule. A lemma of Fatou [300] implies the requested result for a recurrence sequence of integers. Let us give some more details. Assume that u_n is a rational integer for $n \geq 0$ and write

$$u(X) = \sum_{n\geq 0} u_n X^n = \frac{P(X)}{Q(X)} = \frac{p_\ell + p_{\ell-1}X + \cdots + p_0 X^\ell}{q_k + q_{k-1}X + \cdots + q_0 X^k},$$

where the above rational function is irreducible. The coefficients $p_0, \ldots, p_\ell, q_0, \ldots, q_k$ are rational numbers and are wholly determined up to a constant factor. Without loss of generality, we assume that they are integers and that they have no common divisor ≥ 2. This implies that the integer polynomial $Q(X)$ is primitive.

Since $P(X)$ and $Q(X)$ have no common divisor, there exist integer polynomials $A(X)$ and $B(X)$ and a positive integer m such that

$$P(X)A(X) + Q(X)B(X) = m.$$

From $m = Q(X)(A(X)u(X) + B(X))$ and the fact that $Q(X)$ is primitive, we deduce that all the coefficients of the series $A(X)u(X) + B(X)$ are divisible by m. But m is equal to q_k times $A(0)u(0) + B(0)$. This shows that $q_k = \pm 1$. □

THEOREM F.3. *A sequence $(u_n)_{n\geq 0}$ of complex numbers is a recurrence sequence if, and only if, the determinants Δ_n of the Hankel matrices*

$$\mathcal{M}_n = \begin{pmatrix} u_0 & u_1 & \dots & u_n \\ u_1 & u_2 & \dots & u_{n+1} \\ \dots & \dots & \dots & \dots \\ u_n & u_{n+1} & \dots & u_{2n} \end{pmatrix}$$

are zero for every sufficiently large integer n.

PROOF. If the sequence $(u_n)_{n\geq 0}$ satisfies (F.1), then, for every integer $n \geq k+1$, the $(k+1)$th column of the Hankel matrix $\mathcal{M}_n$ is a linear combination of the first k columns. Consequently, $\Delta_n = 0$ for $n \geq k+1$.

Assume now that $\Delta_n = 0$ for every large integer n, and let k be the largest integer such that $\Delta_k = \Delta_{k+i} = 0$ for $i \geq 0$. We have $k \geq 1$, since otherwise $u_n = 0$ for $n \geq 0$. By assumption, the last column of $\mathcal{M}_k$ is a linear combination of the first k columns. This implies that there exist complex numbers $v_0, \dots, v_{k-1}$ such that

$$v_0 u_j + v_1 u_{j+1} + \dots + v_{k-1} u_{j+k-1} + u_{j+k} = 0, \quad 0 \leq j \leq k.$$

For $j \geq 0$, set $L_{j+k} = v_0 u_j + v_1 u_{j+1} + \dots + v_{k-1} u_{j+k-1} + u_{j+k}$. Let m be an integer such that $m > k$ and $L_k = L_{1+k} = \dots = L_{m-1+k} = 0$. Since

$$\mathcal{M}_m = \begin{pmatrix} & & & u_k & \dots & u_m \\ & \mathcal{M}_{k-1} & & \dots & \dots & \dots \\ & & & u_{2k-1} & \dots & u_{k-1+m} \\ u_k & \dots & u_{2k-1} & u_{2k} & \dots & u_{k+m} \\ \dots & \dots & \dots & \dots & \dots & \dots \\ u_m & \dots & u_{k-1+m} & u_{k+m} & \dots & u_{2m} \end{pmatrix},$$

adding to every column C_ℓ with $\ell \geq k+1$ the linear combination $v_0 C_{\ell-k} + v_1 C_{\ell-k+1} + \cdots + v_{k-1} C_{\ell-1}$ gives

$$\Delta_m = \begin{vmatrix} & & & L_k & \dots & L_m \\ & \mathcal{M}_{k-1} & & \dots & \dots & \dots \\ & & & L_{2k-1} & \dots & L_{k-1+m} \\ u_k & \dots & u_{2k-1} & L_{2k} & \dots & L_{k+m} \\ \dots & \dots & \dots & \dots & \dots & \dots \\ u_m & \dots & u_{k-1+m} & L_{k+m} & \dots & L_{2m} \end{vmatrix}$$

$$= \begin{vmatrix} & & & 0 & \dots & \dots & \dots & \dots & 0 \\ & \mathcal{M}_{k-1} & & \dots & \dots & \dots & \dots & \dots & \dots \\ & & & 0 & \dots & \dots & \dots & \dots & 0 \\ u_k & \dots & u_{2k-1} & 0 & & \dots & & 0 & L_{k+m} \\ \dots & \dots & \dots & \vdots & & \iddots & & \iddots & \vdots \\ \dots & \dots & \dots & 0 & & L_{k+m} & & \dots & L_{2m-1} \\ u_m & \dots & u_{k-1+m} & L_{k+m} & & L_{k+m-1} & & \dots & L_{2m} \end{vmatrix}$$

$$= \pm \Delta_{k-1} (L_{k+m})^{m-k+1}.$$

Since $\Delta_m = 0$ and Δ_{k-1} is non-zero, this shows that $L_{k+m} = 0$. Consequently, the sequence $(u_n)_{n \geq 0}$ is a recurrence sequence. □

References

[1] B. Adamczewski, *On the expansion of some exponential periods in an integer base*, Math. Ann. 346 (2010), 107–116. (Cited in Chapter 8.)

[2] B. Adamczewski and J.-P. Allouche, *Reversals and palindromes in continued fractions*, Theoret. Comput. Sci. 380 (2007), 220–237. (Cited in Chapter 7.)

[3] B. Adamczewski and Y. Bugeaud, *On the complexity of algebraic numbers, II. Continued fractions*, Acta Math. 195 (2005), 1–20. (Cited in Chapter 9.)

[4] ———, *On the independence of expansions of algebraic numbers in an integer base*, Bull. London Math. Soc. 39 (2007), 283–289. (Cited in Chapter 8.)

[5] ———, *A short proof of the transcendence of the Thue–Morse continued fractions*, Amer. Math. Monthly 114 (2007), 536–540. (Cited in Chapter 9.)

[6] ———, *Dynamics for β-shifts and Diophantine approximation*, Ergod. Theory Dynam. Systems 27 (2007), 1695–1711. (Cited in Chapters 8 & 9, and Appendix A.)

[7] ———, *On the complexity of algebraic numbers I. Expansions in integer bases*, Ann. of Math. 165 (2007), 547–565. (Cited in Chapter 8.)

[8] ———, *On the Maillet–Baker continued fractions*, J. Reine Angew. Math. 606 (2007), 105–121. (Cited in Chapter 9.)

[9] ———, *Palindromic continued fractions*, Ann. Inst. Fourier (Grenoble) 57 (2007), 1557–1574. (Cited in Chapter 9.)

[10] ———, *Mesures de transcendance et aspects quantitatifs de la méthode de Thue–Siegel–Roth–Schmidt*, Proc. London Math. Soc. 101 (2010), 1–31. (Cited in Chapters 4 & 9.)

[11] ———, *Transcendence measures for continued fractions involving repetitive or symmetric patterns*, J. Europ. Math. Soc. 12 (2010), 883–914. (Cited in Chapter 9.)

[12] ———, *Diophantine approximation and transcendence*. In: Encyclopedia of Mathematics and its Applications 135, Cambridge University Press, Cambridge, 2010, pp. 424–465. (Cited in Chapter 8.)

[13] ———, *Nombres réels de complexité sous-linéaire: mesures d'irrationalité et de transcendance*, J. Reine Angew. Math. 658 (2011), 65–98. (Cited in Chapter 8 and Appendix A.)

[14] B. Adamczewski, Y. Bugeaud and L. Davison, *Continued fractions and transcendental numbers*, Ann. Inst. Fourier (Grenoble) 56 (2006), 2093–2113. (Cited in Chapter 9.)

[15] B. Adamczewski, Y. Bugeaud and F. Luca, *Sur la complexité des nombres algébriques*, C. R. Acad. Sci. Paris 339 (2004), 11–14. (Cited in Chapter 8.)

[16] ———, *On the values of a class of analytic functions at algebraic points*, Acta Arith. 135 (2008), 1–18. (Cited in Chapter 8.)

[17] B. Adamczewski and J. Cassaigne, *Diophantine properties of real numbers generated by finite automata*, Compos. Math. 142 (2006), 1351–1372. (Cited in Chapter 8.)

[18] B. Adamczewski and C. Faverjon, *Chiffres non nuls dans le développement en base entière d'un nombre algébrique irrationnel.* C.R. Acad. Sci. Paris 350 (2012), 1–4. (Cited in Chapter 8.)

[19] B. Adamczewski and N. Rampersad, *On patterns occurring in binary algebraic numbers*, Proc. Amer. Math. Soc. 136 (2008), 3105–3109. (Cited in Chapters 8 & 10.)

[20] B. Adamczewski and T. Rivoal, *Irrationality measures for some automatic real numbers*, Math. Proc. Cambridge Phil. Soc. 147 (2009), 659–678. (Cited in Chapter 8.)

[21] W. W. Adams and J. L. Davison, *A remarkable class of continued fractions*, Proc. Amer. Math. Soc. 65 (1977), 194–198. (Cited in Chapter 7.)

[22] S. D. Adhikari, P. Rath and N. Saradha, *On the sets of uniqueness of a distribution function of* $\{\xi(p/q)^n\}$, Acta Arith. 119 (2005), 307–316. (Cited in Chapter 3.)

[23] R. Adler, M. Keane and M. Smorodinsky, *A construction of a normal number for the continued fraction transformation*, J. Number Theory 13 (1981), 95–105. (Cited in Chapter 9.)

[24] V. N. Agafonov, *Normal sequences and finite automata*, Dokl. Akad. Nauk SSSR 179 (1968), 255–256 (in Russian). English translation in Soviet Math. Dokl. 9 (1968), 324–325. (Cited in Chapter 4.)

[25] Ch. Aistleitner, *On modifying normal numbers*, Unif. Distrib. Theory 6 (2011), 49–58. (Cited in Chapters 4 & 10.)

[26] M. Ajtai, I. Havas and J. Komlós, *Every group admits a bad topology.* In: Studies in Pure Mathematics, 21–34, Birkhäuser, Basel, 1983. (Cited in Chapter 2.)

[27] M. Aka and U. Shapira, *On the evolution of continued fractions in a fixed quadratic field.* Preprint. (Cited in Chapter 9.)

[28] R. K. Akhunzhanov, *On nonnormal numbers*, Mat. Zametki 72 (2002), 150–152 (in Russian). English translation in Math. Notes 72 (2002), 135–137. (Cited in Chapter 7.)

[29] ———, *On the distribution modulo 1 of exponential sequences*, Mat. Zametki 76 (2004), 163–171 (in Russian). English translation in Math. Notes 76 (2004), 153–160. (Cited in Chapter 7.)

[30] R. K. Akhunzhanov and N. G. Moshchevitin, *On the chromatic number of a distance graph associated with a lacunary sequence*, Dokl. Akad. Nauk 397 (2004), 295–296 (in Russian). (Cited in Chapter 2.)

[31] ———, *On the distribution modulo 1 of subexponential sequences*, Mat. Zametki 77 (2005), 803–813 (in Russian). English translation in Math. Notes 77 (2005), 741–750. (Cited in Chapter 2.)

[32] S. Akiyama, *Mahler's Z-number and 3/2 number system*, Unif. Distrib. Theory 3 (2008), 91–99. (Cited in Chapter 3.)

[33] S. Akiyama, Ch. Frougny and J. Sakarovitch, *Powers of rationals modulo 1 and rational base number systems*, Israel J. Math. 168 (2008), 53–91. (Cited in Chapter 3.)

[34] S. Akiyama and Y. Tanigawa, *Salem numbers and uniform distribution modulo 1*, Publ. Math. Debrecen 64 (2004), 329–341. (Cited in Chapter 3.)

[35] S. Albeverio, M. Pratsiovytyi and G. Torbin, *Singular probability distributions and fractal properties of sets of real numbers defined by the asymptotic frequencies of their s-adic digits*, Ukraïn. Mat. Zh. 57 (2005), 1163–1170. English translation in Ukrainian Math. J. 57 (2005), 1361–1370. (Cited in Chapters 4 & 7.)

[36] ———, *Topological and fractal properties of real numbers which are not normal*, Bull. Sci. Math. 129 (2005), 615–630. (Cited in Chapters 4 & 7.)

[37] G. Alkauskas and A. Dubickas, *Prime and composite numbers as integer parts of powers*, Acta Math. Hungar. 105 (2004), 249–256. (Cited in Chapter 3.)

[38] J.-P. Allouche, *Nouveaux résultats de transcendance de réels à développements non aléatoire*, Gaz. Math. 84 (2000), 19–34. (Cited in Chapter 8.)

[39] J.-P. Allouche, A. Arnold, J. Berstel, S. Brlek, W. Jockush, S. Plouffe and B. E. Sagan, *A relative of the Thue–Morse sequence*, Discrete Math. 139 (1995), 455–461. (Cited in Appendix A.)

[40] J.-P. Allouche and M. Cosnard, *The Komornik–Loreti constant is transcendental*, Amer. Math. Monthly (2000), 448–449. (Cited in Chapter 9.)

[41] ———, *Non-integer bases, iteration of continuous real maps, and an arithmetic self-similar set*, Acta Math. Hungar. 91 (2001), 325–332. (Cited in Chapter 3.)

[42] J.-P. Allouche, J. L. Davison, M. Queffélec and L. Q. Zamboni, *Transcendence of Sturmian or morphic continued fractions*, J. Number Theory 91 (2001), 39–66. (Cited in Chapter 9 and Appendix A.)

[43] J.-P. Allouche and A. Glen, *Distribution modulo* 1 *and the lexicographic world*, Ann. Sci. Math. Québec 33 (2009), 125–143. (Cited in Chapter 3.)

[44] ———, *Extremal properties of (epi)sturmian sequences and distribution modulo* 1, Enseign. Math. 56 (2010), 365–401. (Cited in Chapter 3.)

[45] J.-P. Allouche and J. Shallit, *The ubiquitous Prouhet–Thue–Morse sequence*. In: Sequences and their Applications (Singapore, 1998), 1–16, Springer Ser. Discrete Math. Theor. Comput. Sci., Springer, London, 1999. (Cited in Chapter 3 and Appendix A.)

[46] ———, Automatic Sequences: Theory, Applications, Generalizations, Cambridge University Press, Cambridge, 2003. (Cited in Chapters 7, 8 & 10, and Appendix A.)

[47] N. Alon and Yu. Peres, *Uniform dilations*, Geom. Funct. Anal. 2 (1992), 1–28. (Cited in Chapter 8.)

[48] C. Altomare and B. Mance, *Cantor series constructions contrasting two notions of normality*, Monatsh. Math. 164 (2011), 1–22. (Cited in Chapter 9.)

[49] M. Amou, *Approximation to certain transcendental decimal fractions by algebraic numbers*, J. Number Theory 37 (1991), 231–241. (Cited in Chapters 4 & 7.)

[50] M. Amou and Y. Bugeaud, *Expansions in integer bases and exponents of Diophantine approximation*, J. London Math. Soc. 81 (2010), 297–316. (Cited in Chapters 4 & 7.)

[51] J. Auslander and Y. N. Dowker, *On disjointness of dynamical systems*, Math. Proc. Cambridge Philos. Soc. 85 (1979), 477–491. (Cited in Chapter 4.)

[52] D. Badziahin and S. Velani, *Multiplicatively badly approximable numbers and the mixed Littlewood conjecture*, Adv. Math. 228 (2011), 2766–2796. (Cited in Chapter 2.)

[53] D. H. Bailey, J. M. Borwein, R. E. Crandall and C. Pomerance, *On the binary expansions of algebraic numbers*, J. Théor. Nombres Bordeaux 16 (2004), 487–518. (Cited in Chapters 6 & 8.)

[54] D. Bailey, P. Borwein and S. Plouffe, *On the rapid computation of various polylogarithmic constants*, Math. Comp. 66 (1997), 903–913. (Cited in Chapter 5.)

[55] D. H. Bailey and R. E. Crandall, *On the random character of fundamental constant expansions*, Experiment. Math. 10 (2001), 175–190. (Cited in Chapter 5.)

[56] ———, *Random generators and normal numbers*, Experiment. Math. 11 (2002), 527–546. (Cited in Chapter 5.)

[57] D. H. Bailey and M. Misiurewicz, *A strong hot spot theorem*, Proc. Amer. Math. Soc. 134 (2006), 2495–2501. (Cited in Chapters 4 & 5.)

[58] A. Baker, *Continued fractions of transcendental numbers*, Mathematika 9 (1962), 1–8. (Cited in Chapter 9.)

[59] G. Barat, Ch. Frougny and A. Pethő, *A note on linear recurrent Mahler numbers*, Integers 5 (2005), A1, 17 pp. (Cited in Chapter 10.)

[60] G. Barat, R. F. Tichy and R. Tijdeman, *Digital blocks in linear numeration systems*. In: Number Theory in Progress, Vol. 2 (Zakopane-Kościelisko, 1997), 607–631, de Gruyter, Berlin, 1999. (Cited in Chapter 6.)

[61] J. Barral and S. Seuret, *Ubiquity and large intersections properties under digit frequencies constraints*, Math. Proc. Cambridge Philos. Soc. 145 (2008), 527–548. (Cited in Chapter 7.)

[62] L. Barreira and G. Iommi, *Partial quotients of continued fractions and β-expansions*, Nonlinearity 21 (2008), 2211–2219. (Cited in Chapter 9.)

[63] L. Barreira, B. Saussol and J. Schmeling, *Higher-dimensional multifractal analysis*, J. Math. Pures Appl. 81 (2002), 67–91. (Cited in Chapter 7.)

[64] ———, *Distribution of frequencies of digits via multifractal analysis*, J. Number Theory 97 (2002), 410–438. (Cited in Chapter 7.)

[65] L. Barreira and C. Valls, *Asymptotic behavior of distribution of frequencies of digits*, Math. Proc. Cambridge Philos. Soc. 145 (2008), 177–195. (Cited in Chapter 7.)

[66] ———, *Cubic relations between frequencies of digits and Hausdorff dimension*, Michigan Math. J. 60 (2011), 579–602. (Cited in Chapter 7.)

[67] V. Becher, *Turing's note on normal numbers.* In: Alan Turing – his work and impact, (eds. S. Barry Cooper and Jan van Leeuwen), Elsevier, Amsterdam, 2012. (Cited in Chapter 5.)

[68] V. Becher and S. Figueira, *An example of a computable absolutely normal number*, Theoret. Comput. Sci. 270 (2002), 947–958. (Cited in Chapter 5.)

[69] V. Becher, S. Figueira and R. Picchi, *Turing's unpublished algorithm for normal numbers*, Theoret. Comput. Sci. 377 (2007), 126–138. (Cited in Chapter 5.)

[70] J. P. Bell and M. Coons, *Diophantine approximation of Mahler numbers.* Preprint. (Cited in Chapter 8.)

[71] D. Berend, *Multi-invariant sets on tori*, Trans. Amer. Math. Soc. 280 (1983), 509–532. (Cited in Chapters 2 & 8.)

[72] ———, *Dense* (mod 1) *dilated semigroups of algebraic numbers*, J. Number Theory 26 (1987), 246–256. (Cited in Chapter 2.)

[73] D. Berend and M. D. Boshernitzan, *On a result of Mahler on the decimal expansions of* $(n\alpha)$, Acta Arith. 66 (1994), 315–322. (Cited in Chapter 8.)

[74] ———, *Numbers with complicated decimal expansions*, Acta Math. Hungar. 66 (1995), 113–126. (Cited in Chapters 8, 9 & 10.)

[75] D. Berend and Yu. Peres, *Asymptotically dense dilations of sets on the circle*, J. London Math. Soc. 47 (1993), 1–17. (Cited in Chapter 8.)

[76] V. Bergström, *Einige Bemerkungen zur Theorie der diophantischen Approximationen*, Fysiogr. Sllsk. Lund Forhandl. 6 (1936), 113–131. (Cited in Chapter 1.)

[77] M. Bernay, *La dimension de Hausdorff de l'ensemble des nombres r-déterministes*, C. R. Acad. Sci. Paris 280 (1975), A539–A541. (Cited in Chapter 4.)

[78] T. Bernhard, Einfach Kompliziert, Suhrkamp Verlag, 1986.

[79] V. Berthé, C. Holton and L. Q. Zamboni, *Initial powers of Sturmian sequences*, Acta Arith. 122 (2006), 315–347. (Cited in Chapter 8 and Appendix A.)

[80] M.-J. Bertin, A. Decomps-Guilloux, M. Grandet-Hugot, M. Pathiaux-Delefosse and J.-P. Schreiber, Pisot and Salem Numbers, Birkhäuser, Basel, 1992. (Cited in Preface and Chapter 2.)

[81] A. Bertrand, *Développements en base de Pisot et répartition modulo* 1, C. R. Acad. Sci. Paris 285 (1977), A419–A421. (Cited in Chapter 9.)

[82] ———, *Répartition modulo 1 et développement en base θ*, C. R. Acad. Sci. Paris 289 (1979), A1–A4. (Cited in Chapter 9.)

[83] A. Bertrand-Mathis, *Développements en base θ, répartition modulo un de la suite $(x\theta^n)_{n\geq 0}$, langages codés et θ-shift*, Bull. Soc. Math. France 114 (1986), 271–323. (Cited in Chapters 6 & 9.)

[84] ———, *Points génériques de Champernowne sur certains systèmes codes; application aux θ-shifts*, Ergodic Theory Dynam. Systems 8 (1988), 35–51. (Cited in Chapter 9.)

[85] ———, *Nombres normaux dans diverses bases*, Ann. Inst. Fourier (Grenoble) 45 (1995), 1205–1222. (Cited in Chapter 9.)

[86] ———, *Nombres normaux*, J. Théor. Nombres Bordeaux 8 (1996), 397–412. (Cited in Chapter 9.)

[87] ———, *Ensembles normaux relatifs à des matrices non-commutantes*, J. Number Theory 63 (1997), 180–190. (Cited in Chapter 6.)

[88] ———, *Sur les mesures simultanément invariantes pour les transformations $x \to \{\lambda x\}$ et $x \to \{\beta x\}$*, Acta Math. Hungar. 78 (1998), 71–78. (Cited in Chapter 9.)

[89] ———, *Sur les nombres normaux liés à des bases non équivalentes*, Ergodic Theory Dynam. Systems 18 (1998), 89–93. (Cited in Chapter 9.)

[90] ———, *Nombres self normaux*, Bull. Soc. Math. France. To appear. (Cited in Chapter 9.)

[91] A. Bertrand-Mathis and B. Volkmann, *On (ε, k)-normal words in connecting dynamical systems*, Monatsh. Math. 107 (1989), 267–279. (Cited in Chapter 9.)

[92] D. Bertrand, *Theta functions and transcendence*, Ramanujan J. 1 (1997), 339–350. (Cited in Chapter 8.)

[93] A. S. Besicovitch, *The asymptotic distribution of the numerals in the decimal representation of the squares of the natural numbers*, Math. Z. 39 (1935), 146–156. (Cited in Chapters 4 & 5.)

[94] F. Beukers, *Fractional parts of powers of rationals*, Math. Proc. Cambridge Phil. Soc. 90 (1981), 13–20. (Cited in Chapter 3.)

[95] W. A. Beyer, N. Metropolis and J. R. Neergaard, *Statistical study of digits of some square roots of integers in various bases*, Math. Comp. 24 (1970), 455–473. (Cited in Chapter 8.)

[96] ———, *The generalized serial test applied to expansions of some irrational square roots in various bases*, Math. Comp. 24 (1970), 745–747. (Cited in Chapter 8.)

[97] L. Bienvenu, *Qu'est-ce qu'un nombre au hasard? Hasard et calculabilité*. In: Journée Annuelle 2011, pp. 1–26, Société mathématique de France, 2011. (Cited in Chapter 5.)

[98] P. Billingsley, Probability and Measure, Wiley Ser. Probab. Math. Statist., John Wiley & Sons, New York, 1979. (Cited in Chapter 6.)

[99] Yu. F. Bilu, *The many faces of the subspace theorem [after Adamczewski, Bugeaud, Corvaja, Zannier...]*. Séminaire Bourbaki, Vol. 2006/2007, Astérisque No. 317 (2008), Exp. No. 967, vii, 1–38. (Cited in Chapter 8 and Appendix E.)

[100] A. Bisbas, *Normal numbers from infinite convolution measures*, Ergodic Theory Dynam. Systems 23 (2003), 389–393. (Cited in Chapter 6.)

[101] ———, *A Cassels–Schmidt theorem for non-homogeneous Markov chains*, Bull. Sci. Math. 129 (2005), 25–37. (Cited in Chapter 6.)

[102] A. Bisbas and N. Snigireva, *Divergence points and normal numbers*, Monatsh. Math. To appear. (Cited in Chapter 6.)

[103] F. Blanchard, *β-expansions and symbolic dynamics*, Theoret. Comput. Sci. 65 (1989), 131–141. (Cited in Chapter 9.)

[104] ———, *Nonliteral transducers and some problems of normality*, J. Théor. Nombres Bordeaux 5 (1993), 303–321. (Cited in Chapter 4.)

[105] F. Blanchard and J.-M. Dumont, *Generic sequences, transducers and multiplication of normal numbers*, Israel J. Math. 80 (1992), 257–287. (Cited in Chapter 4.)

[106] R. Blecksmith, M. Filaseta and C. Nicol, *A result on the digits of a^n*, Acta Arith. 64 (1993), 331–339. (Cited in Chapter 6.)

[107] C. Bluhm, *On a theorem of Kaufman: Cantor-type construction of linear fractal Salem sets*, Ark. Mat. 36 (1998), 307–316. (Cited in Chapter 7.)

[108] ———, *Liouville numbers, Rajchman measures, and small Cantor sets*, Proc. Amer. Math. Soc. 128 (2000), 2637–2640. (Cited in Chapter 7.)

[109] P. Bohl, *Über ein in der Theorie der säkularen Störungen vorkommendes Problem*, J. Reine Angew. Math. 135 (1909), 189–283. (Cited in Chapter 1.)

[110] P. E. Böhmer, *Über die Transzendenz gewisser dyadischer Brüche*, Math. Ann. 96 (1927), 367–377. (Cited in Chapter 7.)

[111] E. Bombieri and W. Gubler, Heights in Diophantine Geometry, New Math. Monogr. 4, Cambridge University Press, Cambridge, 2006. (Cited in Appendix E.)

[112] E. Bombieri and A. J. van der Poorten, *Continued fractions of algebraic numbers*. In: Computational Algebra and Number Theory (Sydney, 1992), 137–152, Math. Appl., 325, Kluwer Academic Publishers, Dordrecht, 1995. (Cited in Chapter 9.)

[113] É. Borel, *Sur une application d'un théorème de M. Hadamard*, Bull. Sci. Math. 18 (1894), 22–25. (Cited in Chapter 2.)

[114] ———, *Les probabilités dénombrables et leurs applications arithmétiques*, Rend. Circ. Math. Palermo 27 (1909), 247–271. (Cited in Chapter 4.)

[115] ———, Leçons sur la Théorie des Fonctions, seconde édition, Gauthier-Villars, Paris, 1914. (Cited in Chapter 4.)

[116] ———, *Sur les chiffres décimaux de $\sqrt{2}$ et divers problèmes de probabilités en chaîne*, C. R. Acad. Sci. Paris 230 (1950), 591–593. (Cited in Chapter 8.)

[117] J.-P. Borel, *Parties d'ensembles B-normaux*, Manuscripta Math. 62 (1988), 317–335. (Cited in Chapter 4.)

[118] ———, *Suites de longueur minimale associées à un ensemble normal donné*, Israel J. Math. 64 (1988), 229–250. (Cited in Chapter 4.)

[119] ———, *Sur certains ensembles normaux*, Sém. Théor. Nombres Bordeaux 1 (1989), 67–79. (Cited in Chapter 4.)

[120] J. Borwein and D. Bailey, Mathematics by Experiment. Plausible Reasoning in the 21st Century, second edition, A. K. Peters, Ltd, Wellesley, MA, 2008. (Cited in Chapters 4, 5 & 8.)

[121] M. D. Boshernitzan, *Homogeneously distributed sequences and Poincaré sequences of integers of sublacunary growth*, Monatsh. Math. 96 (1983), 173–181. (Cited in Chapter 2.)

[122] ———, *Elementary proof of Furstenberg's Diophantine result*, Proc. Amer. Math. Soc. 122 (1994), 67–70. (Cited in Chapter 2.)

[123] ———, *Density modulo* 1 *of dilations of sublacunary sequences*, Adv. Math. 108 (1994), 104–117. (Cited in Chapter 2.)

[124] W. Bosma, K. Dajani and C. Kraaikamp, *Entropy quotients and correct digits in number-theoretic expansions*, In: Dynamics and Stochastics, 176–188, IMS Lecture Notes Monogr. Ser., 48, Institute of Mathematics and Statistics, Beachwood, OH, 2006. (Cited in Chapters 6 & 9.)

[125] J. Bourgain, E. Lindenstrauss, Ph. Michel and A. Venkatesh, *Some effective results for* $\times a \times b$, Ergodic Theory Dynam. Systems 29 (2009), 1705–1722. (Cited in Chapter 2.)

[126] F. Bowman, Introduction to Bessel functions, Dover Publications Inc., New York, 1958. (Cited in Chapter 3.)

[127] D. Boyd, *Transcendental numbers with badly distributed powers*, Proc. Amer. Math. Soc. 23 (1969), 424–427. (Cited in Chapter 2.)

[128] ———, *Salem numbers of degree four have periodic expansions.* In: Théorie des Nombres (Quebec, PQ, 1987), pp. 57–64, de Gruyter, Boston, MA, 1989. (Cited in Chapter 9.)

[129] ———, *On the beta expansion for Salem numbers of degree* 6, Math. Comp. 65 (1996), 861–875. (Cited in Chapter 9.)

[130] R.P. Brent, A. J. van der Poorten and H. J. J. te Riele, *A comparative study of algorithms for computing continued fractions of algebraic numbers.* In: Algorithmic Number Theory (Talence, 1996), 3547, Lecture Notes in Comput. Sci., 1122, Springer, Berlin, 1996. (Cited in Chapter 9.)

[131] R. Broderick, Y. Bugeaud, L. Fishman, D. Kleinbock and B. Weiss, *Schmidt's game, fractals, and numbers normal to no base*, Math. Res. Lett. 17 (2010), 307–321. (Cited in Chapter 7 and Appendix C.)

[132] R. Broderick, L. Fishman and A. Reich, *Intrinsic approximation on Cantor-like sets, a problem of Mahler*, Moscow J. Comb. Number Theory 1 (2011), 291–300. (Cited in Chapter 7.)

[133] G. Brown, *Normal numbers and dynamical systems*, Probabilistic and stochastic methods in analysis, with applications (Il Ciocco, 1991), 207–216, NATO Adv. Sci. Inst. Ser. C Math. Phys. Sci., 372, Kluwer Academic Publishers, Dordrecht, 1992. (Cited in Chapter 6.)

[134] G. Brown, M. S. Keane, W. Moran and C. E. M. Pearce, *An inequality, with applications to Cantor measures and normal numbers*, Mathematika 35 (1988), 87–94. (Cited in Chapter 6.)

[135] G. Brown and W. Moran, *Schmidt's conjecture on normality for commuting matrices*, Invent. Math. 111 (1993), 449–463. (Cited in Chapter 6.)

[136] G. Brown, W. Moran and C. E. M. Pearce, *Riesz products and normal numbers*, J. London Math. Soc. 32 (1985), 12–18. (Cited in Chapter 6.)

[137] ———, *A decomposition theorem for numbers in which the summands have prescribed normality properties*, J. Number Theory 24 (1986), 259–271. (Cited in Chapter 6.)

[138] ———, *Riesz products, Hausdorff dimension and normal numbers*, Math. Proc. Cambridge Philos. Soc. 101 (1987), 529–540. (Cited in Chapter 6.)

[139] G. Brown, W. Moran and A. D. Pollington, *Normality to noninteger bases*, C. R. Acad. Sci. Paris 316 (1993), 1241–1244. (Cited in Chapter 6.)

[140] ———, *Normality with respect to powers of a base*, Duke Math. J. 88 (1997), 247–265. (Cited in Chapter 6.)

[141] ———, *Schmidt's conjecture on normality for dimension two*, J. Fourier Anal. Appl. 8 (2002), 427–441. (Cited in Chapter 6.)

[142] G. Brown and Q. Yin, *β-expansions and frequency of zero*, Acta Math. Hungar. 84 (1999), 275–291. (Cited in Chapter 9.)

[143] N. G. de Bruijn, *A combinatorial problem*, Nederl. Akad. Wet. Proc. 49 (1946), 758–764. (Cited in Chapter 5.)

[144] Y. Bugeaud, *Nombres de Liouville et nombres normaux*, C. R. Acad. Sci. Paris 335 (2002), 117–120. (Cited in Chapter 7.)

[145] ———, *Linear mod one transformations and the distribution of fractional parts $\{\xi(p/q)^n\}$*, Acta Arith. 114 (2004), 301–311. (Cited in Chapter 3.)

[146] ———, Approximation by Algebraic Numbers. Cambridge Tracts in Math. 160, Cambridge, 2004. (Cited in Chapters 2 & 7 and Appendices C, D & E.)

[147] ———, *On the b-ary expansion of an algebraic number*, Rend. Sem. Mat. Univ. Padova 118 (2007), 217–233. (Cited in Chapters 6 & 8.)

[148] ———, *Diophantine approximation and Cantor sets*, Math. Ann. 341 (2008), 677–684. (Cited in Chapters 7 & 8.)

[149] ———, *An explicit lower bound for the block complexity of an algebraic number*, Atti Accad. Naz. Lincei Cl. Sci. Fis. Mat. Natur. Rend. Lincei (9) Mat. Appl. 19 (2008), 229–235. (Cited in Chapter 8.)

[150] ———, *On the convergents to algebraic numbers.* In: Analytic Number Theory, 133–143, Cambridge University Press, Cambridge, 2009. (Cited in Chapter 9.)

[151] ———, *On the β-expansion of an algebraic number in an algebraic base β*, Integers 9 (2009), 215–226. (Cited in Chapter 9.)

[152] ———, *On sequences $(a_n\xi)_{n\geq1}$ converging modulo* 1, Proc. Amer. Math. Soc. 137 (2009), 2609–2612. (Cited in Chapter 2.)

[153] ———, *Quantitative versions of the Subspace Theorem and applications*, J. Théor. Nombres Bordeaux 23 (2011), 35–57. (Cited in Appendix E.)

[154] ________, *On the irrationality exponent of the Thue–Morse–Mahler numbers.* Ann. Inst. Fourier (Grenoble) 61 (2011). (Cited in Chapter 8.)

[155] ________, *On the expansions of a real number to several integer bases*, Rev. Mat. Iberoam. To appear. (Cited in Chapters 4, 5, 6, 7 & 10.)

[156] ________, *Continued fractions with low complexity: Transcendence measures and quadratic approximation*, Compos. Math. To appear. (Cited in Chapter 9.)

[157] ________, *Variations around a problem of Mahler and Mendès France*, J. Austral. Math. Soc. To appear. (Cited in Chapter 10 and Appendix D.)

[158] ________, *Continued fractions of transcendental numbers.* Preprint. (Cited in Chapter 9.)

[159] Y. Bugeaud, M. Cipu and M. Mignotte, *On the representation of Fibonacci and Lucas numbers in an integer base*, Ann. Sci. Math. Québec. To appear. (Cited in Chapter 6.)

[160] Y. Bugeaud, M. Drmota and B. de Mathan, *On a mixed Littlewood conjecture in Diophantine approximation*, Acta Arith. 128 (2007), 107–124. (Cited in Chapter 2.)

[161] Y. Bugeaud and A. Dubickas, *Fractional parts of powers and Sturmian words*, C.R. Acad. Sci. Paris 341 (2005), 69–74. (Cited in Chapter 3.)

[162] ________, *On a problem of Mahler and Szekeres on approximation by roots of integers*, Mich. Math. J. 56 (2008), 703–715. (Cited in Chapter 2.)

[163] Y. Bugeaud and J.-H. Evertse, *On two notions of complexity of algebraic numbers*, Acta Arith. 133 (2008), 221–250. (Cited in Chapter 8 and Appendix E.)

[164] Y. Bugeaud, A. Haynes and S. Velani, *Metric considerations concerning the mixed Littlewood Conjecture*, Intern. J. Number Theory 7 (2011), 593–609. (Cited in Chapter 2.)

[165] Y. Bugeaud, D. Krieger and J. Shallit, *Morphic and automatic words: maximal blocks and Diophantine approximation*, Acta Arith. 149 (2011), 181–199. (Cited in Chapter 8.)

[166] Y. Bugeaud and F. Luca, *On the period of the continued fraction expansion of* $\sqrt{2^{2n+1}+1}$, Indag. Math. 16 (2005), 21–35. (Cited in Chapter 9.)

[167] Y. Bugeaud and N. Moshchevitin, *Badly approximable numbers and Littlewood-type problems*, Math. Proc. Cambridge Phil. Soc. 150 (2011), 215–226. (Cited in Chapter 2.)

[168] ________, *On fractional parts of powers of real numbers close to* 1, Math. Z. To appear. (Cited in Chapters 2 & 10.)

[169] P. Bundschuh, *Über eine Klasse reeller transzendenter Zahlen mit explicit angebbarer g-adischer und Kettenbruch-Entwicklung*, J. Reine Angew. Math. 318 (1980), 110–119. (Cited in Chapter 7.)

[170] ________, *Generalization of a recent irrationality result of Mahler*, J. Number Theory 19 (1984), 248–253. (Cited in Chapter 10.)

[171] P. Bundschuh and A. Pethő, *Zur Transzendenz gewisser Reihen*, Monatsh. Math. 104 (1987), 199–223. (Cited in Chapter 8.)

[172] P. Bundschuh, P. J.-S. Shiue and X. Y. Yu, *Transcendence and algebraic independence connected with Mahler type numbers*, Publ. Math. Debrecen 56 (2000), 121–130. (Cited in Chapter 10.)

[173] C. S. Calude, Information and Randomness. An Algorithmic Perspective. Texts in Theoret. Comput. Sci., an EATCS series, Springer, Berlin, 2002. (Cited in Chapter 5.)

[174] C. S. Calude, M. J. Dinneen and C.-K. Shu, *Computing a glimpse of randomness*, Experiment. Math. 11 (2002), 361–370. (Cited in Chapter 5.)

[175] C. Calude and H. Jürgensen, *Randomness as an invariant for number representations*. In: Results and Trends in Theoretical Computer Science (Graz, 1994), 44–66, Lecture Notes in Comput. Sci., 812, Springer, Berlin, 1994. (Cited in Chapter 5.)

[176] D. G. Cantor, *On powers of real numbers* (mod 1), Proc. Amer. Math. Soc. 16 (1965), 791–793. (Cited in Chapter 2.)

[177] ———, *On power series with only finitely many coefficients* (mod 1)*: Solution of a problem of Pisot and Salem*, Acta Arith. 34 (1977), 43–55. (Cited in Chapter 2.)

[178] D. Cass, *Integer parts of powers of quadratic units*, Proc. Amer. Math. Soc. 101 (1987), 610–612. (Cited in Chapter 3.)

[179] J. Cassaigne, *Sequences with grouped factors*. In: DLT'97, Developments in Language Theory III, Thessaloniki, Aristotle University of Thessaloniki, 1998, pp. 211–222. (Cited in Chapter 8 and Appendix A.)

[180] J. W. S. Cassels, *On a paper of Niven and Zuckerman*, Pacific J. Math 2 (1952), 555–557. (Cited in Chapter 4.)

[181] ———, An Introduction to Diophantine Approximation. Cambridge Tracts in Math. and Math. Phys., vol. 99, Cambridge University Press, Cambridge, 1957. (Cited in Chapter 2 and Appendix D.)

[182] ———, *On a problem of Steinhaus about normal numbers*, Colloq. Math. 7 (1959), 95–101. (Cited in Chapter 6.)

[183] E. Cesaratto and B. Vallée, *Hausdorff dimension of real numbers with bounded digit averages*, Acta Arith. 125 (2006), 115–162. (Cited in Chapter 7.)

[184] D. G. Champernowne, *The construction of decimals normal in the scale of ten*, J. London Math. Soc. 8 (1933), 254–260. (Cited in Chapter 4.)

[185] K. T. Chang, *A note on normal numbers*, Nanta Math. 9 (1976), 70–72. (Cited in Chapter 4.)

[186] D. P. Chi and D. Y. Kwon, *Sturmian words, β-shifts, and transcendence*, Theoret. Comput. Sci. 321 (2004), 395–404. (Cited in Chapter 9.)

[187] G. H. Choe, T. Hamachi and H. Nakada, *Mod 2 normal numbers and skew products*, Studia Math. 165 (2004), 53–60. (Cited in Chapter 4.)

[188] G. Choquet, *Répartition des nombres $k(3/2)^n$; et ensembles associés*, C. R. Acad. Sci. Paris 290 (1980), A575–A580. (Cited in Chapter 3.)

[189] ———, *Algorithmes adaptés aux suites $(k\theta^n)$ et aux chaînes associées*, C. R. Acad. Sci. Paris 290 (1980), A719–A724. (Cited in Chapter 3.)

[190] ———, *θ-jeux récursifs et application aux suites $(k\theta^n)$; solénoïdes de $T^{\mathbb{Z}}$*, C. R. Acad. Sci. Paris 290 (1980), A863–A868. (Cited in Chapter 3.)

[191] ———, *Construction effective de suites* $(k(3/2)^n)$. *Étude des mesures (3/2)-stables*, C. R. Acad. Sci. Paris 291 (1980), A69–A74. (Cited in Chapter 3.)

[192] ———, *Les fermés (3/2)-stables de T; structure des fermés dénombrables; applications arithmétiques*, C. R. Acad. Sci. Paris 291 (1980), A239–A244. (Cited in Chapter 3.)

[193] ———, *θ-fermés; θ-chaînes et θ-cycles (pour* $\theta = 3/2$*)*, C. R. Acad. Sci. Paris 292 (1981), 5–10. (Cited in Chapters 3 & 9.)

[194] ———, *θ-fermés et dimension de Hausdorff. Conjectures de travail. Arithmétique des θ-cycles (où* $\theta = 3/2$*)*, C. R. Acad. Sci. Paris 292 (1981), 339–344. (Cited in Chapters 3 & 9.)

[195] S. Chowla and M. J. De Leon, *On a conjecture of Littlewood*, Norske Vid. Selsk. Forh. (Trondheim) 41 (1968), 45–47. (Cited in Chapter 3.)

[196] J. Cigler, *Ein gruppentheoretisches Analogon zum Begriff der normalen Zahl*, J. Reine Angew. Math. 206 (1961), 5–8. (Cited in Chapter 6.)

[197] A. Cobham, *On the Hartmanis–Stearns problem for a class of tag machines*. In: Conference Record of 1968 Ninth Annual Symposium on Switching and Automata Theory, Schenectady, NY (1968), 51–60. (Cited in Chapter 8.)

[198] ———, *Uniform tag sequences*, Math. Systems Theory 6 (1972), 164–192. (Cited in Chapter 8.)

[199] H. J. Cohen and F. Supnick, *Concerning real numbers whose powers have nonintegral differences*, Proc. Amer. Math. Soc. 14 (1963), 626–627. (Cited in Chapter 2.)

[200] C. M. Colebrook, *The Hausdorff dimension of certain sets of nonnormal numbers*, Michigan Math. J. 17 (1970), 103–116. (Cited in Chapters 4 & 9.)

[201] C. M. Colebrook and J. H. B. Kemperman, *On non-normal numbers*, Indag. Math. 30 (1968), 1–11. (Cited in Chapters 6 & 10.)

[202] K. J. Compton, *On rich words*. In: Combinatorics on Words (Waterloo, Ont., 1982), 39–61, Academic Press, Toronto, ON, 1983. (Cited in Chapter 4.)

[203] A. H. Copeland and P. Erdős, *Note on normal numbers*, Bull. Amer. Math. Soc. 52 (1946), 857–860. (Cited in Chapter 4.)

[204] J. Coquet, *Ensembles normaux et équirépartition complète*, Acta Arith. 39 (1981), 49–52. (Cited in Chapter 4.)

[205] J. G. van der Corput, *Diophantische Ungleichungen. I. Zur Gleichverteilung Modulo Eins*, Acta Math. 56 (1931), 373–456. (Cited in Chapter 1.)

[206] ———, *Sur la discrépance modulo un. I, II, III*, Proc. Akad. Wet. Amsterdam 42 (1939), 476–486, 554–565, 713–722. (Cited in Chapter 1.)

[207] P. Corvaja and U. Zannier, *Some new applications of the subspace theorem*, Compositio Math. 131 (2002), 319–340. (Cited in Chapter 8.)

[208] ———, *On the rational approximations to the powers of an algebraic number: Solution of two problems of Mahler and Mendès France*, Acta Math. 193 (2004), 175–191. (Cited in Chapter 9.)

[209] ———, *On the length of the continued fraction for values of quotients of power sums*, J. Théor. Nombres Bordeaux 17 (2005), 737–748. (Cited in Chapter 9.)

[210] E. Curry, *Multidimensional radix representations and hot spot theorem*. In: Tapas in Experimental Mathematics, 181190, Contemp. Math., 457, American Mathematical Society, Providence, RI, 2008. (Cited in Chapter 4.)

[211] K. Dajani and A. Fieldsteel, *Equipartition of interval partitions and an application to number theory*, Proc. Amer. Math. Soc. 129 (2001), 3453–3460. (Cited in Chapter 6.)

[212] K. Dajani and C. Kraaikamp, Ergodic Theory of Numbers, Carus Math Monogr., 29, Mathematical Association of America, Washington, DC, 2002, x + 190 pp. (Cited in Chapter 9 and Appendices C & D.)

[213] L. V. Danilov, *Certain classes of transcendental numbers*, Mat. Zametki 12 (1972), 149–154 (in Russian). English translation in Math. Notes 12 (1972), 524–527. (Cited in Chapter 7.)

[214] H. Davenport, *A remark on continued fractions*, Michigan Math. J. 11 (1964), 343–344. (Cited in Chapter 9.)

[215] H. Davenport and P. Erdős, *Note on normal decimals*, Canadian J. Math. 4 (1952), 58–63. (Cited in Chapter 4.)

[216] H. Davenport, P. Erdős and W. J. LeVeque, *On Weyl's criterion for uniform distribution*, Michigan Math. J. 10 (1963), 311–314. (Cited in Chapter 1.)

[217] H. Davenport and K. F. Roth, *Rational approximations to algebraic numbers*, Mathematika 2 (1955), 160–167. (Cited in Appendix E.)

[218] J. L. Davison, *A series and its associated continued fraction*, Proc. Amer. Math. Soc. 63 (1977), 29–32. (Cited in Chapter 7.)

[219] J.-M. De Koninck and I. Kátai, *On a problem of normal numbers raised by Igor Shparlinski*, Bull. Austr. Math. Soc. 84 (2011), 337–349. (Cited in Chapter 4.)

[220] ———, *Normal numbers created from primes and polynomials*. Preprint. (Cited in Chapter 4.)

[221] ———, *Construction of normal numbers by classified prime divisors of integers*, Funct. Approx. Comment. Math. 45 (2011), 231–253. (Cited in Chapter 4.)

[222] ———, *Some new methods for constructing normal numbers*. Preprint. (Cited in Chapter 4.)

[223] A. Decomps-Guilloux and M. Grandet-Hugot, *Nouvelles caractérisations des nombres de Pisot et de Salem*. In: Groupe d'Étude en Théorie Analytique des Nombres, 1ère–2ème années: 1984–1985, Exp No. 31, 19 pp., Secrétariat Math., Paris, 1985. (Cited in Chapter 2.)

[224] ———, *Nouvelles caractérisations des nombres de Pisot et de Salem*, Acta Arith. 50 (1988), 155–170. (Cited in Chapter 2.)

[225] M. Dekking, *Transcendance du nombre de Thue–Morse*, C. R. Acad. Sci. Paris 285 (1977), A157–A160. (Cited in Chapter 8.)

[226] M. Denker and K. F. Krämer, *Upper and lower class results for subsequences of the Champernowne number*. In: Ergodic Theory and Related

Topics, III (Güstrow, 1990), 83–89, Lecture Notes in Math., 1514, Springer, Berlin, 1992. (Cited in Chapter 4.)

[227] C. Doche, M. Mendès France and J.-J. Ruch, *Equidistribution modulo 1 and Salem numbers*, Funct. Approx. Comment. Math. 39 (2008), 261–271. (Cited in Chapter 3.)

[228] Y. Dodge and G. Melfi, *Random number generators and rare events in the continued fraction of π*, J. Stat. Comput. Simul. 75 (2005), 189–197. (Cited in Chapter 9.)

[229] D. Doty, J. H. Lutz and S. Nandakumar, *Finite-state dimension and real arithmetic*, Inform. and Comput. 205 (2007), 1640–1651. (Cited in Chapter 4.)

[230] R. G. Downey and D. R. Hirschfeldt, Algorithmic Randomness and Complexity. Theory and Applications of Computability, Springer, New York, 2010. (Cited in Chapter 5.)

[231] F. Dress, *Intersections d'ensembles normaux*, J. Number Theory 2 (1970), 352–362. (Cited in Chapter 4.)

[232] M. Drmota and R. Tichy, Sequences, Discrepancies and Applications. Lecture Notes in Math. 1651, Springer, Berlin, 1997. (Cited in Preface and Chapters 1 & 4.)

[233] A. Dubickas, *A lower bound for the quantity $\|(3/2)^k\|$*, Russian Math. Surveys 45 (1990), 163–164. (Cited in Chapter 3.)

[234] ———, *A note on powers of Pisot numbers*, Publ. Math. Debrecen 56 (2000), 141–144. (Cited in Chapter 3.)

[235] ———, *Integer parts of powers of Pisot and Salem numbers*, Arch. Math. (Basel) 79 (2002), 252–257. (Cited in Chapter 3.)

[236] ———, *Sequences with infinitely many composite numbers*, In: Analytic and Probabilistic Methods in Number Theory (Palanga, 2001), 57–60, TEV, Vilnius, 2002. (Cited in Chapter 3.)

[237] ———, *There are infinitely many limit points of the fractional parts of powers*, Proc. Indian Acad. Sci. Math. Sci. 115 (2005), 4391–4397. (Cited in Chapter 2.)

[238] ———, *Arithmetical properties of powers of algebraic numbers*, Bull. London Math. Soc. 38 (2006), 70–80. (Cited in Chapter 3.)

[239] ———, *On the fractional parts of the natural powers of a fixed number*, Sibirsk. Mat. Zh. 47 (2006), 1071–1075. English translation in Siberian Math. J. 47 (2006), 879–882. (Cited in Chapter 2.)

[240] ———, *On the distance from a rational power to the nearest integer*, J. Number Theory 117 (2006), 222–239. (Cited in Chapter 3.)

[241] ———, *On the limit points of the fractional parts of powers of Pisot numbers*, Arch. Math. (Brno) 42 (2006), 151–158. (Cited in Chapter 3.)

[242] ———, *Even and odd integral parts of powers of a real number*, Glasg. Math. J. 48 (2006), 331–336. (Cited in Chapter 3.)

[243] ———, *On the fractional parts of lacunary sequences*, Math. Scand. 99 (2006), 136–146. (Cited in Chapter 2.)

[244] ———, *Arithmetical properties of linear recurrent sequences*, J. Number Theory 122 (2007), 142–150. (Cited in Chapter 3.)

[245] ———, *On a sequence related to that of Thue–Morse and its applications*, Discrete Math. 307 (2007), 1082–1093. (Cited in Chapter 3.)

[246] ———, *On the powers of some transcendental numbers*, Bull. Austral. Math. Soc. 76 (2007), 433–440. (Cited in Chapter 2.)

[247] ———, *An approximation by lacunary sequence of vectors*, Combin. Probab. Comput. 17 (2008), 339–345. (Cited in Chapter 2.)

[248] ———, *On the powers of 3/2 and other rational numbers*, Math. Nachr. 281 (2008), 951–958. (Cited in Chapter 3.)

[249] ———, *On the limit points of $(a_n\xi)_{n=1}^{\infty}$ mod 1 for slowly increasing integer sequences $(a_n)_{n=1}^{\infty}$*, Proc. Amer. Math. Soc. 137 (2009), 449–456. (Cited in Chapter 2.)

[250] ———, *An approximation property of lacunary sequences*, Israel J. Math. 170 (2009), 95–111. (Cited in Chapters 3 & 10.)

[251] ———, *On the fractional parts of rational powers*, Int. J. Number Theory 5 (2009), 1037–1048. (Cited in Chapter 3.)

[252] ———, *Prime and composite integers close to powers of a number*, Monatsh. Math. 156 (2009), 271–284. (Cited in Chapter 3.)

[253] ———, *Distribution of some sequences modulo 1*. In: Number Theory and Applications. Proc. of the Intern. Conf. on Number Theory and Cryptography (eds. S. D. Adhikari and B. Ramakrishnan), Hindustan Book Agency, India (2009), 33–41. (Cited in Chapter 2.)

[254] ———, *Powers of a rational number modulo* 1 *cannot lie in a small interval*, Acta Arith. 137 (2009), 233–239. (Cited in Chapter 3.)

[255] ———, *On integer sequences generated by linear maps*, Glasgow Math. J. 51 (2009), 243–252. (Cited in Chapter 9.)

[256] ———, *Squares and cubes in Sturmian sequences*, RAIRO Theoret Inform. and Appl. 43 (2009), 615–624. (Cited in Appendix A.)

[257] ———, *On the universality of a sequence of powers modulo 1*. In: Analytic Number Theory and Related Areas, October 17–19, 2007 (ed. Koichi Kawada), RIMS Kokyuroku 1665, RIMS, Kyoto University, Kyoto, Japan (2009), 1–4. (Cited in Chapter 2.)

[258] ———, *Binary words with a given Diophantine exponent*, Theor. Computer Science 410 (2009), 5191–5195. (Cited in Chapter 7.)

[259] ———, *Powers of rational numbers modulo 1 lying in short intervals*, Results in Math. 57 (2010), 23–31. (Cited in Chapter 3.)

[260] ———, *On the distribution of powers of a complex number*, Archiv Math. 95 (2010), 151–160. (Cited in Chapter 3.)

[261] ———, *The lonely runner problem for many runners*, Glas. Mat. 46 (2011), 25–30. (Cited in Chapter 2.)

[262] ———, *On β-expansions of unity for rational and transcendental numbers β*, Math. Slovaca 61 (2011), 705–716. (Cited in Chapter 9.)

[263] A. Dubickas and J. Jankauskas, *On the intersection of infinite geometric and arithmetic progressions*, Bull. Braz. Math. Soc. 41 (2010), 551–566. (Cited in Chapter 2.)

[264] A. Dubickas and M. J. Mossinghoff, *Lower bounds for Z-numbers*, Math. Comp. 78 (2009), 1837–1851. (Cited in Chapter 3.)

[265] A. Dubickas and A. Novikas, *Integer parts of powers of rational numbers*, Math. Z. 251 (2005), 635–648. (Cited in Chapter 3.)

[266] J.-M. Dumont and A. Thomas, *Une modification multiplicative des nombres g normaux*, Ann. Fac. Sci. Toulouse Math. 8 (1986/87), 367–373. (Cited in Chapter 4.)

[267] ———, *Modifications de nombres normaux par des transducteurs*, Acta Arith. 68 (1994), 153–170. (Cited in Chapter 4.)

[268] A. Durand, *Sets with large intersection and ubiquity*, Math. Proc. Cambridge Philos. Soc. 144 (2008), 119–144. (Cited in Chapter 7.)

[269] A. Durner, *Distribution measures and Hausdorff dimensions*, Forum Math. 9 (1997), 687–705. (Cited in Chapters 4 & 9.)

[270] J. Dutka, *The square root of* 2 *to* 1,000,000 *decimals*, Math. Comp. 25 (1971), 927–930. (Cited in Chapter 8.)

[271] D. Duverney, Ke. Nishioka, Ku. Nishioka and I. Shiokawa, *Transcendence of Jacobi's theta series*, Proc. Japan Acad. Ser. A Math. Sci. 72 (1996), 202–203. (Cited in Chapter 8.)

[272] H. G. Eggleston, *The fractional dimension of a set defined by decimal properties*, Quart. J. Math. Oxford Ser. 20 (1949), 31–36. (Cited in Chapter 7.)

[273] ———, *Sets of fractional dimension which occur in some problems in number theory*, Proc. London Math. Soc. 54 (1951), 42–93. (Cited in Chapter 7.)

[274] H. Ehlich, *Die positiven Lösungen der Gleichung* $y^a - [y^a] = y^b - [y^b] = y^c - [y^c]$, Math. Z. 76 (1961), 1–4. (Cited in Chapter 2.)

[275] A. Ehrenfeucht, K.P. Lee and G. Rozenberg, *Subword complexities of various classes of deterministic developmental languages without interactions*, Theor. Comput. Sci. 1 (1975), 59–75. (Cited in Chapter 10.)

[276] M. Einsiedler, L. Fishman and U. Shapira, *Diophantine approximation on fractals*, Geom. Funct. Anal. 21 (2011), 14–35. (Cited in Chapters 7, 9 & 10.)

[277] M. Einsiedler and D. Kleinbock, *Measure rigidity and p-adic Littlewood-type problems*, Compositio Math. 143 (2007), 689–702. (Cited in Chapter 2.)

[278] S.I. El-Zanati and W.R. R. Transue, *On dynamics of certain Cantor sets*, J. Number Theory 36 (1990), 246–253. (Cited in Chapters 4 & 6.)

[279] P. Erdős, *Problems and results on Diophantine approximations. II.* In: Répartition Modulo 1 (Actes Colloq., Marseille-Luminy, 1974), pp. 89–99, Lecture Notes in Math., Vol. 475, Springer, Berlin, 1975. (Cited in Chapter 2.)

[280] ———, *Some unconventional problems in number theory*, Math. Mag. 52 (1979), 67–70. (Cited in Chapter 6.)

[281] P. Erdős and S. J. Taylor, *On the set of points of convergence of a lacunary trigonometric series and the equidistribution properties of related sequences*, Proc. London Math. Soc. 7 (1957), 598–615. (Cited in Chapter 2.)

[282] L. Euler, *De fractionibus continuis*, Commentarii Acad. Sci. Imperiali Petropolitanae 9 (1737). (Cited in Appendix D.)

[283] G. Everest, A. van der Poorten, I. Shparlinski and T. Ward, Recurrence Sequences, Math. Surveys Monogr., 104, American Mathematical Society, Providence, RI, 2003. (Cited in Chapter 9.)

[284] J.-H. Evertse and H.P. Schlickewei, *A quantitative version of the Absolute Subspace Theorem*, J. Reine Angew. Math. 548 (2002), 21–127. (Cited in Appendix E.)

[285] C. Faivre, *On decimal and continued fraction expansions of a real number*, Acta Arith. 82 (1997), 119–128. (Cited in Chapter 9.)

[286] ———, *A central limit theorem related to decimal and continued fraction expansion*, Arch. Math. (Basel) 70 (1998), 455–463. (Cited in Chapter 9.)

[287] ———, *On calculating a continued fraction expansion from a decimal expansion*, Acta Sci. Math. (Szeged) 67 (2001), 505–519. (Cited in Chapter 9.)

[288] K. Falconer, The Geometry of Fractal Sets, Cambridge Tracts in Mathematics 85, Cambridge University Press, Cambridge, 1985. (Cited in Appendix C.)

[289] ———, Fractal Geometry: Mathematical Foundations and Applications, John Wiley & Sons, Chichester, 1990. (Cited in Chapter 7 and Appendix C.)

[290] ———, *Sets with large intersection properties*, J. London Math. Soc. 49 (1994), 267–280. (Cited in Chapter 7.)

[291] B. Faller and C.-E. Pfister, *A point is normal for almost all maps* $\beta x + \alpha$ mod 1 *or generalized* β*-transformations*, Ergodic Theory Dynam. Systems 29 (2009), 1529–1547. (Cited in Chapter 9.)

[292] A.-H. Fan, L. Liao and J.-H. Ma, *On the frequency of partial quotients of regular continued fractions*, Math. Proc. Cambridge Philos. Soc. 148 (2010), 179–192. (Cited in Chapter 9.)

[293] B. Farhi, *Sur l'ensemble exceptionnel de Weyl*, C. R. Math. Acad. Sci. Paris 343 (2006), 441–445. (Cited in Chapter 2.)

[294] D. Färm, *Applications of large intersection classes to non-typical points of expanding interval maps*, Licentiate Theses in Mathematical Sciences, University of Lund, 2009. (Cited in Chapter 7.)

[295] ———, *Simultaneously non-dense orbits under different expanding maps*, Dynamical Syst. 25 (2010), 531–545. (Cited in Chapter 9.)

[296] ———, *Simultaneously non-convergent frequencies of words in different expansions*, Monatsh. Math. 162 (2011), 409–427. (Cited in Chapter 7.)

[297] D. Färm and T. Persson, *Large intersection classes on fractals*, Nonlinearity 24 (2011), 1291–1309. (Cited in Chapter 7.)

[298] ———, *Non-typical points for* β*-shifts*. Preprint. (Cited in Chapter 9.)

[299] D. Färm, T. Persson and J. Schmeling, *Dimension of countable intersections of some sets arising in expansions in non-integer bases*, Fundamenta Math. 209 (2010), 157–176. (Cited in Chapter 9.)

[300] P. Fatou, *Séries trigonométriques et séries de Taylor*, Acta Math. 30 (1906), 335–400. (Cited in Chapter 2 and Appendix F.)

[301] J. Feldman, *A generalization of a result of R. Lyons about measures on* $[0,1)$, Israel J. Math. 81 (1993), 281–287. (Cited in Chapters 2 & 10.)

[302] J. Feldman and M. Smorodinsky, *Normal numbers from independent processes*, Ergodic Theory Dynam. Systems 12 (1992), 707–712. (Cited in Chapter 6.)

[303] N. I. Feldman and Yu. V. Nesterenko, Number Theory. IV. Transcendental Numbers, Encyclopaedia Math. Sci. 44, Springer, Berlin, 1998. (Cited in Chapter 8.)

[304] D.-J. Feng, *On the topology of polynomials with bounded integer coefficients.* Preprint. (Cited in Chapter 2.)

[305] S. Ferenczi and Ch. Mauduit, *Transcendence of numbers with a low complexity expansion*, J. Number Theory 67 (1997), 146–161. (Cited in Chapter 8 and Appendix A.)

[306] L. Fishman, *Schmidt's game on fractals*, Israel J. Math. 171 (2009), 77–92. (Cited in Chapter 7.)

[307] L. Flatto, *Z-numbers and* β*-transformations.* In: Symbolic Dynamics and its Applications, (ed. P. Walters), Contemp. Math. 135 (1992), American Mathematical Society, Providence, RI, 181–202. (Cited in Chapter 3.)

[308] L. Flatto, J. C. Lagarias and A. D. Pollington, *On the range of fractional parts* $\{\xi(p/q)^n\}$, Acta Arith. 70 (1995), 125–147. (Cited in Chapters 2 & 3.)

[309] C. Flye Sainte-Marie, *Solution au problème 48*, L'intermédiaire des mathématiciens 1 (1894), 107–110. (Cited in Chapter 5.)

[310] N. P. Fogg, Substitutions in Dynamics, Arithmetics and Combinatorics, (eds. V. Berthé, S. Ferenczi, C. Mauduit and A. Siegel), Lecture Notes in Mathematics, 1794, Springer, Berlin, 2002. (Cited in Appendix A.)

[311] W. Forman and H. N. Shapiro, *An arithmetic property of certain rational powers*, Comm. Pure Appl. Math. 20 (1967), 561–573. (Cited in Chapter 3.)

[312] J. N. Franklin, *Deterministic simulation of random processes*, Math. Comp. 17 (1963), 28–59. (Cited in Chapter 1.)

[313] O. Frostman, *Potentiel d'équilibre et capacité des ensembles avec quelques applications à la théorie des fonctions*, Meddel. Lunds Univ. Math. Sem. 3 (1935), 1–118. (Cited in Appendix C.)

[314] M. Fuchs, *Digital expansion of exponential sequences*, J. Théor. Nombres Bordeaux 14 (2002), 477–487. (Cited in Chapter 6.)

[315] H. Furstenberg, *Disjointness in ergodic theory, minimal sets, and a problem in Diophantine approximation*, Math. Systems Theory 1 (1967), 1–49. (Cited in Chapters 2 & 10.)

[316] ———, *Intersections of Cantor sets and transversality of semigroups.* In: Problems in Analysis (Sympos. Salomon Bochner, Princeton University, Princeton, NJ, 1969), p. 4159. Princeton University Press, Princeton, NJ, 1970. (Cited in Chapter 10.)

[317] I. Gál and L. Gál, *The discrepancy of the sequence* $\{(2^n x)\}$, Indag. Math. 26 (1964), 129–143. (Cited in Chapter 4.)

[318] J. Galambos, Representations of Real Numbers by Infinite Series, Lecture Notes in Math., Vol. 502, Springer, Berlin, 1976. (Cited in Chapter 9.)

[319] A. Gelfond, *On fractional parts of linear combinations of polynomials and exponential functions*, Rec. Math. [Mat. Sbornik] N.S. 9 (1941), 721–726 (in Russian). (Cited in Chapter 2.)

[320] I. Gheorghiciuc, *The subword complexity of a class of infinite binary words*, Adv. Appl. Math. 39 (2007), 237–259. (Cited in Chapter 8.)

[321] S. Glasner, *Almost periodic sets and measures on the torus*, Israel J. Math. 32 (1979), 161–172. (Cited in Chapter 8.)

[322] M. Goldstern, J. Schmeling and R. Winkler, *Metric, fractal dimensional and Baire results on the distribution of subsequences*, Math. Nachr. 219 (2000), 97–108. (Cited in Chapter 2.)

[323] ———, *Further Baire results on the distribution of subsequences*, Unif. Distrib. Theory 2 (2007), 127–149. (Cited in Chapter 2.)

[324] I. J. Good, *Normal recurring decimals*, J. London Math. Soc. 21 (1946), 167–169. (Cited in Chapter 5.)

[325] G. S. Goodman, *Statistical independence and normal numbers: an aftermath to Mark Kac's Carus Monograph*, Amer. Math. Monthly 106 (1999), 112–126. (Cited in Chapter 4.)

[326] A. Gorodnik and S. Kadyrov, *Algebraic numbers, hyperbolicity, and density modulo one*. Preprint. (Cited in Chapter 2.)

[327] P. J. Grabner, *On digit expansions with respect to second order linear recurring sequences*. In: Number-theoretic Analysis (Vienna, 1988–89), 58–64, Lecture Notes in Math., 1452, Springer, Berlin, 1990. (Cited in Chapter 9.)

[328] R. L. Graham and H. O. Pollak, *Note on a nonlinear recurrence related to $\sqrt{2}$*, Math. Mag. 43 (1970), 143–145. (Cited in Chapter 8.)

[329] G. Grisel, *Length of continued fractions in principal quadratic fields*, Acta Arith. 85 (1998), 35–49. (Cited in Chapter 9.)

[330] X. Gu, J.H. Lutz and P. Moser, *Dimensions of Copeland–Erdős sequences*, Inform. and Comput. 205 (2007), 1317–1333. (Cited in Chapter 4.)

[331] L. Habsieger, *Explicit lower bounds for* $\|(3/2)^k\|$, Acta Arith. 106 (2003), 299–309. (Cited in Chapter 3.)

[332] H. A. Hanson, *Some relations between various types of normality of numbers*, Canadian J. Math. 6 (1954), 477–485. (Cited in Chapter 4.)

[333] G. H. Hardy, *A problem of Diophantine approximation*, J. Indian Math. Soc. 11 (1919), 162–166. (Cited in Chapters 2 & 10.)

[334] G. H. Hardy and E. M. Wright, An Introduction to the Theory of Numbers, 5th edition, Clarendon Press, Oxford, 1979. (Cited in Chapters 1, 3 & 4 and Appendix D.)

[335] G. Harman, Metric Number Theory, LMS Monographs New Series, vol. 18, Clarendon Press, Oxford, 1998. (Cited in Preface.)

[336] S. Harrap and A. Haynes, *The mixed Littlewood conjecture for pseudo-absolute values*. Preprint. (Cited in Chapter 2.)

[337] M. Hata, *Note on the fractional parts of $\lambda\theta^n$*, Acta Arith. 120 (2005), 153–157. (Cited in Chapter 2.)

[338] ———, *A lower estimate for* $||e^n||$, J. Number Theory 130 (2010), 1685–1704. (Cited in Chapter 2.)

[339] F. Hausdorff, *Dimension und äusseres Maß*, Math. Ann. 79 (1919), 157–179. (Cited in Appendix C.)

[340] G. Helmberg, *On subsequences of normal sequences.* In: Proceedings of the Conf. on Ergodic Theory and Related Topics, II (Georgenthal, 1986), 70–71, Teubner-Texte Math., 94, Teubner, Leipzig, 1987. (Cited in Chapter 4.)

[341] H. Helson and J.-P. Kahane, *A Fourier method in Diophantine problems*, J. Analyse Math. 15 (1965), 245–262. (Cited in Chapter 2.)

[342] P. Hertling, *Disjunctive* ω*-words and real numbers*, J. UCS 2 (1996), 549–568. (Cited in Chapters 4, 5, 6 & 10.)

[343] ———, *The Kolmogorov complexity of real numbers*, Theoret. Comput. Sci. 284 (2002), 455–466. (Cited in Chapter 5.)

[344] ———, *Simply normal numbers to different bases*, J. UCS 8 (2002), 235–242. (Cited in Chapter 4.)

[345] D. Hilbert, *Beweis für die Darstellbarkeit der ganzen Zahlen durch eine feste Anzahl* n^{ter} *Potenzen (Waringsches Problem)*, Math. Ann. 67 (1909), 281–300. (Cited in Chapter 3.)

[346] M. Hochman and P. Shmerkin, *Local entropy averages and projections of fractal measures*, Ann. of Math. To appear. (Cited in Chapters 2 & 10.)

[347] F. Hofbauer, β*-shifts have unique maximal measure*, Monatsh. Math. 85 (1978), 189–198. (Cited in Chapter 9.)

[348] B. Host, *Nombres normaux, entropie, translations*, Israel J. Math. 91 (1995), 419–428. (Cited in Chapters 2 & 10.)

[349] B. Host, *Some results of uniform distribution in the multidimensional torus*, Ergodic Theory Dynam. Systems 20 (2000), 439–452. (Cited in Chapters 2 & 10.)

[350] J. Hyde, V. Laschos, L. Olsen, I. Petrykiewicz and A. Shaw, *Iterated Cesàro averages, frequencies of digits, and Baire Category*, Acta Arith. 144 (2010), 287–293. (Cited in Chapter 4.)

[351] M. Iosifescu and C. Kraaikamp, Metrical theory of continued fractions, Mathematics and its Applications 547, Kluwer Academic Publishers, Dordrecht, 2002. (Cited in Appendix D.)

[352] S. Ito and I. Shiokawa, *A construction of* β*-normal sequences*, J. Math. Soc. Japan 27 (1975), 20–23. (Cited in Chapter 9.)

[353] V. Jarník, *Zur metrischen Theorie der diophantischen Approximationen*, Prace Mat.-Fiz. 36 (1928/29), 91–106. (Cited in Chapter 7.)

[354] J. L. Jensen *On lower bounded orbits of the times-q map*, Unif. Distrib. Theory 6 (2011), 157–175. (Cited in Chapter 7.)

[355] A. S. A. Johnson, *Measures on the circle invariant under multiplication by a nonlacunary subsemigroup of the integers*, Israel J. Math. 77 (1992), 211–240. (Cited in Chapters 2 & 10.)

[356] A. Johnson and D. J. Rudolph, *Convergence under* $\times_q$ *of* $\times_p$ *invariant measures on the circle*, Adv. Math. 115 (1995), 117–140. (Cited in Chapters 2 & 10.)

[357] S.-M. Jung and B. Volkmann, *Remarks on a paper of Wagner*, J. Number Theory 56 (1996), 329–335. (Cited in Chapters 5 & 6.)

[358] J.-P. Kahane and R. Salem, *Distribution modulo 1 and sets of uniqueness*, Bull. Amer. Math. Soc. 70 (1964), 259–261. (Cited in Chapter 7.)

[359] T. Kamae, *Subsequences of normal sequences*, Israel J. Math. 16 (1973), 121–149. (Cited in Chapters 4 & 10.)

[360] ———, *Cyclic extensions of odometer transformations and spectral disjointness*, Israel J. Math. 59 (1987), 41–63. (Cited in Chapter 6.)

[361] ———, *Number-theoretic problems involving two independent bases.* In: Number Theory and Cryptography (Sydney, 1989), 196–203, London Math. Soc. Lecture Note Ser., 154, Cambridge University Press, Cambridge, 1990. (Cited in Chapter 3.)

[362] T. Kamae and B. Weiss, *Normal numbers and selection rules*, Israel J. Math. 21 (1975), 101–110. (Cited in Chapter 4.)

[363] H. Kaneko, *Distribution of geometric sequences modulo 1*, Results Math. 52 (2008), 91–109. (Cited in Chapters 2 & 3.)

[364] ———, *Limit points of fractional parts of geometric sequences*, Unif. Distrib. Theory 4 (2009), 1–37. (Cited in Chapter 3.)

[365] ———, *On normal numbers and powers of algebraic numbers*, Integers 10 (2010), 31–64. (Cited in Chapter 3.)

[366] ———, *On the binary digits of algebraic numbers*, J. Austral. Math. Soc. 89 (2010), 233–244. (Cited in Chapter 8.)

[367] ———, *Algebraic independence of real numbers with low density of nonzero digits.* Preprint. (Cited in Chapter 8.)

[368] ———, *On the number of digit changes in base-b expansions of algebraic numbers.* Preprint. (Cited in Chapter 8.)

[369] H. Kano, *A remark on Wagner's ring of normal numbers*, Arch. Math. (Basel) 60 (1993), 46–50. (Cited in Chapter 6.)

[370] ———, *General constructions of normal numbers of Korobov type*, Osaka J. Math. 30 (1993), 909–919. (Cited in Chapter 5.)

[371] H. Kano and I. Shiokawa, *Rings of normal and nonnormal numbers*, Israel J. Math. 84 (1993), 403–416. (Cited in Chapter 6.)

[372] J. Karhumäki and J. O. Shallit, *Polynomial versus exponential growth in repetition-free binary words*, J. Combin. Theory, Ser. A 105 (2004), 335–347. (Cited in Appendix A.)

[373] Y. Katznelson, *Chromatic numbers of Cayley graphs on $\mathbb{Z}$ and recurrence*, Combinatorica 21 (2001), 211–219. (Cited in Chapter 2.)

[374] R. Kaufman, *Continued fractions and Fourier transforms*, Mathematika 27 (1980), 262–267. (Cited in Chapter 7.)

[375] ———, *On the theorem of Jarník and Besicovitch*, Acta Arith. 39 (1981), 265–267. (Cited in Chapter 7.)

[376] A. S. Kechris, Classical Descriptive Set Theory, Springer, Berlin, 1995. (Cited in Chapter 4.)

[377] A. J. Kempner, *On transcendental numbers*, Trans. Amer. Math. Soc. 17 (1916), 476–482. (Cited in Chapter 8.)

[378] R. E. Kennedy and C. Cooper, *A generalization of a result by Narkiewicz concerning large digits of powers*, Univ. Beograd. Publ. Elektrotehn. Fak. Ser. Mat. 11 (2000), 36–40. (Cited in Chapter 6.)

[379] A. Ya. Khintchine, *Über eine Klasse linearer diophantischer Approximationen*, Rendiconti Circ. Mat. Palermo 50 (1926), 170–195. (Cited in Chapter 2.)

[380] ———, Continued fractions. The University of Chicago Press, Chicago, IL, 1964. (Cited in Chapter 9 and Appendix D.)

[381] H. Ki and T. Linton, *Normal numbers and subsets of* $\mathbb{N}$ *with given densities*, Fund. Math. 144 (1994), 163–179. (Cited in Chapter 4.)

[382] P. Kiss and S. Molnár, *On distribution of linear recurrences modulo* 1, Studia Sci. Math. Hungar. 17 (1982), 113–127. (Cited in Chapter 3.)

[383] D. Kleinbock, E. Lindenstrauss and B. Weiss, *On fractal measures and Diophantine approximation*, Selecta Math. 10 (2004), 479–523. (Cited in Chapter 7.)

[384] D. Kleinbock and B. Weiss, *Badly approximable vectors on fractals*, Israel J. Math. 149 (2005), 137–170. (Cited in Chapter 7.)

[385] V. Knichal, *Dyadische Entwicklungen und Hausdorffsches Mass*, Časopis Mat. Fys., Praha, 65 (1936), 195–210. (Cited in Chapter 7.)

[386] M. J. Knight, *An 'ocean of zeroes' proof that a certain non-Liouville number is transcendental*, Amer. Math. Monthly 98 (1991), 947–949. (Cited in Chapter 8.)

[387] D. E. Knuth, The Art of Computer Programming, Vol. 2, Seminumerical Algorithms, Second edition. Addison-Wesley Ser. in Comput. Sci. Inform. Process., Addison-Wesley Publishing Co., Reading, MA, 1981. (Cited in Chapter 5.)

[388] J. F. Koksma, *Ein mengentheoretischer Satz über die Gleichverteilung modulo Eins*, Compositio Math. 2 (1935), 250–258. (Cited in Chapter 1.)

[389] ———, Diophantische Approximationen, Ergebnisse d. Math. u. ihrer Grenzgebiete, vol. 4, Springer, Berlin, 1936. (Cited in Preface, Chapters 1 & 4.)

[390] ———, *Über die Mahlersche Klasseneinteilung der transzendenten Zahlen und die Approximation komplexer Zahlen durch algebraische Zahlen*, Monatsh. Math. Phys. 48 (1939), 176–189. (Cited in Appendix E.)

[391] ———, *Sur la théorie métrique des approximations diophantiques*, Indag. Math. 7 (1945), 54–70. (Cited in Chapters 2 & 10.)

[392] A. N. Kolmogorov, *Three approaches to the definition of the concept "quantity of information"*, Problemy Peredači Informacii 1 (1965), 3–11 (in Russian). (Cited in Chapter 5.)

[393] V. Komornik, *Expansions in noninteger bases*, Integers 11B (2011), A9, 30 pp. (Cited in Chapter 9.)

[394] I. Környei, *On a theorem of Pisot*, Publ. Math. Debrecen 34 (1987), 169–179. (Cited in Chapter 2.)

[395] A. N. Korobov, *Continued fractions of some normal numbers*, Mat. Zametki 47 (1990), 28–33, 158 (in Russian). English translation in Math. Notes 47 (1990), 128–132. (Cited in Chapters 5 & 7.)

[396] N. M. Korobov, *Concerning some questions of uniform distribution*, Izv. Akad. Nauk SSSR. Ser. Mat. 14 (1950), 215–238 (in Russian). (Cited in Chapters 1 & 5.)

[397] ———, *Normal periodic systems and their applications to the estimation of sums of fractional parts*, Izv. Akad. Nauk SSSR. Ser. Mat. 15 (1951), 17–46 (in Russian). English translation in Amer. Math. Soc. Translations 4 (1962), 31–58. (Cited in Chapter 5.)

[398] ———, *Numbers with bounded quotient and their applications to questions of Diophantine approximation*, Izv. Akad. Nauk SSSR. Ser. Mat. 19 (1955), 361–380 (in Russian). (Cited in Chapter 6.)

[399] ———, *On completely uniform distribution and conjunctly normal numbers*, Izv. Akad. Nauk SSSR. Ser. Mat. 20 (1956), 649–660 (in Russian). (Cited in Chapter 1.)

[400] ———, *Distribution of fractional parts of exponential functions*, Vestnik Moskov. Univ. Ser. I Mat. Meh. 21 (1966), 42–46 (in Russian). (Cited in Chapter 4.)

[401] ———, *The distribution of digits in periodic fractions*, Mat. Sb. (N.S.) 89 (131) (1972), 654–670, 672. English translation in Math. USSR-Sb. 18 (1972), 659–676. (Cited in Chapter 5.)

[402] ———, Exponential Sums and their Applications. Translated from the 1989 Russian original by Yu. N. Shakhov. Mathematics and its Applications (Soviet Series), 80, Kluwer Academic Publishers Group, Dordrecht, 1992. (Cited in Chapters 1, 5 & 6.)

[403] N. M. Korobov and A. G. Postnikov, *Some general theorems on the uniform distribution of fractional parts*, Doklady Akad. Nauk SSSR 84 (1952), 217–220 (in Russian). (Cited in Chapter 1.)

[404] O. V. Kotova, *Continuum cardinality of the set of solutions of a class of equations that contain the function of the frequency of ternary digits of a number*, Ukraïn. Mat. Zh. 60 (2008), 1414–1421. English translation in Ukrainian Math. J. 60 (2008), 1650–1659. (Cited in Chapter 7.)

[405] B. Kra, *A generalization of Furstenberg's Diophantine theorem*, Proc Amer. Math. Soc. 127 (1999), 1951–1956. (Cited in Chapter 2.)

[406] C. Kraaikamp and H. Nakada, *On normal numbers for continued fractions*, Ergodic Theory Dynam. Systems 20 (2000), 1405–1421. (Cited in Chapter 9.)

[407] ———, *On a problem of Schweiger concerning normal numbers*, J. Number Theory 86 (2001), 330–340. (Cited in Chapter 6.)

[408] S. Kristensen, *Approximating numbers with missing digits by algebraic numbers*, Proc. Edinb. Math. Soc. 49 (2006), 657–666. (Cited in Chapter 7.)

[409] S. Kristensen, R. Thorn and S. Velani, *Diophantine approximation and badly approximable sets*, Adv. Math. 203 (2006), 132–169. (Cited in Chapter 7.)

[410] L. Kronecker, *Näherungsweise ganzzahlige Auflösung linearer Gleichungen*, Berl. Ber. (1884), 1179–1193, 1271–1299. (Cited in Chapter 1.)

[411] L. Kuipers and H. Niederreiter, *Uniform distribution of sequences*. Wiley–Interscience, New York, 1974. (Cited in Preface and Chapter 1.)

[412] M. F. Kulikova, *A construction problem connected with the distribution of fractional parts of the exponential function*, Dokl. Akad. Nauk SSSR 143 (1962), 522–524 (in Russian). (Cited in Chapters 2 & 4.)

[413] ———, *Construction of a number α whose fractional parts $\{\alpha g^\nu\}$ are rapidly and uniformly distributed*, Dokl. Akad. Nauk SSSR 143 (1962), 782–784 (in Russian). (Cited in Chapter 4.)

[414] J. Kwapisz, *A dynamical proof of Pisot's theorem*, Canad. Math. Bull. 49 (2006), 108–112. (Cited in Chapter 2.)

[415] D. Y. Kwon, *The orbit of a β-transformation cannot lie in a small interval*, J. Korean Math. Soc. To appear. (Cited in Chapters 3 & 9.)

[416] R. Labarca and C. G. Moreira, *Essential dynamics for Lorenz maps on the real line and the lexicographical world*, Ann. Inst. H. Poincaré Anal. Non Linéaire 23 (2006), 683–694. (Cited in Chapter 7.)

[417] J. C. Lagarias, *On the normality of arithmetical constants*, Experiment. Math. 10 (2001), 355–368. (Cited in Chapter 5.)

[418] ———, *Ternary expansions of powers of* 2, J. London Math. Soc. 79 (2009), 562–588. (Cited in Chapter 6.)

[419] J. L. Lagrange, *Additions au mémoire sur la résolution des équations numériques*, Mém. Berl. 24 (1770). (Cited in Appendix D.)

[420] S. Lang and H. Trotter, *Continued fractions for some algebraic numbers*, J. Reine Angew. Math. 255 (1972), 112–134; addendum, ibid. 267 (1974), 219–220. (Cited in Chapter 9.)

[421] H. Lebesgue, *Sur certaines démonstrations d'existence*, Bull. Soc. Math. France 45 (1917), 132–144. (Cited in Chapter 5.)

[422] A.-M. Legendre, Théorie des Nombres, troisième édition, Tome 1, Paris, 1830. (Cited in Appendix D.)

[423] E. Lehrer, *The game of normal numbers*, Math. Oper. Res. 29 (2004), 259–265. (Cited in Chapter 4.)

[424] M. A. Lerma, *Construction of a number greater than one whose powers are uniformly distributed modulo one.* Preprint. See `http://www.math.northwestern.edu/ mlerma/papers/`. (Cited in Chapter 2.)

[425] J. Levesley, C. Salp and S. L. Velani, *On a problem of K. Mahler: Diophantine approximation and Cantor sets*, Math. Ann. 338 (2007), 97–118. (Cited in Chapter 7.)

[426] M. B. Levin, *The uniform distribution of the sequence $\{\alpha\lambda^x\}$*, Mat. Sb. (N.S.) 98(140) (1975), 207–222, 333 (in Russian). (Cited in Chapters 1 & 4.)

[427] ———, *The distribution of the fractional parts of the exponential function*, Soviet Math. (Iz. VUZ) 21 (1977), 41–47. (Cited in Chapter 4.)

[428] ———, *On absolutely normal numbers*, Vestnik Moskov. Univ. Ser. I Mat. Mekh. (1979), 31–37, 87 (in Russian). English translation in Moscow Univ. Math. Bull. 34 (1979), 32–39. (Cited in Chapters 4 & 5.)

[429] ———, *On the complete uniform distribution of the fractional parts of the exponential function*, Trudy Sem. Petrovsk. 7 (1981), 245–256 (in Russian). (Cited in Chapter 2.)

[430] ———, *Effectivization of the Koksma theorem*, Mat. Zametki 47 (1990), 163–166 (in Russian). (Cited in Chapter 2.)

[431] ———, *Jointly absolutely normal numbers*, Mat. Zametki 48 (1990), 61–71 (in Russian). English translation in Math. Notes 48 (1990), 1213–1220. (Cited in Chapters 2, 4 & 6.)

[432] ———, *On the discrepancy of Markov-normal sequences*, J. Théor. Nombres Bordeaux 8 (1996), 413–428. (Cited in Chapter 7.)

[433] ———, *On the discrepancy estimate of normal numbers*, Acta Arith. 88 (1999), 99–111. (Cited in Chapters 4 & 10.)

[434] ———, *On normal lattice configurations and simultaneously normal numbers*, J. Théor. Nombres Bordeaux 13 (2001), 483–527. (Cited in Chapter 6.)

[435] M. B. Levin and I. E. Shparlinskii, *Uniform distribution of fractional parts of recurrent sequences*, Uspekhi Mat. Nauk 34 (1979), 203–204. (Cited in Chapter 4.)

[436] M. B. Levin and M. Smorodinsky, *A $\mathbb{Z}^d$ generalization of the Davenport–Erdős construction of normal numbers*, Colloq. Math. 84/85 (2000), 431–441. (Cited in Chapter 4.)

[437] ———, *Explicit construction of normal lattice configurations*, Colloq. Math. 102 (2005), 33–47. (Cited in Chapter 4.)

[438] S. Leys, Le Bonheur des Petits Poissons, J.-C. Lattès, 2008.

[439] B. Li and Y.-C. Chen, *Chaotic and topological properties of β-transformations*, J. Math. Anal. Appl. 383 (2011), 585–596. (Cited in Chapter 9.)

[440] B. Li and J. Wu, *Beta-expansion and continued fraction expansion*, J. Math. Anal. Appl. 339 (2008), 1322–1331. (Cited in Chapter 9.)

[441] W. Li and F. M. Dekking, *Hausdorff dimension of subsets of Moran fractals with prescribed group frequency of their codings*, Nonlinearity 16 (2003), 187–199. (Cited in Chapter 7.)

[442] W. Li, L. Olsen and Z. Wen, *Hausdorff and packing dimensions of subsets of Moran fractals with prescribed mixed group frequency of their codings*, Aequationes Math. 77 (2009), 171–185. (Cited in Chapter 7.)

[443] L. Liao and W. Steiner, *Dynamical properties of the negative beta transformation*, Ergodic Theory Dynam. Systems. To appear. (Cited in Chapter 9.)

[444] E. Lindenstrauss, *p-adic foliation and equidistribution*, Israel J. Math. 122 (2001), 29–42. (Cited in Chapters 2 & 10.)

[445] J. Liouville, *Remarques relatives à des classes très-étendues de quantités dont la valeur n'est ni algébrique, ni même réductible à des irrationnelles algébriques*, C. R. Acad. Sci. Paris 18 (1844), 883–885. (Cited in Appendix E.)

[446] ———, *Nouvelle démonstration d'un théorème sur les irrationnelles algébriques*, C. R. Acad. Sci. Paris 18 (1844), 910–911. (Cited in Appendix E.)

[447] J. E. Littlewood, Some Problems in Real and Complex Analysis. D. C. Heath and Co. Raytheon Education Co., Lexington, MA, 1968. (Cited in Chapter 2.)

[448] G. Lochs, *Vergleich der Genauigkeit von Dezimalbruch und Kettenbruch*, Abh. Math. Sem. Univ. Hamburg 27 (1964), 142–144. (Cited in Chapter 9.)

[449] C. T. Long, *Note on normal numbers*, Pacific J. Math. 7 (1957), 1163–1165. (Cited in Chapter 4.)

[450] M. Lothaire, Algebraic Combinatorics on Words, Encyclopaedia Math. Appl. 90, Cambridge University Press, Cambridge, 2002. (Cited in Appendix A.)

[451] F. Luca, *On a question of G. Kuba*, Arch. Math. (Basel) 74 (2000), 269–275. (Cited in Chapter 3.)

[452] ———, *.121221222... is not quadratic*, Rev. Mat. Comput. 18 (2005), 353–362. (Cited in Chapter 8.)

[453] R. Lyons, *Characterizations of measures whose Fourier–Stieltjes transforms vanish at infinity*, Bull. Amer. Math. Soc. 10 (1984), 93–96. (Cited in Chapter 7.)

[454] ———, *The measure of non-normal sets*, Invent. Math. 83 (1986), 605–616. (Cited in Chapter 7.)

[455] C. McMullen, *Uniformly Diophantine numbers in a fixed real quadratic field*, Compos. Math. 145 (2009), 827–844. (Cited in Chapter 9.)

[456] ———, *Winning sets, quasiconformal maps and Diophantine approximation*, Geom. Funct. Anal. 20 (2010), 726–740. (Cited in Chapter 7.)

[457] M. G. Madritsch, *A note on normal numbers in matrix number systems*, Math. Pannon. 18 (2007), 219–227. (Cited in Chapter 4.)

[458] ———, *Generating normal numbers over Gaussian integers*, Acta Arith. 135 (2008), 63–90. (Cited in Chapter 4.)

[459] M. G. Madritsch, J. M. Thuswaldner and R. F. Tichy, *Normality of numbers generated by the values of entire functions*, J. Number Theory 128 (2008), 1127–1145. (Cited in Chapter 4.)

[460] K. Mahler, *Über das Verschwinden von Potenzreihen mehrerer Veränderlichen in speziellen Punktfolgen*, Math. Ann. 103 (1930), 573–587. (Cited in Chapter 8.)

[461] ———, *Zur Approximation der Exponentialfunktionen und des Logarithmus. I, II*, J. Reine Angew. Math. 166 (1932), 118–150. (Cited in Appendix E.)

[462] ———, *Über die Dezimalbruchentwicklung gewisser Irrationalzahlen*, Mathematica 4B (Zutphen) (1937), 15 pp. (Cited in Chapters 4 & 8.)

[463] ———, *Arithmetische Eigenschaften einer Klasse von Dezimalbrüchen*, Proc. Akad. v. Wetensch. 40 (1937), 421–428. (Cited in Chapter 4.)

[464] ———, *On the approximation of logarithms of algebraic numbers*, Philos. Trans. Roy. Soc. London 245 (1953), 371–398. (Cited in Chapter 2.)

[465] ———, *On the fractional parts of the powers of a rational number, II*, Mathematika 4 (1957), 122–124. (Cited in Chapters 3 & 9.)

[466] ———, Lectures on Diophantine approximation, Part 1: g-adic Numbers and Roth's theorem, University of Notre Dame, Ann Arbor, MI, 1961. (Cited in Appendix E.)

[467] ———, *Applications of some formulas by Hermite to the approximation of exponentials and logarithms*, Math. Ann. 168 (1967), 200–227. (Cited in Chapter 2.)

[468] ———, *An unsolved problem on the powers of* 3/2, J. Austral. Math. Soc. 8 (1968), 313–321. (Cited in Chapters 3 & 10.)

[469] ———, *Arithmetical properties of the digits of the multiples of an irrational number*, Bull. Austral. Math. Soc. 8 (1973), 191–203. (Cited in Chapter 8.)

[470] ———, *On a class of transcendental decimal fractions*, Comm. Pure Appl. Math. 29 (1976), 717–725. (Cited in Chapter 4.)

[471] ———, *On some irrational decimal fractions*, J. Number Theory 13 (1981), 268–269. (Cited in Chapter 10.)

[472] ———, *Fifty years as a mathematician*, J. Number Theory 14 (1982), 121–155. (Cited in Chapter 4.)

[473] ———, *Some suggestions for further research*, Bull. Austral. Math. Soc. 29 (1984), 101–108. (Cited in Chapter 8.)

[474] K. Mahler and G. Szekeres, *On the approximation of real numbers by roots of integers*, Acta Arith. 12 (1966/1967), 315–320. (Cited in Chapter 2.)

[475] E. Maillet, Introduction à la Théorie des Nombres Transcendants et des Propriétés Arithmétiques des Fonctions, Gauthier-Villars, Paris, 1906. (Cited in Chapter 9.)

[476] B. Mance, *Construction of normal numbers with respect to the Q-Cantor series expansion for certain Q*, Acta Arith. To appear. (Cited in Chapter 9.)

[477] G. Margulis, *Problems and conjectures in rigidity theory*, In: Mathematics: Frontiers and Perspectives, 161–174, American Mathematical Society, Providence, RI, 2000. (Cited in Chapter 10.)

[478] G. Martin, *Absolutely abnormal numbers*, Amer. Math. Monthly 108 (2001), 746–754. (Cited in Chapter 5.)

[479] ———, *The unreasonable effectualness of continued function expansions*, J. Aust. Math. Soc. 77 (2004), 305–319. (Cited in Chapter 9.)

[480] P. Martin-Löf, *The definition of random sequences*, Information and Control 9 (1966), 602–619. (Cited in Chapter 5.)

[481] B. de Mathan, *Sur un problème de densité modulo* 1, C. R. Acad. Sci. Paris 287 (1978), A277–A279. (Cited in Chapter 2.)

[482] ———, *Numbers contravening a condition in density modulo* 1, Acta Math. Acad. Sci. Hungar. 36 (1980), 237–241. (Cited in Chapter 2.)

[483] B. de Mathan and O. Teulié, *Problèmes diophantiens simultanés*, Monatsh. Math. 143 (2004), 229–245. (Cited in Chapters 2 & 10.)

[484] P. Mattila, Geometry of Sets and Measures in Euclidean Spaces. Cambridge University Press, Cambridge, 1995. (Cited in Appendix C.)

[485] Ch. Mauduit, *Automates finis et ensembles normaux*, Ann. Inst. Fourier (Grenoble) 36 (1986), 1–25. (Cited in Chapter 4.)

[486] ———, *Caractérisation des ensembles normaux substitutifs*, Invent. Math. 95 (1989), 133–147. (Cited in Chapter 4.)

[487] ———, *Problem session dedicated to Gérard Rauzy*. In: Dynamical Systems (Luminy-Marseille, 1998), 1–19, World Scientific Publishing, River Edge, NJ, 2000. (Cited in Chapter 10.)

[488] ———, *Multiplicative properties of the Thue–Morse sequence*, Period. Math. Hungar. 43 (2001), 137–153. (Cited in Appendix A.)

[489] Ch. Mauduit and C. G. Moreira, *Complexity of infinite sequences with zero entropy*, Acta Arith. 142 (2010), 331–346. (Cited in Chapters 4 & 9.)

[490] ———, *Generalized Hausdorff dimensions of sets of real numbers with zero entropy expansion*, Ergodic Theory Dynam. Systems. To appear. (Cited in Chapters 4 & 9.)

[491] ———, *Generalized fractal dimensions and complexity of infinite sequences with positive entropy*. Preprint. (Cited in Chapter 4.)

[492] J. E. Maxfield, *Sums and products of normal numbers*, Amer. Math. Monthly 59 (1952), 98. (Cited in Chapter 4.)

[493] ———, *A short proof of Pillai's theorem on normal numbers*, Pacific J. Math. 2 (1952), 23–24. (Cited in Chapter 4.)

[494] ———, *Normal k-tuples*, Pacific J. Math. 3 (1953), 189–196. (Cited in Chapters 4, 5 & 6.)

[495] D. Meiri, *Entropy and uniform distribution of orbits in T^d*, Israel J. Math. 105 (1998), 155–183. (Cited in Chapters 2 & 10.)

[496] M. Mendès France, *Nombres normaux et fonctions pseudo-aléatoires*, Ann. Inst. Fourier (Grenoble) 13 (1963), 91–104. (Cited in Chapter 4.)

[497] ———, *A set of nonnormal numbers*, Pacific J. Math. 15 (1965), 1165–1170. (Cited in Chapter 4.)

[498] ———, *Nombres normaux. Applications aux fonctions pseudo-aléatoires*, J. Analyse Math. 20 (1967), 1–56. (Cited in Chapters 4 & 10.)

[499] ———, *Deux remarques concernant l'équirépartition des suites*, Acta Arith. 14 (1967/1968), 163–167. (Cited in Chapter 4.)

[500] ———, *Nombres transcendants et ensembles normaux*, Acta Arith. 15 (1968/1969), 189–192. (Cited in Chapter 4.)

[501] ———, *Sur les fractions continues limitées*, Acta Arith. 23 (1973), 207–215. (Cited in Chapter 9 and Appendix D.)

[502] ———, *Les ensembles de Bésineau*, Séminaire Delange-Pisot-Poitou (15e année: 1973/74), Théorie des nombres, Fasc. 1, Exp. No. 7, 6 pp. Secrétariat Mathématique, Paris, 1975. (Cited in Chapters 6 & 10.)

[503] ———, *On a theorem of Davenport concerning continued fractions*, Mathematika 23 (1976), 136–141. (Cited in Chapter 9.)

[504] ———, *Sur les décimales des nombres algébriques réels*, Séminaire de Théorie des Nombres de Bordeaux, 1979–1980, Exp. No. 28, 7 pp., Univ. Bordeaux I, Talence, 1980. (Cited in Chapter 10.)

[505] ———, *Nombres algébriques et théorie des automates*, Enseign. Math. 26 (1980), 193–199. (Cited in Chapters 8 & 10.)

[506] ———, *Remarks on problems on finite and periodic continued fractions*, Enseign. Math. 39 (1993), 249–257. (Cited in Chapters 9 & 10.)

[507] W. Merkle and J. Reimann, *Selection functions that do not preserve normality*, Theory Comput. Syst. 39 (2006), 685–697. (Cited in Chapter 4.)

[508] Y. Meyer, *Nombres algébriques, nombres transcendants et équirépartition modulo* 1, Acta Arith. 16 (1969/1970), 347–350. (Cited in Chapter 4.)

[509] ———, Nombres de Pisot, Nombres de Salem et Analyse Harmonique, Lecture Notes in Mathematics, Vol. 117, Springer, Berlin, 1970. (Cited in Chapter 4.)

[510] ———, *Nombres transcendants et partition modulo* 1. In: Colloque de Théorie des Nombres (Univ. Bordeaux, Bordeaux, 1969), pp. 143–149. Bull. Soc. Math. France, Mem. 25, Soc. Math. France, Paris, 1971. (Cited in Chapter 4.)

[511] ———, Algebraic Numbers and Harmonic Analysis, North-Holland Mathematical Library, Vol. 2, North-Holland Publishing Co., Amsterdam, 1972. (Cited in Chapter 3.)

[512] M. Mignotte, *Approximations rationnelles de π et quelques autres nombres*, Bull. Soc. Math. France 37 (1974), 121–132. (Cited in Chapter 2.)

[513] ———, *An application of W. Schmidt's theorem: transcendental numbers and golden number*, Fibonacci Quart. 15 (1977), 15–16. (Cited in Chapter 8.)

[514] ———, *A characterization of integers*, Amer. Math. Monthly 84 (1977), 278–281. (Cited in Chapter 2.)

[515] ———, *Sur les entiers qui s'crivent simplement en diffrentes bases*, European J. Combin. 9 (1988), 307–316. (Cited in Chapter 6.)

[516] R. von Mises, *Über Zahlenfolgen, die ein kollektiv-ähnliches Verhalten zeigen*, Math. Ann. 108 (1933), 757–772. (Cited in Chapter 4.)

[517] H. M. Montgomery, Ten Lectures on the Interface Between Analytic Number Theory and Harmonic Analysis, American Mathematical Society, Providence, RI, 1994. (Cited in Chapter 7.)

[518] W. Moran and C. E. M. Pearce, *Discrepancy results for normal numbers*. In: Miniconferences on Harmonic Analysis and Operator Algebras (Canberra, 1987), 203–210, Proc. Centre Math. Anal. Austral. Nat Univ, 16,. Australian National University, Canberra, 1988. (Cited in Chapter 6.)

[519] W. Moran and A. D. Pollington, *Normality to non-integer bases*. In: Analytic Number Theory and Related Topics (Tokyo, 1991), 109–117, World Scientific Publishing, River Edge, NJ, 1993. (Cited in Chapter 6.)

[520] ———, *Metrical results on normality to distinct bases*, J. Number Theory 54 (1995), 180–189. (Cited in Chapter 6.)

[521] ———, *A problem in multi-index normality*. In: Proceedings of the Conference in Honor of Jean-Pierre Kahane (Orsay, 1993), J. Fourier Anal. Appl. 1995, Special Issue, 455–466. (Cited in Chapter 6.)

[522] ———, *The discrimination theorem for normality to non-integer bases*, Israel J. Math. 100 (1997), 339–347. (Cited in Chapter 6.)

[523] M. Morse, *Recurrent geodesics on a surface of negative curvature*, Trans. Amer. Math. Soc. 22 (1921), 84–100. (Cited in Appendix A.)

[524] M. Morse and G. A. Hedlund, *Symbolic dynamics*, Amer. J. Math. 60 (1938), 815–866. (Cited in Appendix A.)

[525] ———, *Symbolic dynamics II*, Amer. J. Math. 62 (1940), 1–42. (Cited in Appendix A.)

[526] N. G. Moshchevitin, *On sublacunary sequences and winning sets*, Mat. Zametki 78 (2005), 634–637 (in Russian). English translation in Math. Notes 78 (2005), 592–596. (Cited in Chapters 2 & 7.)

[527] ———, *Density modulo 1 of lacunary and sublacunary sequences: application of Peres–Schlag's construction*, Fundam. Prikl. Mat. 16 (2010), 117–138 (in Russian). (Cited in Chapter 2.)

[528] ———, *A note on badly approximable affine forms and winning sets*, Moscow Math. J. 11 (2011), 129–137. (Cited in Chapter 7.)

[529] N. G. Moshchevitin and I. D. Shkredov, *On the Pyatetskiĭ–Shapiro criterion for normality*, Mat. Zametki 73 (2003), 577–589 (in Russian). English translation in Math. Notes 73 (2003), 539–550. (Cited in Chapter 4.)

[530] R. H. Muhutdinov, *A test for the nonuniform distribution of the fractional parts of the exponential function*, Izv. Akad. Nauk UzSSR Ser. Fiz.-Mat. Nauk 1975, 23–26, 104 (in Russian). (Cited in Chapter 4.)

[531] K. Nagasaka, *On the Hausdorff dimension of sets missing statistical laws*, Proc. Inst. Statist. Math. 25 (1978), 1–9 (in Japanese). (Cited in Chapter 6.)

[532] ———, *La dimension de Hausdorff de certains ensembles dans* $[0,1]$, Proc. Japan Acad. Ser. A Math. Sci. 54 (1978), 109–112. (Cited in Chapter 6.)

[533] ———, *Nonnormal numbers to different bases and their Hausdorff dimension*, Tsukuba J. Math. 10 (1986), 89–99. (Cited in Chapter 6.)

[534] Y. Nakai and I. Shiokawa, *A class of normal numbers*, Japan. J. Math. (N.S.) 16 (1990), 17–29. (Cited in Chapter 4.)

[535] ———, *A class of normal numbers. II.* In: Number Theory and Cryptography (Sydney, 1989), 204–210, London Math. Soc. Lecture Note Ser., 154, Cambridge University Press, Cambridge, 1990. (Cited in Chapter 4.)

[536] ———, *Discrepancy estimates for a class of normal numbers*, Acta Arith. 62 (1992), 271–284. (Cited in Chapter 4.)

[537] ———, *Normality of numbers generated by the values of polynomials at primes*, Acta Arith. 81 (1997), 345–356. (Cited in Chapter 4.)

[538] W. Narkiewicz, *A note on a paper of H. Gupta concerning powers of 2 and 3*, Univ. Beograd Publ. Elecktrotehn. Fak. Ser. Mat. Fiz. 678–715 (1980), 173–174. (Cited in Chapter 6.)

[539] H. Niederreiter, *On an irrationality theorem of Mahler and Bundschuh*, J. Number Theory 24 (1986), 197–199. (Cited in Chapter 10.)

[540] A. Nies, Computability and Randomness, Oxford Logic Guides, 51, Oxford University Press, Oxford, 2009. (Cited in Chapter 5.)

[541] J. Nilsson, *On numbers badly approximable by dyadic rationals*, Israel J. Math. 171 (2009), 93–110. (Cited in Chapter 7.)

[542] ———, *The fine structure of dyadically badly approximable numbers*. Preprint. (Cited in Chapter 7.)

[543] I. Niven, Irrational Numbers, The Carus Mathematical Monographs, No. 11, The Mathematical Association of America. Distributed by John Wiley and Sons, Inc., New York, 1956. (Cited in Preface.)

[544] I. Niven and H.S. Zuckerman, *On the definition of normal numbers*, Pacific J. Math. 1 (1951), 103–109. (Cited in Chapter 4.)

[545] L. Olsen, *Extremely non-normal numbers*, Math. Proc. Cambridge Philos. Soc. 137 (2004), 43–53. (Cited in Chapter 4.)

[546] ———, *Applications of multifractal divergence points to sets of numbers defined by their N-adic expansion*, Math. Proc. Cambridge Philos. Soc. 136 (2004), 139–165. (Cited in Chapter 7.)

[547] ———, *A generalization of a result by W. Li and F. Dekking on the Hausdorff dimension of subsets of self-similar sets with prescribed group frequency of their codings*, Aequationes Math. 72 (2006), 10–26. (Cited in Chapter 7.)

[548] ———, *Frequencies of digits, divergence points, and Schmidt games*, Chaos Solitons Fractals 40 (2009), 2222–2232. (Cited in Chapter 7.)

[549] ———, *Hausdorff and packing dimensions of non-normal tuples of numbers: Non-linearity and divergence points*, Bull. Sci. Math. 134 (2010), 64–96. (Cited in Chapter 7.)

[550] J.C. Oxtoby, *The Banach–Mazur game and Banach category theorem*, Annals of Mathematics Studies, no. 39, pp. 159–163, Princeton University Press, Princeton, NJ, 1957. (Cited in Chapter 7.)

[551] F. Parreau and M. Queffélec, *M_0 measures for the Walsh system*, J. Fourier Anal. Appl. 15 (2009), 502–514. (Cited in Chapter 7.)

[552] W. Parry, *On the β-expansions of real numbers*, Acta Math. Acad. Sci. Hung. 11 (1960), 401–416. (Cited in Chapter 9.)

[553] R.P. Pass, *On the partial quotients of algebraic integers*, J. Number Theory 11 (1979), 14–15. (Cited in Chapter 9.)

[554] C.E. M. Pearce and M.S. Keane, *On normal numbers*, J. Austral. Math. Soc. Ser. A 32 (1982), 79–87. (Cited in Chapter 6.)

[555] D. Pellegrino, *On normal numbers*, Proyecciones 25 (2006), 19–30. (Cited in Chapter 4.)

[556] Yu. Peres and W. Schlag, *Two Erdős problems on lacunary sequences: chromatic numbers and Diophantine approximations*, Bull. London Math. Soc. 42 (2010), 295–300. (Cited in Chapter 2.)

[557] O. Perron, Die Lehre von den Ketterbrüchen, Teubner, Leipzig, 1929. (Cited in Appendix D.)

[558] T. Persson and J. Schmeling, *Dyadic Diophantine approximation and Katok's horseshoe approximation*, Acta Arith. 132 (2008), 205–230. (Cited in Chapter 9.)

[559] I.I. Piatetski-Shapiro, *On the laws of distribution of the fractional parts of an exponential function*, Izv. Akad. Nauk SSSR Ser. Mat. 15 (1951), 47–52 (in Russian). (Cited in Chapter 4.)

[560] ———, *On the distribution of the fractional parts of the exponential function*, Moskov. Gos. Ped. Inst. Uč. Zap. 108 (1957), 317–322 (in Russian). (Cited in Chapter 4.)

[561] S.S. Pillai, *On normal numbers*, Proc. Indian Acad. Sci. 10 (1939), 13–15. (Cited in Chapters 4 & 10.)

[562] ———, *On normal numbers*, Proc. Indian Acad. Sci. 12 (1940), 179–184. (Cited in Chapter 4.)

[563] Ch. Pisot, *Sur une propriété caractéristique de certains entiers algébriques*, C.R. Acad. Sci. Paris 202 (1936), 892–894. (Cited in Chapter 2.)

[564] ———, *Sur la répartition modulo 1 des puissances successives d'un même nombre*, C.R. Acad. Sci. Paris 204 (1937), 312–314. (Cited in Chapter 2.)

[565] ———, *La répartition modulo 1 et les nombres algébriques*, Ann. Sc. Norm. Sup. Pisa 7 (1938), 205–248. (Cited in Chapter 2.)

[566] ———, *Répartition (mod 1) des puissances successives des nombres réels,* Comment. Math. Helv. 19 (1946), 153–160. (Cited in Chapter 2.)

[567] Ch. Pisot and R. Salem, *Distribution modulo 1 of the powers of real numbers larger than 1*, Compositio Math. 16 (1964), 164–168. (Cited in Chapters 2 & 3.)

[568] A. D. Pollington, Interval constructions in the theory of numbers, Thesis, University of London, 1978. (Cited in Chapter 3.)

[569] ———, *The Hausdorff dimension of a set of non-normal well approximable numbers*, Number Theory, Carbondale (1979), Lectures Notes in Math. 751, Springer-Verlag, Berlin, 1979, 256–264. (Cited in Chapter 7.)

[570] ———, *On the density of sequence $n_k\xi$*, Illinois J. Math. 23 (1979), 511–515. (Cited in Chapter 2.)

[571] ———, *The Hausdorff dimension of certain sets related to sequences which are not dense mod* 1, Quart. J. Math. Oxford Ser. (2) 31 (1980), 351–361. (Cited in Chapter 2.)

[572] ———, *Progressions arithmétiques généralisées et le problème des* $(\frac{3}{2})^n$, C.R. Acad. Sci. Paris 292 (1981), 383–384. (Cited in Chapter 3.)

[573] ———, *The Hausdorff dimension of a set of normal numbers*, Pacific J. Math. 95 (1981), 193–204. (Cited in Chapter 6.)

[574] ———, *Sur les suites $k\theta^n$*, C.R. Acad. Sci. Paris 296 (1983), 941–943. (Cited in Chapter 3.)

[575] ———, *The Hausdorff dimension of a set of normal numbers. II*, J. Austral. Math. Soc. Ser. A 44 (1988), 259–264. (Cited in Chapter 6.)

[576] A. D. Pollington and S. L. Velani, *Metric Diophantine approximation and 'absolutely friendly' measures*, Selecta Math. 11 (2005), 297–307. (Cited in Chapter 7.)

[577] A. M. Polosuev, *Uniform distribution of a system of functions representing a solution of linear first order difference equations*, Dokl. Akad. Nauk SSSR 123 (1958), 405–406 (in Russian). (Cited in Chapter 4.)

[578] ———, *On the uniformity of the distribution of a system of functions forming the solution of a system of linear finite-difference equations of first order*, Vestnik Moskov. Univ. Ser. I Mat. Meh. (1960), no. 5, 29–39 and no. 6, 20–25 (in Russian). (Cited in Chapter 6.)

[579] A. J. van der Poorten, *Some problems of recurrent interest.* In: Topics in Classical Number Theory, Vol. I, II (Budapest, 1981), 1265–1294, Colloq. Math. Soc. Jnos Bolyai, 34, North-Holland, Amsterdam, 1984. (Cited in Chapter 9.)

[580] ———, *An introduction to continued fractions*, Diophantine Analysis (Kensington, 1985), 99–138, London Math. Soc. Lecture Note

Ser. 109, Cambridge University Press, Cambridge, 1986. (Cited in Appendix D.)

[581] E. C. Posner, *Diophantine problems involving powers modulo one*, Illinois J. Math. 6 (1962), 251–263. (Cited in Chapter 2.)

[582] A. G. Postnikov, *A criterion for testing the uniform distribution of an exponential function in the complex domain*, Vestnik Leningrad. Univ. 12 (1957), 81–88 (in Russian). (Cited in Chapter 4.)

[583] ———, *Arithmetic modeling of random processes*, Trudy Mat. Inst. Steklov. 57 (1960), 84 pp. (in Russian). English translation in Selected Translations in Mathematical Statistics and Probability, Vol. 13, American Mathematical Society, Providence, RI, 1973. (Cited in Chapters 1, 5, 6 & 9.)

[584] ———, *Ergodic problems in the theory of congruences and of Diophantine approximations*, Trudy Mat. Inst. Steklov 82 (1966), 3–112 (in Russian). English translation in Proceedings of the Steklov Institute of Mathematics, 82, American Mathematical Society, Providence, RI, 1967. (Cited in Chapters 4 & 5.)

[585] A. G. Postnikov and I. I. Pyateckiĭ, *A Markov-sequence of symbols and a normal continued fraction*, Izv. Akad. Nauk SSSR. Ser. Mat. 21 (1957), 729–746 (in Russian). (Cited in Chapters 6 & 9.)

[586] L. P. Postnikova, *A quantitative form of Borel's problem*, Acta Arith. 21 (1972), 235–250 (in Russian). (Cited in Chapter 4.)

[587] Y. Pourchet. Private letter to Mendès France, 1972. (Cited in Chapter 9.)

[588] L. N. Pushkin, *A metric variant of the Cassels–Schmidt theorem*, Mat. Zametki 46 (1989), 60–66, 124 (in Russian). English translation in Math. Notes 46 (1989), 538–542. (Cited in Chapter 6.)

[589] ———, *Vectors that are Borel normal on a manifold in* $\mathbb{R}^n$, Teor. Veroyatnost. i Primenen. 36 (1991), 372–376 (in Russian). English translation in Theory Probab. Appl. 36 (1991), 391–395. (Cited in Chapter 5.)

[590] ———, *Ergodic properties of sets defined by frequencies of numbers*, Teor. Veroyatnost. i Primenen. 41 (1996), 672–677 (in Russian). English translation in Theory Probab. Appl. 41 (1996), 593–597. (Cited in Chapter 6.)

[591] ———, *Small digitwise perturbations of a number make it normal to unrelated bases*, Lobachevskii J. Math. 11 (2002), 22–25. (Cited in Chapter 6.)

[592] ———, *On the behavior of the spectrum of the limit frequencies of digits under perturbations of a real number*, Mat. Zametki 86 (2009), 884–891 (in Russian). English translation in Math. Notes 86 (2009), 824–830. (Cited in Chapter 6.)

[593] L. N. Pushkin and E. Sh. Rakhmatullina, *On the category of numerical sets defined by frequencies of numbers*. In: Studies in Information Science, No. 4, 95–98, Izd. 'Otechestvo', Kazan, 2002 (in Russian). (Cited in Chapter 4.)

[594] M. Queffélec, *Transcendance des fractions continues de Thue–Morse*, J. Number Theory 73 (1998), 201–211. (Cited in Chapter 9.)

[595] ———, *Approximations diophantiennes des nombres sturmiens*, J. Théor. Nombres Bordeaux 14 (2002), 613–628. (Cited in Chapter 8.)

[596] ———, *Fourier analysis and continued fractions.* In: European Women in Mathematics—Marseille 2003, 47–57, CWI Tract, 135, Centrum Wisk. Inform., Amsterdam, 2005. (Cited in Chapter 7.)

[597] ———, *Old and new results on normality.* In: Dynamics and Stochastics, 225–236, IMS Lecture Notes Monogr. Ser., 48, Institute of Mathematics and Statistics, Beachwood, OH, 2006. (Cited in Chapters 6 & 10.)

[598] ———, *An introduction to Littlewood's conjecture.* In: Dynamical Systems and Diophantine Approximation, (eds. Yann Bugeaud, Françoise Dal'Bo, Cornelia Druţu), Séminaires et Congrès 19, pp. 129–152, Société Mathématique de France, 2009. (Cited in Chapter 2.)

[599] M. Queffélec and O. Ramaré, *Analyse de Fourier des fractions continues à quotients restreints*, Enseign. Math. 49 (2003), 335–356. (Cited in Chapter 7.)

[600] P. Rath, *Two exceptional classes of real numbers*, Funct. Approx. Comment. Math. 38 (2008), 81–91. (Cited in Chapter 3.)

[601] G. Rauzy, *Suites partiellement recurrentes (Applications la répartition modulo 1 et aux propriétés arithmétiques des fonctions analytiques)*, Ann. Inst. Fourier (Grenoble) 16 (1966), 159–234. (Cited in Chapter 2.)

[602] ———, *Caractérisation des ensembles normaux*, Bull. Soc. Math. France 98 (1970), 401–414. (Cited in Chapter 4.)

[603] ———, *Normalité de* $\mathbb{Q}^*$, Acta Arith. 19 (1971), 43–47. (Cited in Chapter 4.)

[604] ———, *Nombres normaux et processus déterministes*, Acta Arith. 29 (1976), 211–225. (Cited in Chapters 4 & 10.)

[605] ———, Propriétés Statistiques de Suites Arithmétiques, Le Mathématicien, No. 15, Collection SUP, Presses Universitaires de France, Paris, 1976, 133 pp. (Cited in Preface and Chapter 1.)

[606] L. Rédei, *Zu einem Approximationssatz von Koksma*, Math. Z. 48 (1942), 500–502. (Cited in Chapter 2.)

[607] ———, *Über eine diophantische Approximation im bereich der algebraischen Zahlen*, Math. Naturwiss. Anz. Ungar. Akad. Wiss. 61 (1942), 460–470. (Cited in Chapter 2.)

[608] A. Rényi, *Representations for real numbers and their ergodic properties*, Acta Math. Acad. Sci. Hung. 8 (1957), 477–493. (Cited in Chapter 9.)

[609] R. D. Richtmyer, M. Devaney and N. Metropolis, *Continued fraction expansions of algebraic numbers*, Numer. Math. 4 (1962), 68–84. (Cited in Chapter 9.)

[610] D. Ridout, *Rational approximations to algebraic numbers*, Mathematika 4 (1957), 125–131. (Cited in Appendix E.)

[611] T. Rivoal, *On the bits counting function of real numbers*, J. Austral. Math. Soc. 85 (2008), 95–111. (Cited in Chapters 8 & 10.)

[612] I. Rochev, *On distribution of fractional parts of linear forms*, Fundam. Prikl. Mat. 16 (2010), 123–137 (in Russian). (Cited in Chapter 2.)

[613] C. A. Rogers, Hausdorff Measures, Cambridge University Press, Cambridge, 1970. (Cited in Appendix C.)

[614] K. F. Roth, *Rational approximations to algebraic numbers*, Mathematika 2 (1955), 1–20; corrigendum, 168. (Cited in Appendix E.)

[615] D. J. Rudolph, $\times 2$ *and* $\times 3$ *invariant measures and entropy*, Ergodic Theory Dynam. Systems 10 (1990), 395–406. (Cited in Chapters 2 & 10.)

[616] Yu. N. Šahov, *The construction of a sequence of signs that is normal in the sense of Markov*, Moskov. Gos. Ped. Inst. Učen. Zap. No. 375 (1971), 143–155 (in Russian). (Cited in Chapter 4.)

[617] ———, *Certain estimates in the construction of Bernoulli-normal sequences of symbols*, Mat. Zametki 10 (1971), 501–510 (in Russian). (Cited in Chapter 4.)

[618] T. Šalát, *A remark on normal numbers*, Rev. Roumaine Math. Pures Appl. 11 (1966), 53–56. (Cited in Chapter 4.)

[619] R. Salem, Algebraic Numbers and Fourier Analysis, D. C. Heath and Co., Boston, MA, 1963. (Cited in Preface and Chapter 2.)

[620] L. Schaeffer and J. Shallit, *The critical exponent is computable for automatic sequences.* Preprint. (Cited in Chapter 8.)

[621] J. Schiffer, *Discrepancy of normal numbers*, Acta Arith. 47 (1986), 175–186. (Cited in Chapter 4.)

[622] A. Schinzel, *On the reduced length of a polynomial with real coefficients*, Funct. Approx. Comment. Math. 35 (2006), 271–306. (Cited in Chapter 3.)

[623] ———, *On the reduced length of a polynomial with real coefficients. II*, Funct. Approx. Comment. Math. 37 (2007), 445–459. (Cited in Chapter 3.)

[624] H. P. Schlickewei, *On products of special linear forms with algebraic coefficients*, Acta Arith. 31 (1976), 389–398. (Cited in Appendix E.)

[625] J. Schmeling, *Symbolic dynamics for β-shifts and self-normal numbers*, Ergodic Theory Dynam. Systems 17 (1997), 675–694. (Cited in Chapter 9.)

[626] K. Schmidt, *On periodic expansions of Pisot and Salem numbers*, Bull. London Math. Soc. 12 (1980), 269–278. (Cited in Chapter 9.)

[627] W. M. Schmidt, *On normal numbers*, Pacific J. Math. 10 (1960), 661–672. (Cited in Chapters 6 & 10.)

[628] ———, *Über die Normalität von Zahlen zu verschiedenen Basen*, Acta Arith. 7 (1961/1962), 299–309. (Cited in Chapters 5 & 6.)

[629] ———, *Normalität bezüglich Matrizen*, J. reine angew. Math. 214/215 (1964), 227–260. (Cited in Chapter 6.)

[630] ———, *Über simultane Approximation algebraischer Zahlen durch Rationale*, Acta Math. 114 (1965), 159–206. (Cited in Chapter 8.)

[631] ———, *On badly approximable numbers and certain games*, Trans. Amer. Math. Soc. 123 (1966), 178–199. (Cited in Chapter 7.)

[632] ———, *Simultaneous approximations to algebraic numbers by rationals*, Acta Math. 125 (1970), 189–201. (Cited in Appendix E.)

[633] ———, *Norm form equations*, Ann. of Math. 96 (1972), 526–551. (Cited in Appendix E.)

[634] ———, *Irregularities of distribution. VII*, Acta Arith. 21 (1972), 45–50. (Cited in Chapter 1.)

[635] ———, Diophantine Approximation, Lecture Notes in Math. 785, Springer, Berlin, 1980. (Cited in Preface, Chapter 7, and Appendices D & E.)

[636] ———, *The subspace theorem in Diophantine approximation*, Compositio Math. 69 (1989), 121–173. (Cited in Appendix E.)

[637] Th. Schneider, Einführung in die transzendenten Zahlen, Springer, Berlina, 1957. (Cited in Chapters 8 & 9.)

[638] C.-P. Schnorr und H. Stimm, *Endliche Automaten und Zufallsfolgen*, Acta Informat. 1 (1971/72), 345–359. (Cited in Chapter 4.)

[639] A. Scremin, *On the period of the continued fraction for values of the square root of power sums*, Acta Arith. 123 (2006), 297–312. (Cited in Chapter 9.)

[640] F. Schweiger, *Normalität bezüglich zahlentheoretischer Transformationen*, J. Number Theory 1 (1969), 390–397. (Cited in Chapter 6.)

[641] ———, Ergodic Theory of Fibred Systems and Metric Number Theory, Oxford Science Publications, The Clarendon Press, New York, 1995. (Cited in Appendix D.)

[642] F. Schweiger and T. Šalát, *Some sets of sequences of positive integers and normal numbers*, Rev. Roumaine Math. Pures Appl. 26 (1981), 1255–1264. (Cited in Chapter 4.)

[643] H. G. Senge and E. G. Straus, PV*-numbers and sets of multiplicity*, Period. Math. Hungar. 3 (1973), 93–100. (Cited in Chapter 6.)

[644] J. O. Shallit, *Simple continued fractions for some irrational numbers*, J. Number Theory 11 (1979), 209–217. (Cited in Chapter 7.)

[645] ———, *Simple continued fractions for some irrational numbers, II*, J. Number Theory 14 (1982), 228–231. (Cited in Chapter 7.)

[646] ———, *Real numbers with bounded partial quotients: a survey*, Enseign. Math. 38 (1992), 151–187. (Cited in Chapter 9.)

[647] Z. Shan, *A note on irrationality of some numbers*, J. Number Theory 25 (1987), 211–212. (Cited in Chapter 10.)

[648] D. Sherwell, *Note on normal numbers and uniform distribution of sequences*, Ann. Univ. Sci. Budapest. Eötvös Sect. Math. 43 (2000), 117–123. (Cited in Chapter 4.)

[649] I. Shiokawa, *A remark on a theorem of Copeland–Erdős*, Proc. Japan Acad. 50 (1974), 273–276. (Cited in Chapter 4.)

[650] I. Shiokawa and S. Uchiyama, *On some properties of the dyadic Champernowne numbers*, Acta Math. Acad. Sci. Hungar. 26 (1975), 9–27. (Cited in Chapter 4.)

[651] T. N. Shorey and R. Tijdeman, Exponential Diophantine Equations, Cambridge Tracts in Mathematics, 87, Cambridge University Press, Cambridge, 1986. (Cited in Appendix F.)

[652] ———, *Irrationality criteria for numbers of Mahler's type*. In: Analytic Number Theory (Kyoto, 1996), 343–351, London Math. Soc. Lecture Note Ser., 247, Cambridge University Press, Cambridge, 1997. (Cited in Chapter 10.)

[653] M. W. Sierpiński, *Sur la valeur asymptotique d'une certaine somme*, Krakau Anz. A (1910), 9–11. (Cited in Chapter 1.)

[654] ———, *Démonstration élémentaire du théorème de M. Borel sur les nombres absolument normaux et détermination effective d'un tel nombre*, Bull. Soc. Math. France 45 (1917), 127–132. (Cited in Chapters 4 & 5.)

[655] K. Sigmund, *Normal and quasiregular points for automorphisms of the torus*, Math. Systems Theory 8 (1974/75), 251–255. (Cited in Chapter 6.)

[656] J. G. Simonsen, *Beta-shifts, their languages, and computability*, Theory Comput. Syst. 48 (2011), 297–318. (Cited in Chapter 9.)

[657] J. Slivka and N. C. Severo, *Measures of sets decomposing the simply normal numbers in the unit interval*, Mathematika 41 (1994), 164–172. (Cited in Chapter 4.)

[658] M. Smorodinsky and B. Weiss, *Normal sequences for Markov shifts and intrinsically ergodic subshifts*, Israel J. Math. 59 (1987), 225–233. (Cited in Chapter 9.)

[659] J. L. Spears and J. E. Maxfield, *Further examples of normal numbers*. Publ. Math. Debrecen 16 (1969), 119–127. (Cited in Chapter 4.)

[660] L. Staiger, *The Kolmogorov complexity of real numbers*, Theoret. Comput. Sci. 284 (2002), 455–466. (Cited in Chapter 5.)

[661] L. P. Starčenko, *Construction of sequences jointly normal with a given one*, Izv. Akad. Nauk SSSR. Ser. Mat. 22 (1958), 757–770; erratum 23 (1958) 635–636 (in Russian). (Cited in Chapters 1, 5 & 6.)

[662] ———, *The construction of a completely uniformly distributed sequence*, Dokl. Akad. Nauk SSSR 129 (1959), 519–521 (in Russian). (Cited in Chapter 1.)

[663] H. Steinhaus, Problem 144, The New Scottish book, Wrocław 1946–1958. (Cited in Chapter 6.)

[664] C. L. Stewart, *On the representation of an integer in two different bases*, J. Reine Angew. Math. 319 (1980), 63–72. (Cited in Chapter 6.)

[665] T. Stoll, *On a problem of Erdős and Graham concerning digits*, Acta Arith. 125 (2006), 89–100. (Cited in Chapter 8.)

[666] ———, *A fancy way to obtain the binary digits of* $759250125\sqrt{2}$, Amer. Math. Monthly 117 (2010), 611–617. (Cited in Chapter 8.)

[667] R. G. Stoneham, *The reciprocals of integral powers of primes and normal numbers*, Proc. Amer. Math. Soc. 15 (1964), 200–208. (Cited in Chapter 5.)

[668] ———, *A study of* 60,000 *digits of the transcendental "e"*, Amer. Math. Monthly 72 (1965), 483–500. (Cited in Chapter 8.)

[669] ———, *On* (j, ε)*-normality in the rational fractions*, Acta Arith. 16 (1969/1970), 221–237. (Cited in Chapter 5.)

[670] ———, *A general arithmetic construction of transcendental non-Liouville normal numbers from rational fractions*, Acta Arith. 16 (1969/1970), 239–253. (Cited in Chapter 5.)

[671] ———, *On absolute* (j, ε)*-normality in the rational fractions with applications to normal numbers*, Acta Arith. 22 (1972/73), 277–286. (Cited in Chapter 5.)

[672] ———, *On the uniform ε-distribution of residues within the periods of rational fractions with applications to normal numbers*, Acta Arith. 22 (1973), 371–389. (Cited in Chapter 5.)

[673] ———, *Some further results concerning (j, ε)-normality in the rationals*, Acta Arith. 26 (1974/75), 83–96. (Cited in Chapter 5.)

[674] ———, *Normal recurring decimals, normal periodic systems (j, ε)-normality, and normal numbers*, Acta Arith. 28 (1975/76), 349–361. (Cited in Chapter 5.)

[675] ———, *On a sequence of (j, ε)-normal approximations to $\pi/4$ and the Brouwer conjecture*, Acta Arith. 42 (1983), 265–279. (Cited in Chapter 5.)

[676] O. Strauch, *On distribution functions of* $\xi(3/2)^n$ mod 1, Acta Arith. 81 (1997), 25–35. (Cited in Chapter 3.)

[677] O. Strauch and R. Nair. *Unsolved problems.* Available at `http://www.boku.ac.at/MATH/udt/`. (Cited in Chapter 10.)

[678] O. Strauch and Š. Porubský, Distribution of Sequences: A Sampler, Schriftenreihe der Slowakischen Akademie der Wissenschaften, 1, Peter Lang, Frankfurt am Main, 2005. (Cited in Preface and Chapter 10.)

[679] E. Strzelecki, *On sequences* $\{\xi t_n\}(\text{mod} 1)$, Canad. Math. Bull. 18 (1975), 727–738. (Cited in Chapter 2.)

[680] F. Supnick, H. J. Cohen, and J. F. Keston, *On the powers of a real number reduced modulo one*, Trans. Amer. Math. Soc. 94 (1960), 244–257. (Cited in Chapter 2.)

[681] P. Szüsz and B. Volkmann, *On numbers containing each block infinitely often*, J. Reine Angew. Math. 339 (1983), 199–206. (Cited in Chapter 8.)

[682] ———, *On numbers with given digit distributions*, Arch. Math. (Basel) 52 (1989), 237–244. (Cited in Chapter 4.)

[683] ———, *A combinatorial method for constructing normal numbers*, Forum Math. 6 (1994), 399–414. (Cited in Chapter 4.)

[684] A. Thue, *Über Annäherungswerte algebraischer Zahlen*, J. Reine Angew. Math. 135 (1909), 284–305. (Cited in Appendix E.)

[685] ———, *Über die gegenseitige Lage gleicher Teile gewisser Zeichenreihen*, Kra. Vidensk Selsk. Skr. Mat. Nat. Kl. 2 (1912) No. 1; reprinted in Selected Mathematical Papers of Axel Thue, T. Nagell, ed., Universitetsforlaget, Oslo, 1977, pp. 413–478. (Cited in Chapter 8 and Appendix A.)

[686] ———, *Über eine Eigenschaft, die keine transcendente Grössen haben kann*, Kra. Vidensk Selsk. Skr. Mat. Nat. Kl. 2 (1912) No. 20, 15 pp.; reprinted in Selected Mathematical Papers of Axel Thue, T. Nagell, ed., Universitetsforlaget, Oslo, 1977, pp. 479–491. (Cited in Chapter 2.)

[687] R. Tijdeman, *Note on Mahler's 3/2-problem*, Norske Vid. Selsk. Skr. 16 (1972), 1–4. (Cited in Chapters 2 & 3.)

[688] G. Troi and U. Zannier, *Note on the density constant in the distribution of self-numbers. II*, Boll. Unione Mat. Ital. Sez. B Artic. Ric. Mat. (8) 2 (1999), 397–399. (Cited in Chapter 8.)

[689] A. Turing, *On computable numbers, with an application to the Entscheidungsproblem*, Proc. Lond. Math. Soc. 42 (1936), 230–265. (Cited in Chapter 5.)

[690] ———, *A note on normal numbers*. In: Collected Works of A. M. Turing, Pure Mathematics, (ed. J. L. Britton), North-Holland, Amsterdam, 1992, pp. 117–119. (Cited in Chapter 5.)

[691] E. Ugalde, *An alternative construction of normal numbers*, J. Théor. Nombres Bordeaux 12 (2000), 165–177. (Cited in Chapter 5.)

[692] ———, *A constructible set of normals with positive measure*. In: Dynamical Systems (Luminy-Marseille, 1998), 220–228, World Scientific Publishing, River Edge, NJ, 2000. (Cited in Chapter 5.)

[693] R. Urban, *On density modulo 1 of some expressions containing algebraic integers*, Acta Arith. 127 (2007), 217–229. Corrigendum in Acta Arith. 141 (2010), 209–210. (Cited in Chapter 2.)

[694] ———, *Sequences of algebraic integers and density modulo 1*, J. Théor. Nombres Bordeaux 19 (2007), 755–762. (Cited in Chapter 2.)

[695] ———, *Sequences of algebraic numbers and density modulo 1*, Publ. Math. Debrecen 72 (2008), 141–154. (Cited in Chapter 2.)

[696] ———, *A survey of results on density modulo 1 of double sequences containing algebraic numbers*, Acta Math. Univ. Ostrav. 16 (2008), 31–43. (Cited in Chapter 2.)

[697] ———, *Algebraic numbers and density modulo 1*, J. Number Theory 128 (2008), 645–661. Corrigendum in J. Number Theory 129 (2009), 2879–288. (Cited in Chapter 2.)

[698] ———, *Multiplicatively independent integers and dense modulo 1 sets of sums*, Unif. Distrib. Theory 4 (2009), 27–33. (Cited in Chapter 2.)

[699] ———, *A note on density modulo 1 of certain sets of sums*, Funct. Approx. Comment. Math. 42 (2010), 29–35. (Cited in Chapter 2.)

[700] M. Urbański, *Diophantine approximation and self-conformal measures*, J. Number Theory 110 (2005), 219–235. (Cited in Chapter 7.)

[701] ———, *Diophantine approximation for conformal measures of one-dimensional iterated function systems*, Compos. Math. 141 (2005), 869–886. (Cited in Chapter 7.)

[702] R. C. Vaughan and T. D. Wooley, *Waring's problem: a survey*. In: Number Theory for the Millennium, III (Urbana, IL, 2000), 301–340, A. K. Peters, Natick, MA, 2002. (Cited in Chapter 3.)

[703] P. Veerman, *Symbolic dynamics and rotation numbers*, Phys. A 134 (1986), 543–576. (Cited in Chapter 3.)

[704] ———, *Symbolic dynamics of order-preserving orbits*, Phys. D 29 (1987), 191–201. (Cited in Chapter 3.)

[705] A. Venkatesh, *The work of Einsiedler, Katok and Lindenstrauss on the Littlewood conjecture*, Bull. Amer. Math. Soc. 45 (2008), 117–134. (Cited in Chapter 2.)

[706] T. Vijayarhagavan, *On decimals of irrational numbers*, Proc. Indian Acad. Sci., Sect. A. 12 (1940), 20. (Cited in Chapter 3.)

[707] ———, *On the fractional parts of the powers of a number. I*, J. London Math. Soc. 15 (1940), 159–160. (Cited in Chapter 2.)

[708] ———, *On the fractional parts of the powers of a number. II*, Proc. Cambridge Philos. Soc. 37 (1941), 349–357. (Cited in Chapters 2 & 3.)

[709] ———, *On the fractional parts of the powers of a number. III*, J. London Math. Soc. 17 (1942), 137–138. (Cited in Chapter 2.)

[710] ———, *On the fractional parts of powers of a number. IV*, J. Indian Math. Soc. (N.S.) 12, (1948), 33–39. (Cited in Chapter 2.)

[711] J. Ville, Étude Critique de la Notion de Collectif, Monographie des Probabilités, Gauthier-Villars, Paris, 1939. (Cited in Chapter 4.)

[712] B. Volkmann, *Über Hausdorffsche Dimensionen von Mengen die durch Ziffereigenschaften charakterisiert sind. III*, Math. Z. 59 (1953), 259–270. (Cited in Chapter 7.)

[713] ———, *Über Hausdorffsche Dimensionen von Mengen die durch Ziffereigenschaften charakterisiert sind. V*, Math. Z. 65 (1956), 389–413. (Cited in Chapter 7.)

[714] ———, *Über Hausdorffsche Dimensionen von Mengen, die durch Ziffereigenschaften charakterisiert sind. VI*, Math. Z. 68 (1958), 439–449. (Cited in Chapters 4 & 7.)

[715] ———, *Gewinnmengen*, Arch. Math. 10 (1959), 235–240. (Cited in Chapter 7.)

[716] ———, *On non-normal numbers*, Compositio Math. 16 (1964), 186–190. (Cited in Chapter 10.)

[717] ———, *Über extreme Anormalität bei Ziffernentwicklungen*, Math. Ann. 190 (1970/1971), 149–153. (Cited in Chapter 6.)

[718] ———, *Verallgemeinerung eines Satzes von Maxfield*, J. Reine Angew. Math. 271 (1974), 203–213. (Cited in Chapter 4.)

[719] ———, *On modifying constructed normal numbers*, Ann. Fac. Sci. Toulouse Math. 1 (1979), 269–285. (Cited in Chapter 4.)

[720] ———, *On the Cassels–Schmidt theorem. I*, Bull. Sci. Math. 108 (1984), 321–336. (Cited in Chapter 6.)

[721] ———, *On the Cassels–Schmidt theorem. II*, Bull. Sci. Math. 109 (1985), 209–223. (Cited in Chapter 6.)

[722] G. Wagner, *On rings of numbers which are normal to one base but non-normal to another*, J. Number Theory 54 (1995), 211–231. (Cited in Chapter 5 & 6.)

[723] M. Waldschmidt, *Linear independence measures for logarithms of algebraic numbers*. In: Diophantine Approximation (Cetraro, 2000), 250–344, Lecture Notes in Math., 1819, Springer, Berlin, 2003. (Cited in Chapter 10.)

[724] ———, *Words and transcendence*. In: Analytic Number Theory, 449–470, Cambridge University Press, Cambridge, 2009. (Cited in Chapter 10.)

[725] D.D. Wall, Normal numbers. PhD thesis, University of California, Berkeley, CA, 1949. (Cited in Chapter 4.)

[726] P. Walters, Ergodic Theory – Introductory Lectures, Lecture Notes in Mathematics, Vol. 458, Springer, Berlin, 1975. (Cited in Chapter 1 and Appendix C.)

[727] T. C. Watson, *An improved characterization of normal sets and some counter-examples*, Israel J. Math. 109 (1999), 173–179. (Cited in Chapter 4.)

[728] B. Weiss, *Almost no points on a Cantor set are very well approximable*, R. Soc. Lond. Proc. Ser. A Math. Phys. Eng. Sci. 457 (2001), 949–952. (Cited in Chapter 7.)

[729] ———, *Dynamics on parameter spaces: submanifold and fractal subset questions*, Rigidity in Dynamics and Geometry (Cambridge, 2000), 425–440, Springer, Berlin, 2002. (Cited in Chapter 7.)

[730] B. Weiss, Single Orbit Dynamics, CBMS Regional Conference Series in Mathematics, 95, American Mathematical Society, Providence, RI, 2000. (Cited in Chapters 2 & 4.)

[731] H. Weyl, *Über die Gibbssche Erscheinung und verwandte Konvergenzphänomene*, Rend. Circ. Mat. Palermo 30 (1910), 377–407. (Cited in Chapter 1.)

[732] ———, *Über ein Problem aus dem Gebiete der diophantischen Approximationen*, Gött. Nachr. (1914), 234–244. (Cited in Chapter 1.)

[733] ———, *Über die Gleichverteilung von Zahlen mod. Eins*, Math. Ann 77 (1916), 313–352. (Cited in Chapter 1.)

[734] F. Wielonsky, *Hermite–Padé approximants to exponential functions and an inequality of Mahler*, J. Number Theory 74 (1999), 230–249. (Cited in Chapter 2.)

[735] J. M. Wills, *Zwei Sätze über inhomogene diophantische Approximation von Irrationalzahlen*, Monatsh. Math. 71 (1967), 263–269. (Cited in Chapter 2.)

[736] S. M. J. Wilson, *Limit points in the Lagrange spectrum of a quadratic field*, Bull. Soc. Math. France 108 (1980), 137–141. (Cited in Chapter 9.)

[737] R. Winkler, *Some constructive examples in uniform distribution on finite sets and normal numbers*, Anz. Österreich. Akad. Wiss. Math.-Natur. Kl. 126 (1989), 1–8. (Cited in Chapter 4.)

[738] T. H. Wolff, Lectures on Harmonic Analysis, University Lecture Series, 29, American Mathematical Society, Providence, RI, 2003. (Cited in Chapter 7.)

[739] J. Wu, *Continued fraction and decimal expansions of an irrational number*, Adv. Math. 206 (2006), 684–694. (Cited in Chapter 9.)

[740] ———, *An iterated logarithm law related to decimal and continued fraction expansions*, Monatsh. Math. 153 (2008), 83–87. (Cited in Chapter 9.)

[741] T. Zaïmi, *On integer and fractional parts of powers of Salem numbers*, Arch. Math. (Basel) 87 (2006), 124–128. (Cited in Chapter 3.)

[742] ———, *An arithmetical property of powers of Salem numbers*, J. Number Theory 120 (2006), 179–191. (Cited in Chapter 3.)

[743] ———, *On the spectra of Pisot numbers*, Glasgow Math. J. 54 (2012), 127–132. (Cited in Chapter 2.)

[744] ———, *Comments on the distribution modulo one of powers of Pisot and Salem numbers*, Publ. Math. Debrecen. To appear. (Cited in Chapter 2.)

[745] A. Zame, *The distribution of sequences modulo* 1, Canad. J. Math. 19 (1967), 697–709. (Cited in Chapter 2.)

[746] ———, *On normal sets of numbers*, Acta Arith. 20 (1972), 147–154. (Cited in Chapter 4.)

[747] U. Zannier, Some Applications of Diophantine Approximation to Diophantine Equations, Editrice Forum, Udine, 2003. (Cited in Chapter 9 and Appendix E.)

[748] Y. C. Zhu, *On a problem of Mahler concerning the transcendence of a decimal fraction*, Acta Math. Sinica 24 (1981), 246–253 (in Chinese). (Cited in Chapter 4.)

[749] ———, *A note of irrationality measure of Mahler's transcendental fractions*, Acta Math. Sinica 43 (2000), 1–8 (in Chinese). (Cited in Chapter 4.)

[750] Y. C. Zhu, L. X. Wang, and G. S. Xu, *On the transcendence of a class of series*, Kexue Tongbao (English Edition) 25 (1980), 1–6. (Cited in Chapter 8.)

[751] W. Zudilin, *A new lower bound for* $\|(3/2)^k\|$, J. Théor. Nombres Bordeaux 19 (2007), 311–323. (Cited in Chapter 3.)

Index

For EU product safety concerns, contact us at Calle de José Abascal, 56–1°,
28003 Madrid, Spain or eugpsr@cambridge.org.

www.ingramcontent.com/pod-product-compliance
Lightning Source LLC
LaVergne TN
LVHW102008210926
848483LV00010B/160

DIE REIHE
Archivbilder

WADERN

VERZEICHNISS der Märkte in Wadern für das Jahr 1880.

1.	Montag vor Lichtmeß	am 26. Januar
2.	Zweiter Montag in den Fasten	23. Februar
3.	„ „ im März	8. März
4.	„ „ nach Oster-Sonntag	5. April
5.	Montag vor Pfingsten	10. Mai
6.	Dritter Montag nach Pfingsten	31. Mai
7.	Montag vor Laurentius	9. August
8.	„ „ Mathäus	20. September
9.	„ „ Allerheiligen	25. October
10.	„ „ Nikolaus	29. November

Dienstags in jeder Woche wird Schweinemarkt abgehalten.

Wadern, den 27. November 1879.

Der Bürgermeister,

Friedrich.

J. F. Ziegler'sche Buchdruckerei in Merzig.

Erster Waderner Marktkalender für das Jahr 1880.

DIE REIHE

Archivbilder

WADERN

Autorenteam der Stadt Wadern

SUTTON VERLAG

Sutton Verlag GmbH
Hochheimer Straße 59
99094 Erfurt
www.suttonverlag.de

ISBN: 978-3-89702-254-6

Druck: Books on Demand GmbH, Norderstedt, Deutschland

Waderner Frauen präsentieren, vermutlich anläßlich eines Erntedankfestes Mitte der dreißiger Jahre, nach alten Überlieferungen erstellte Hochwälder Frauentrachten.

Inhaltsverzeichnis

Vorwort

Mit der Herausgabe dieses historischen Bildbandes der Stadt Wadern soll ein Teil der wechselvollen Geschichte unserer Kommune im Bild dokumentiert werden.

Das Buch hält Erinnerungen an unsere Stadt wach, während gleichzeitig der Wandel im Laufe einer Epoche in historischen Fotografien und Ansichtskarten anschaulich dokumentiert wird: ein Wandel vom ehedem ländlich-idyllischen Marktflecken im Kernort Wadern zu einem Zentrum für Handel, Gewerbe, Dienstleistungen und Verwaltung.

Das Werk bildet einen wesentlichen Beitrag zu unserer Heimatgeschichte. Die alten Fotografien eröffnen eine neue Sicht auf Entwicklung und Veränderung des Erscheinungsbildes und der Lebensverhältnisse in der heutigen Stadt Wadern mit ihren 14 Stadtteilen. Damit ist das Buch Album, Chronik und Nachschlagewerk in einem.

Dem Verlag und dem Autorenteam sowie allen, die für den historischen Bildband der Stadt Wadern Fotografien und Postkarten überließen, gilt mein herzlicher Dank.

Allen Leserinnen und Lesern wünsche ich viel Spaß bei ihrem Spaziergang durch Waderns vergangene und doch oft noch lebendige Zeiten.

Wadern, im Juni 2000

Fredi Dewald
Bürgermeister der Stadt Wadern

Bildnachweis

Archiv Fotohaus Lohrig: Seiten 12 u., 13 o., 16, 21, 24, 39 o., 52 o., 65 o., 72 u., 87 u., 95 o., 101, 104, 109 o., 113 o., 116 o., 118 o., 125; Archiv Karl-Heinz Müller: Seiten 10/11, 12 o., 13 u., 14, 15, 18/19, 20, 63, 79 u., 82, 84 o., 89 o., 92/93, 94, 96, 97, 98/99, 100, 102, 103, 106/107, 108, 109 o., 110, 111, 112, 116 u., 118 u., 120 o., 124, 126/127; Archiv Stadt Wadern: Seiten 9, 88; Archiv Richard Scheid: Seiten 68/69, 90 o., 123; Michael Biesel: Seiten 54/55; Johanna Birtel: Seiten 45 o., 122 u.; Lorenz Birtel: Seite 61 o.; Hugo Dörr: Seiten 30 o., 31 o., 38 u., 44 o., 57, 60 u., 86 o., 115, 122 o.; Friedrich Ebert: Seiten 4, 17, 40 o., 58, 59 o., 91; Anton Faber: Seiten 40 u., 89 u.; Heimat- und Geschichtsfreunde Büschfeld: Seiten 32 u., 119; Günther Junk: Seiten 29, 33 u., 42, 61 u., 62, 66 u., 70 u., 73 o., 76, 77, 84 u., 90 u., 120 u., 121; Christoph Kaub: Seiten 32 o., 37, 39 u., 41, 51, 56 o., 59 u., 71 o., 73 u., 87 o.; Richard Klein: Seiten 26 u., 28 u., 38 u., 45 u., 74/75, 83, 95 u., 105 o., 113 u.; Paul Koch: Seiten 60 o., 117; Kolpingkapelle Steinberg: Seiten 53 u., 66 o.; Landesarchiv Saarbrücken: Seite 2; Alban Lauer: Seite 79 o.; Arnold Lauer (U): Seiten 44 u., 46, 47, 48/49, 50; H. Lauer: Seiten 22/23, 25, 26 o., 27; Hermann Leidinger: Seiten 67 o., 70 o.; Walter Lohrig: Seiten 43 o., 71 u.; Grete Meyer-Thome: Seiten 36 u., 43 u., 64, 78, 80/81, 85 o.; Josef Molter: Seiten 28 o., 114; Anneliese Mörsdorf: Seiten 67 u., 72 o.; Helmut Müller: Seiten 35, 36 o.; Musikverein „Lyra" Noswendel; Seite 65 o.; Christa Peifer: Seiten 31 u., 33 o., 34, 52 u., 85 u., 96 u.; Gerd Simon: Seiten 30 u., 53 o., 56 u., 105 u.

Einleitung

Sie werden verdutzt dreingeschaut haben, die Hochwälder Bauern, als Graf Joseph Anton von Öttingen-Sötern im Jahre 1758 aus dem fernen Nördlinger Ries nach Wadern übersiedelte, um forthin in seiner Herrschaftsdependance Dagstuhl Hof zu halten. Doch auch dem Grafen wird nicht sonderlich wohl zumute gewesen sein, als er seinen Stammsitz Schloß Hohenbaldern verließ, um den Schulden, die ihm sein Vater hinterlassen hatte, zu entkommen. Für Wadern allerdings war dies ein Glücksfall, denn seitdem entwickelte sich das Dorf zum Zentrum des Hochwaldes. Wadern erhielt die Marktrechte, Handwerker und Händler wurden angesiedelt. Die wirtschaftlich schwache Region erlebte seitdem einen bescheidenen Aufschwung, der die Grundlage für die heutige Bedeutung der Stadt Wadern bildet.

Die Geschichte beginnt jedoch schon früher: Nacheinander besiedelten der keltische Stamm der Treverer, die Römer und die Franken die Täler der Wadrill, Prims und Löster. Die erste urkundliche Erwähnung eines Stadtteils findet sich für Büschfeld in einer Grenzbeschreibung aus dem Jahre 802. Waderns Ersterwähnung folgte um 950 in Verbindung mit der Abtei Mettlach. Geprägt wurde die territoriale Entwicklung durch die Errichtung der Herrschaft Dagstuhl, die sich mit der Burg Dagstuhl einen eigenen Sitz erbaute. Die durch Erbfolge zersplitterte Herrschaft wurde 1635 von Kurfürst Philipp Christoph von Sötern, dem Bischof von Trier, zu einer Familienbesitzung wieder zusammengeführt, bevor sie 1698 in den Besitz der Grafen von Öttingen überging, die damit zu Herren von Dagstuhl wurden.

Seit der Grafenzeit entwickelte sich Wadern kontinuierlich bis zur Verleihung der Stadtrechte im Jahre 1978. Nach der französischen Revolution wurde der Ort Bürgermeisterei im französischen Saar-Departement Trier. Diese Funktion behielt er auch zur preußischen Zeit. Durch die Abtrennung des Saargebietes und der damit verbundenen Trennung des Kreises Merzig in Stammkreis und Restkreis Wadern erhielt der Ort im Jahre 1920 eine weitere Aufwertung: In Wadern wurden zahlreiche Ämter und Verwaltungseinrichtungen angesiedelt. Erst 1946 wurden die formal noch getrennten Kreise zum heutigen Landkreis Merzig-Wadern wiedervereinigt.

Mit der Gebietsreform des Jahres 1974 entstand eine neue Gemeindestruktur. Wadern wurde zum Sitz der Amtsverwaltung für nunmehr 14 vorher eigenständige Gemeinden. Ein weiterer großer Sprung in der Entwicklung folgte 1978: Die junge Gemeinde erhielt die Stadtrechte und nimmt seither die Funktion eines Mittelzentrums im nördlichen Saarland wahr. Knapp 17500 Menschen leben in den Stadtteilen und genießen die kurzen Wege, denn die junge Stadt bietet alles, was man zum Leben braucht.

Wenn dieser Bildband auch nur Fotografien aus der Zeit von der Jahrhundertwende bis in die sechziger Jahre zeigt, so sind doch viele Spuren der Geschichte früherer Jahrhunderte in den Aufnahmen und Ansichten erkennbar. Dem von Graf Joseph Anton erbauten Schloß Dagstuhl, dem heutigen Wahrzeichen der Stadt, ist ein eigenes Kapitel gewidmet. Ebenso beschäftigt sich der Band mit der Landwirtschaft, die lange Zeit die entscheidende Einnahmequelle für die Menschen der Region bildete, bevor es Arbeit in den Gruben und Hütten des Saartals gab. Ein Kapitel beschäftigt sich mit den Wirren der Weltkriege und mit dem Restkreis Merzig-Wadern. Aber auch der Alltag der Menschen der Region kommt nicht zu kurz, der gleichsam harte Arbeit und vielgestaltige Lebensfreude in Vergangenheit und Gegenwart beinhaltet. Die Gesichter der Menschen in den Vordergrund zu stellen, war die erklärte Absicht des Autorenteams, denn nur so kann die allgemeine Historie zu einer persönlichen und erfahrbaren Geschichte werden. Die Eltern, Großeltern und Urgroßeltern der heutigen Stadtbewohner sind die Hauptdarsteller des Buches.

Ein fotografischer Chronist der Jahrhundertwende sei an dieser Stelle besonders erwähnt. Der Fotograf August Fuisting hat mit großer Meisterschaft und Liebe zum Detail Aufnahmen gemacht und einen beachtlichen Fundus an hochwertigen historischen Ansichten hinterlassen. Ohne seine Arbeit zu Beginn des 20. Jahrhunderts wäre dieser Bildband nicht realisierbar gewesen.

Die Recherche der Bilder und deren Inhalte war nicht einfach. Das Autorenteam war auf Unterstützung von vielen Seiten angewiesen. All jenen, die zur Entstehung dieses Bildbandes beigetragen haben, sei ganz herzlich gedankt. Insbesondere gilt dies für die vielen Sammler und Privatpersonen, die ihre Archive und Schuhkartons öffneten, und die Vorstandsmitglieder des Vereins für Heimatkunde Wadern für die inhaltliche Beratung. Sicherlich werden Sie als Betrachter noch viele Lücken und den ein oder anderen Fehler in diesem Buch entdecken. Lassen Sie dies Ansporn für die Vervollständigung eines städtischen Bildarchives sein. Und vielleicht werden noch so viele neue Ansichten zu Tage gefördert, daß wir uns mit einem zweiten Bildband beschäftigen können.

Das Autorenteam der Stadt Wadern
Hanns Peter Ebert und Wolfgang Birtel

1

Schloß Dagstuhl

Graf Joseph Anton von Öttingen-Sötern (1721–1778), Erbauer von Schloß Dagstuhl.

Zwischen 1760 und 1762 entstand Schloß Dagstuhl im Tal der Löster als Stammsitz des Grafen Joseph Anton von Öttingen Sötern, nachdem sich der Wiederaufbau der Burg Dagstuhl als

zu teuer erwiesen hatte. Das Bild zeigt das Schloß vor dem Umbau um 1900 mit Blick auf die Schlosskapelle.

Nachdem die letzte Gräfin Maria Antonia von Zeil-Wurzach im Zuge der Französischen Revolution geflohen war, übernahm die Familie de Lasalle von Louisenthal im Jahre 1806 das Schloß. Als verdienter Offizier der französischen Armee siedelte der Baron aus der fernen Karibik in den Hochwald über. Das Bild zeigt die Familie de Lasalle bei Schloß Dagstuhl um 1875.

Die Familie de Lasalle, im Hintergrund das Löstertal. Gut zu erkennen ist die Gartenanlage und das Brauhaus links. Graf Joseph Anton hatte im Jahre 1762 die erste Brauerei in der Herrschaft errichtet. Als zweiter von links ist das Familienoberhaupt Rudolf de Lasalle von Louisenthal zu erkennen, der von 1869 bis 1875 Landrat des Kreises Merzig war.

Die letzte Baronin Reza de Lasalle von Louisenthal ließ im Jahre 1905 den so markanten Turmteil des Schlosses im neugotischen Stil zwischen Kapelle und Haupthaus erbauen. Das Bild zeigt die gefährlichen Arbeiten in vollem Gange.

Das Schloß erstrahlt nach dem Umbau des Turmes in neuem Glanze. Im Rahmen der Umbaumaßnahmen wurde auch die Familiengruft der Familie de Lasalle, welche rechts im Bild zu sehen ist, erweitert.

Der Altar der barocken Schloßkapelle um die Jahrhundertwende. Die Kapelle wurde um 1763 von Graf Joseph Anton erbaut, der den Innenraum mit einem Altar des süddeutschen Spätbarocks aus seiner Heimat ausstattete. Vor dem Altar kniet der letzte Baron, Theo de Lasalle von Louisenthal, der im Jahre 1959 verstarb.

Blick von der Bannmühle in das Löstertal Richtung Schloß Dagstuhl, um 1900. Die Bannmühle gehörte zu den Schloßanlagen und existierte vermutlich schon vor Schloß Dagstuhl. Ein Nachfolgebau trägt heute noch den Namen Bannmühle.

Auch zu den Zeiten der Familie de Lasalle von Louisenthal wurde auf dem Gut Schloß Dagstuhl intensiv Landwirtschaft betrieben. Die Arbeiter der Bannmühle haben sich um 1900 zum Fototermin vor der Mühle aufgestellt. Einige von ihnen haben die für diese Zeit typischen Tonpfeifen im Mund.

Das mächtige Hofhaus bei Schloß Dagstuhl, um 1892 von der Lösterbrücke aus gesehen. Das Gebäude wurde um 1920 abgerissen. Der Bereich ist heute mit einem Zaun eingefriedet und umschließt den Vorplatz des Schlosses.

Der Zunftbrief von 1774, den Graf Joseph Anton für die in der Herrschaft angesiedelten Handwerker ausstellte, ist ein wichtiges Dokument der Stadtgeschichte. Die Bildnisse zeigen links Schloß Dagstuhl und rechts die Fahrt des Grafen mit Gefolge nach Wadern an einem Markttag. Der Zunftbrief ist heute im Heimatmuseum ausgestellt.

2

Der Waderner Marktplatz

Ansicht des Marktplatzes mit „Hauptwache“ des gräflichen Militärs auf der Nordseite, um 1780.

Der Marktplatz im Jahre 1908, aus Richtung der Kirche gesehen: Die ehemalige „Hauptwache“ auf der Nordseite wurde Mitte des 19. Jahrhunderts von Dompropst Nikolaus München erworben und um ein Geschoß erhöht. Nach dem Tod des Dompropstes erbte die Familie des Waderner Tuchfabrikanten Lauer das Schlößchen, welches im Volksmund „Lauers Schlößchen“ genannt wurde.

Der Marktplatz in Richtung Pfarrkirche Allerheiligen, vor 1900. Der Marktbrunnen weist noch die barocke Laterne über dem Kapitell der Brunnensäule auf. Auch hier hat der Fotograf Fuisting die Marktszene besonders „choreografiert“.

Viehmarkt in Wadern, um 1910. Die Waderner Viehmärkte gehörten seit der Mitte des 19. Jahrhunderts zu den wichtigsten im Regierungsbezirk Trier.

Viehmarkt in Wadern im Jahre 1893. Der seit 1815 übliche Schweinemarkt wurde an Dienstagen abgehalten. Auf Kuh- und Pferdegespannen wurden Ferkel und Läuferschweine angeliefert. Bis 1900 war die Anzahl der Schweinemärkte auf 52 pro Jahr angestiegen.

Die Markttage waren natürlich auch immer ein gesellschaftliches Ereignis. Man traf sich, handelte und feierte dann den erfolgreichen Geschäftsabschluß in den Gaststätten rund um den Markt. Bis heute findet jeweils am letzten Mittwoch im Monat ein Krammarkt statt, welcher auf die dem Ort 1765 verliehenen Marktrechte zurückgeht.

Laurentius-Kirmes auf dem Marktplatz, um 1900. Besonders auffällig auf diesem Kirmesbild sind die vielen Kinder vor dem Karussell. Damit auch die Erwachsenen am Abend den Kirmesfreuden

frönen konnten, sorgten viele Petroleumlampen in der Karussellmitte für die notwendige Helligkeit.

Der Marktbrunnen ist bis heute ein beliebter Treffpunkt in Wadern. Um die Jahrhundertwende haben sich ehrenwerte Waderner Bürger zum Fototermin eingefunden. Auch die Hunde sind für das Foto speziell in Pose gebracht worden.

Treffpunkt Marktbrunnen: Knapp 30 Jahre später hat sich das Bild kaum verändert, nur das Automobil hat Einzug gehalten.

Detailaufnahme an der Ostseite des Marktplatzes. Der wunderschöne Balkon an der „Gaststätte Brücker“ ist in den dreißiger Jahren noch vorhanden. Schon zu Zeiten des Grafen Joseph Anton soll es hier ein Gasthaus gegeben haben.

Der Marktplatz in den dreißiger Jahren, vom Kirchturm aus gesehen. Der Platz zeigt sich weiterhin wenig verändert. Einzig das „Lauers Schlößchen“ ist mit Efeu bewachsen, und der Marktbrunnen hat wieder eine Laterne. Gut zu erkennen ist auch die schnurgerade Anlage der Oberstraße.

Der Marktplatz von der Ecke Oberstraße/Gartenfeldstraße in den dreißiger Jahren. An der ehemaligen Wirtschaft „Nikolaus München“ in der Mitte des Fotos sind die Zeichen der Zeit zu erkennen: Die NSDAP hat hier ihre Zentrale eingerichtet.

Der Marktbrunnen wird wegen des verstärkten Verkehrs Ende der vierziger Jahre abgebaut und für viele Jahre „eingemottet".

Blick auf den Marktplatz, 1957. Das „Lauers Schlößchen" wurde 1945 zerstört. Auch der Brunnen ist bereits abgebaut. Der Platz hat sein Gesicht grundlegend verändert.

Der Marktbrunnen wird in den sechziger Jahren am Kopfende des Marktplatzes wieder aufgebaut. Der Lehrer und Heimatkundler Hans Schwendler, im Bildvordergrund zu sehen, ist als Initiator maßgeblich daran beteiligt.

Der Marktplatz, Anfang der siebziger Jahre. Die Kreissparkasse wurde neben dem Rathaus neu gebaut, der Marktplatz dient zum Parken. Von der barocken Marktplatzanlage ist nur noch wenig übriggeblieben. Die letzte entscheidende Umgestaltung erlebte der Marktplatz dann im Jahre 1987.

3

Die Landwirtschaft als Lebensgrundlage

Johann Görgen aus Morscholz mit seinem gemischten Kuh- und Pferdegespann bei der Heimfahrt nach geleisteter Feldarbeit, 1953.

Nach der Jahrhundertwende wurde die Arbeit in der Landwirtschaft immer mehr mechanisiert. Um 1918 ist hier eine dampfbetriebene Dreschmaschine in Nunkirchen im Einsatz. Tatkräftig helfen auch der Dorflehrer Loste, links hinter dem Sack, und der „lange Briefträger" Demmer, dritter von rechts, mit.

Eine ähnliche Dreschmaschine steht um 1910 in Wadern auf dem Feld. Das Schwungrad der Dampfmaschine und der Treibriemen, der mit der Dreschmaschine verbunden ist, sind gut zu sehen. Bis in die fünfziger Jahre gehörte die schweißtreibende Arbeit an der Dreschmaschine zur jährlichen Ernte.

Jahrhundertelang war die Hochwaldregion von der Landwirtschaft geprägt. Hier wird gerade eine Pause bei der Feldarbeit eingelegt. Auch das Ochsengespann kann sich einmal ausruhen. Aufnahme aus den dreißiger Jahren.

Das Bauernhaus Wecker in Bardenbach, um 1920. Die Familie Wecker betrieb zu dieser Zeit noch Landwirtschaft. Das Haus entspricht der typischen Bauweise der Region: Das Trierische Eindachhaus hat einen Wohn- und einen Scheunentrakt.

Getreidemähen auf dem Waderner Mühlenberg in den dreißiger Jahren: Auf der Mähmaschine sitzen Nikolaus Molter (links) und Heinrich Molter.

Blick auf den Dreschplatz „Auf dem Wacken" in Büschfeld, 1935. In der Mitte ist die Dreschmaschine zu erkennen, zu der das geschnittene Getreide mit Leiterwagen angeliefert wird.

Maria Kasper vom Dösterhof ist mit ihrem Pferd Max, Gespann und Milchkannen unterwegs zu ihren Kunden in Wadern. Das Foto ist in den vierziger Jahren auf dem Weg von Altland nach Lockweiler aufgenommen worden.

Die Morscholzer Johann Müller und Alfred Meyer sind um 1930 mit ihrem massiven Eichenstamm in der Waderner Kräwigstraße unterwegs zum Sägewerk Kirsten. Die Waldwirtschaft ist seit jeher in der Hochwaldregion von besonderer Bedeutung.

Die Feldarbeiter mußten auch ihre Technik in Ordnung halten. Hier wird in der Nähe des Dösterhofes bei Altland in den dreißiger Jahren ein defekter Schlauch geflickt.

In den dreißiger Jahren kam auch schon schweres Gerät in der Landwirtschaft zum Einsatz. Diese Raupe hatte ihr Einsatzgebiet rund um den Dösterhof.

Auf der Flur „Im Stockland“ in Morscholz ist in den sechziger Jahren Kartoffelerntezeit wie in vielen Jahrzehnten zuvor. Die Ardenner Stute „Bella“ zieht den Pflug, der die Kartoffeln aus der Erde hebt.

Die Kartoffelernte ist eingefahren und wird verladen. Von links sind Helmut Müller, Peter Hoffmann und Peter Hoffmann junior zu sehen.

Der Fotograf August Fuisting, diesmal mit einem „landwirtschaftlichen Motiv“ selbst im Bild. Um 1911 hat es sich die Familie Fuisting auf Roß und Pflug bequem gemacht. Gespannbesitzer Matthias Rech, genannt „Reß Matz“, hat die Zügel fest im Griff.

Mit der Technisierung der Landwirtschaft begann auch der Landmaschinenhandel. Die erste landwirtschaftliche Ausstellung in Nunkirchen fand im Jahre 1923 statt. Organisiert hatte sie Elektromeister Mathias Thome, der seine neuesten Gerätschaften vorstellte.

4

Grenzland – Kriegsland

Mit der Abtrennung des Saargebietes vom Deutschen Reich entstand im Jahre 1920 der Restkreis Merzig-Wadern, dessen Verwaltung ihren Sitz in Wadern hatte. Diesen Gutschein unterzeichnete der Restkreisverwalter Ferdinand Graf von Spee, der die Aufgaben eines Landrates wahrnahm.

Als der Erste Weltkrieg begann, erfaßte der Sturm der Kriegsbegeisterung auch die Hochwaldregion. Hier haben sich Reservisten aus Wadern zum Gruppenfoto aufgestellt.

Ins Fotoatelier haben sich hier Reservisten aus Nunkirchen begeben. Mit aufgepflanztem Bajonett blicken sie entschlossen in eine ungewisse Zukunft.

Der Aufmarsch der deutschen Truppen in Richtung Frankreich lief verstärkt über die Eisenbahnanlagen. Am 5. August 1914 besetzte eine Gruppe Reservisten den strategisch wichtigen Bahnhof Wadern.

Mit dem fortschreitenden Verlauf des Krieges wurden kriegswichtige Rohstoffe knapp. Vielerorts wurden die Kirchenglocken aus den Türmen geholt und zu Rüstungszwecken eingeschmolzen. Hier haben sich die Waderner Kinder um die Glocken der Pfarrkirche Allerheiligen versammelt, die gerade abtransportiert werden.

Der Krieg ist verloren. Die deutschen Truppen sind auf dem Rückzug und machen in Wadern auf dem Marktplatz Station. Am 21. November 1918 wird eine „Gulaschkanone“ auf dem Marktplatz „in Stellung gebracht“. Für die Waderner Kinder war dies natürlich eine große Attraktion.

Im November 1918 zieht ein langer Zug von Soldaten durch Wadern am Postamt vorbei in Richtung Heimat. Einige Bürger haben sich an der Straße versammelt, um die Soldaten aufzumuntern.

Die Burgruine Dagstuhl krönt diesen Gutschein aus dem Jahre 1921. Der Gutschein diente als Ersatz für Kleingeld, da nach dem Krieg nicht genug Metallgeld vorhanden war.

Der Wahlspruch der Hochwälder aus der Zeit als Restkreis, mit dem trotz der Isolation der Restkreiskommunen die besondere Beziehung zum Deutschen Reich ausgedrückt wurde.

In den späten zwanziger Jahren stieg die Arbeitslosigkeit im strukturschwachen Hochwald in schwindelerregenden Höhen. Zahlreiche Arbeitsbeschaffungsmaßnahmen sorgten für eine gewisse Entspannung der brisanten Lage. Zwischen Wadrill und Steinberg sind Morscholzer Männer damit beschäftigt, Schotter für den Straßenbau zu klopfen.

Diese Notstandsarbeiter aus Morscholz sind mit der Ausstockung von Gebüschen und Lohhecken zwischen Morscholz und Konfeld beschäftigt. Im ersten Erntejahr sollen auf den neuen Ackerflächen ganz besondere Erträge erreicht worden sein.

Notstandsarbeiter im Jahre 1929 beim Bau der neuen Verbindungsstraße von Noswendel nach Dagstuhl. Bei der Arbeit sind Noswendeler Männer (Peter Meier, Johann Köhl, Aufseher Johann Haas, Albert Becker, Jakob Nickels, Richard Kläser, Alois Schneider, Peter Finkler, Nikolaus Görgen, Peter Sauer und zwei unbekannte Arbeiter) zu sehen.

Die Grenze zwischen französischem Saargebiet und dem Deutschen Reich war Schauplatz intensiver Schmugglertätigkeit. Ein trauriger Höhepunkt der Auseinandersetzungen zwischen Zöllnern und Schmugglern war die Ermordung des deutschen Zöllners Augustin, der Pfingsten 1926 in Nunkirchen beerdigt wurde.

Der Zweite Weltkrieg hat gerade begonnen, die ersten Gefallenen werden Ende 1939 auf dem ehemaligen Soldatenfriedhof in Nunkirchen an der Weiskircherstraße zu Grabe getragen. In den fünfziger Jahren wurden die Toten auf den Soldatenfriedhof in Weiskirchen verlegt.

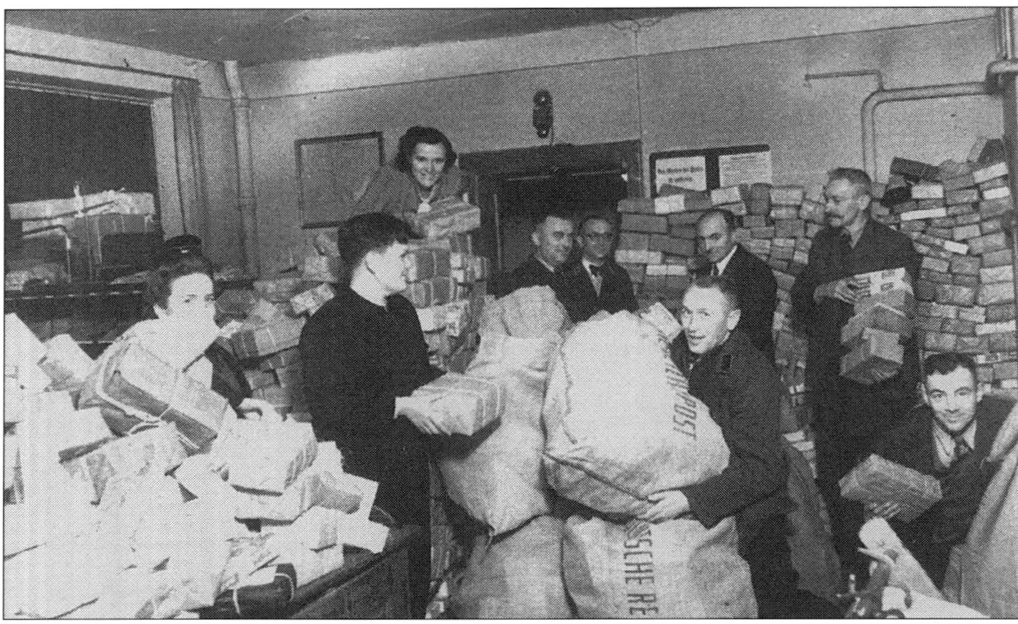

Zum Jahreswechsel 1939/40 sind die Mitarbeiter des Postamtes Wadern beim Verpacken der Feldpost noch bester Laune. Das wahre Gesicht des Krieges sollte sich erst noch zeigen. Von links nach rechts: Josefine Brausch, Rudi Wilkin, Maria Meyer („Zerfersch Maria“), Jakob Kaub, Postmeister Eligius Becker, Sebastian Wilkin, Herr Wagner, Herr Klöser, Karl Gimmler.

Im Verlauf des Krieges wurden immer mehr Frauen in der industriellen Produktion eingesetzt. Die damalige Belegschaft der bekannten Tuchfabrik Lauer hat sich im Jahre 1942 vor der Fabrik ablichten lassen.

Blick in die Produktionshalle der Tuchfabrik Lauer. Die Fabrik wurde bereits 1843 gegründet und hatte ihren letzten Standort bis in die sechziger Jahre im Wadrilltal in Wadern. Sie erlebte während des Zweiten Weltkrieges einen letzten Produktionshöhepunkt. Heute erinnert noch der Straßennahme „Fabrikweg" an sie.

Zu Beginn des Zweiten Weltkrieges zog 1939 ein Flüchtlingstreck durch die Waderner Unterstraße.

Am 18. März 1945 war es soweit: Die amerikanischen Truppen hatten Wadern erreicht. Die Infanteristen der 10. Division liegen im Wadrilltal in Stellung. In der Ferne ist der Waderner Kirchturm zu sehen.

Der Waderner Marktplatz am 19. März 1945. Die letzten Kriegstage haben schwere Schäden hinterlassen. Das „Lauers Schlößchen“ liegt in Schutt und Asche und wird danach nicht wieder aufgebaut. Wadern ist von amerikanischen Truppen besetzt, welche die Bewohner aus ihren Häusern holen.

In der Unterstraße werden von amerikanischen Kettenfahrzeugen die Panzersperren beseitigt. Im Vordergrund tragen Ordensschwestern, die von Merzig nach Wadern evakuiert worden waren, ihr verbliebenes Hab und Gut mit sich.

Die Tochter des damaligen Waderner Bürgermeisters Decku ist als Rotkreuzschwester im Einsatz und unterhält sich mit amerikanischen Soldaten. Die Bewohner der Waderner Oberstraße

werden gerade aus ihren Häusern geholt.

Der Vormarsch der amerikanischen Truppen wurde häufig durch gesprengte Brücken behindert. Hier wird von amerikanischen Pionieren über die Wadrill bei Wedern eine Behelfsbrücke installiert, kurz bevor Wadern eingenommen wird.

Die zerstörte Primsbrücke bei Krettnich: Das im Jahre 1942 vom Waderner Bürgermeister angeschaffte Feuerwehrfahrzeug ist von den Amerikanern requiriert worden und durchquert die Prims. Nur der Anhänger und die Spritze verblieben in Wadern.

5

Feiern und Brauchtum

Der Festzug im Rahmen der landwirtschaftlichen Ausstellung im Jahre 1930 formiert sich in der heutigen Egon-Reinert-Straße.

Das große Fest im Jahreskalender: die Kirmes. Um die Jahrhundertwende ist auf dem kleinen Markt vor dem Öttinger Schlößchen ein Karussell aufgebaut. Gut zu erkennen sind die Holzpferde auf dem Karussell, die Leute gingen auf die „Reiterbahn".

Ziehungsbuben aus Lockweiler-Krettnich im Jahre 1938. Die Aufnahme ist in der Gastwirtschaft Molter in Lockweiler entstanden. Einzige Frau auf dem Bild ist Maria Augustin-Molter. Viele der Männer sind im Zweiten Weltkrieg gefallen

Das wichtigste Fest im Familienkreis war die Hochzeit. Peter Simon und Katharina Schruden heirateten am 30. September 1911. Die ganze Familie hat sich für das Foto vor dem Elternhaus in Wadern aufgestellt.

Im Jahre 1935 feiert die Familie Gimmler in Wedern Hochzeit. Beim Feiern werden sie musikalisch vom katholischen Gesellen- und Musikverein Steinberg unterstützt.

Alois Arend und Magdalena Mersdorf heirateten am 30. Mai 1921. Fotograf Fuisting hat die

ganze Familie in Szene gesetzt.

Zur Erntedankfeier in Wadern, Ende der dreißiger Jahre. Die Nutzung der Feier für nationalsozialistische Propagandazwecke ist unübersehbar.

Eine Erntedankfeier in Wadern, Mitte der fünfziger Jahre: Die Erntekrone wird an der Spitze des Festzuges durch den Ort getragen.

Zur Maifeier, Anfang der fünfziger Jahre in Nunkirchen. An der Spitze des Zuges fährt die Maikönigin Hilde Salm. Auf dem Kutschbock sitzt Franz Lauer.

Bei der Maifeier zogen auch die Belegschaften der Betriebe mit. Die Mitarbeiter der Firma Müller werden rechts in der ersten Reihe angeführt vom Seniorchef Franz Müller. Auf dem Transparent im Hintergrund wird „Die Saargruben dem Saarvolk" gefordert, ein typisches Anliegen vor der Rückgliederung des Saarlandes an die Bundesrepublik.

Das Bild zeigt das Titelbild des Programmheftes der landwirtschaftlichen Ausstellung im Jahre 1930.

In den zwanziger und dreißiger Jahren gehörten die landwirtschaftlichen Ausstellungen mit ihren Festzügen zu den Veranstaltungshöhepunkten in der Region. Vor dem „Lauer Schlößchen" in Wadern hat sich hier ein Viez-Kelterwagen mit Begleitung aufgestellt, der am Festzug in Nunkirchen am 21. September 1924 teilnahm.

Zur landwirtschaftlichen Ausstellung des Jahres 1930 wurde in der „Großwies" im Wadrilltal ein großes Festzelt aufgebaut.

An kirchlichen Festen war das ganze Dorf beteiligt. Die Dorfkinder von Rathen erwarten kniend vermutlich die Fronleichnamsprozession oder die Segnung des Wegekreuzes im Hintergrund. Die Aufnahme stammt etwa aus den zwanziger Jahren.

Kommunion im Jahre 1935 in Nunkirchen. Vor dem Portal der Pfarrkirche Herz-Jesu haben die Jungen aus Nunkirchen, Büschfeld und Michelbach Aufstellung genommen.

Im September 1949 wird am Kirmesmontag in Lockweiler der Hammel ausgetanzt. Die Metzger Lorenz Birtel, links im Bild, und Alois Brust haben das geschmückte Tier fest im Griff.

Das Austanzen des Hammels zur Kirmes war in allen Dörfern üblich. Hier haben sich in den dreißiger Jahren die Morscholzer mit Musikanten zum Zug durch das Dorf versammelt.

Straußtanz in Lockweiler um 1950: Günther Sauer hält den Strauß empor, die unverheirateten Paare folgen ihm durch das Dorf. Der Strauß wird reihum getragen. Der Strauß mit den Würsten gehört dem Paar, der ihn im Moment des Klingelns eines Weckers in der Hand hält.

6

Vereinsleben

Einer der zahlreichen Handwerkervereine aus dem Hochwald im Jahre 1903. Die einzelnen Gewerke von Schreiner bis Maurer sind gut zu erkennen.

Der Mandolinenverein war die Keimzelle des heutigen Musikvereins Nunkirchen. In der Mitte ist ein besonderes Instrument zu erkennen, das in manchen Gegenden des Saarlandes als „Teufelsgeige“ bekannt war und Streich- und Rhythmusinstrument zugleich ist. Aufnahme um 1900.

Diese Nunkircher Musikanten begießen die Jahrhundertwende. Auf dem Faß sitzend ist Mathias Thome zu sehen. Die Familien Thome und Weyand bildeten in den Anfangstagen den harten Kern des Vereins.

Der Musikverein Noswendel im Jahre 1910. Gegründet wurde der Verein 1895. Bis heute ist die Blasmusik ein fester Bestandteil des Vereinswesens und aus dem Dorfleben nicht wegzudenken.

Für einige Jahre war die Kapelle einer Militäreinheit aus Straßburg zu Gast in Wadern und führte, zusammen mit dem Bannschütz ganz rechts, die Rosenmontagsumzüge der 1897 gegründeten Karnevalsgesellschaft an. Die Narrenschar hat sich in der Kräwigstraße aufgestellt.

Der Vorstand des katholischen Musik- und Gesellenvereins Steinberg mit Pastor Stinner im Jahre 1933. Der Verein war der Vorläufer der heutigen Kolpingkapelle Steinberg.

Diese Morscholzer Freizeitmusiker spielen in den dreißiger Jahren auf der Straße auf. Dirigent „Schaplin" gibt den Takt vor, am Schlagzeug ist Peter Schneider, „Häntz Pittschi" genannt.

Der Noswendeler Handwerkerverein hat sich in den dreißiger Jahren mit Mitgliedern des Musikvereins zum Fototermin vor der Pfarrkirche eingefunden, die in dieser Zeit gebaut wurde.

Im Sonntagsanzug präsentieren sich Lockweilerer Handwerker und Bürger um 1914 vor einem Dorfgasthaus.

Der Berg- und Hüttenarbeiterverein Nunkirchen hat sich um 1900 zum Gruppenbild vor dem Gasthaus Spang versammelt. An der Zahl der Mitglieder wird deutlich, wie wichtig die Arbeit

in den Gruben und Hütten für die Menschen der Hochwaldregion war.

Mitglieder des Berg- und Hüttenarbeitervereins Noswendel im verdienten Ruhestand. Die Aufnahme von 1955 zeigt von links Peter Meier, Johann Ruf, Michel Schillo, Mathias Meier, Adam Schömer und Mathias Köhl.

Die Morscholzer Feuerwehr um 1930. Bei Johann Görgen, genannt „Scheitchie", hat es gebrannt. An der Feuerspritze Johann Meier, genannt „Schäresch Johann".

Die Waderner Feuerwehr hat sich zum Gruppenbild in den dreißiger Jahren aufgestellt. Noch heute gibt es in den 14 Stadtteilen zehn eigenständige Löschbezirke.

Die freiwillige Feuerwehr Noswendel in ihrer Gründungszeit 1934. Die Wehr hat sich unter Leitung von Wehrführer Nikolaus Finkler, rechts im Bild, vor dem Haus Ehm in Noswendel-Roth aufgestellt.

Der Männergesangverein „Eintracht Lockweiler-Krettnich" vor dem Ersten Weltkrieg mit der Vereinsfahne in Krettnich.

Während der Zeit des Nationalsozialismus wurde auch das Vereinsleben gleichgeschaltet. Mitglieder des „Bundes deutscher Mädchen" treffen sich zum Ausflug auf dem Waderner Marktplatz.

Mit den gesellschaftlichen Veränderungen im Zuge der Industrialisierung entwickelte sich ein reges Vereinsleben, zu dem auch Sportvereine gehörten. In verschiedenen Dörfern gab und gibt es zum Teil bis heute sehr aktive Radfahrvereine, wie hier in Morscholz in den dreißiger Jahren.

Einer der ältesten Sportvereine in der Region ist der „Turn- und Sportverein Wadern", der im Jahre 1909 gegründet wurde. Das Bild zeigt die Fußballmannschaft um 1925: Von links Vorstandsmitglied Franz Grefig sen., Toni Rech, Mathias Kaspar, ein Unbekannter, Theo Dubois, Jakob Kaspar, Hanni Braun, Josef Molter, Willi Bröcker, Johann Thome, ein Unbekannter und Vorstandsmitglied Peter Blasius.

In den zwanziger und dreißiger Jahren gab es in vielen Dörfern aktive Theatervereinigungen. Der Kirchenchor Wadern führte im Jahre 1933 im Kinosaal Dubois Passionsspiele auf, an denen fast 100 Akteure mitwirkten. Chorleiter Peter Frenzer, vierter von rechts, hatte die musikalische

Leitung. Die Aufführung ging über die Bühne, kurz bevor die Nationalsozialisten religiöse Veranstaltungen außerhalb der Kirche untersagten.

Sehr bekannt waren auch die „Tell-Festspiele“ in Lockweiler. Auf einer eigens angelegten Bühne im Primstal wurden die Festspiele im Jahre 1934 aufgeführt. Die Reichsbahn legte sogar einen Behelfsbahnsteig für die eingesetzten Sonderzüge aus dem Saargebiet an.

Die Frauen des Morscholzer Theatervereins begeisterten ihre Zuschauer in den zwanziger Jahren mit dem Theaterstück „Genoveva“.

7

Personen und Persönlichkeiten

Ältere Morscholzer werden sich vielleicht noch an den alten Kindergarten, auch Spielschule genannt, erinnern. Die Aufnahme zeigt eine Kindergartengruppe um 1938.

Angeln am Seelerbach in Nunkirchen: „Tom Sawyer und Huckleberry Finn“, alias Walter und Nicki Thome, beim Fischefangen im Jahre 1920.

Walter und Nicki Thome ein paar Jahre früher. Nicki trägt ein für die damalige Zeit typisches Jungenkleidchen. Aufnahme von 1914.

Auf der Hauptstraße in Bardenbach sind 1927 die Kinder mit dem Handkarren auf dem Weg zur Schule.

Mädchen in der Kochschule. Die Aufnahme entstand vermutlich in den zwanziger Jahren in Lockweiler.

Die Mädchenklasse der Volksschule Nunkirchen mit Lehrerin Fräulein Schäfer im Jahre 1902. Die Gesichter der Mädchen lassen erahnen, daß die Kindheit um 1900 im Hochwald

keine reine Freude war.

Reza, Freiin von Liebig, Witwe von Heinrich Albert de Lasalle von Louisenthal und spätere Gattin des Freiherrn Vittinghoff-Schell, mit ihren Kindern in einem Porträt von August Fuisting um1900. Wie anders sahen die Kinder der Bauern, Bergleute und Handwerker aus Nunkirchen aus!

Blick in einen Klassensaal der Volksschule Wadern in der Oberstraße. In den dreißiger Jahren wurden mehrere Altersstufen gleichzeitig unterrichtet, und es gab schon gemischte Mädchen- und Jungenklassen.

Pause in der Volksschule Wadern in den dreißiger Jahren. Es herrscht reges Treiben auf dem Schulhof der Volksschule, deren weiterführenden Klassen später das heutige Hochwald-Gymnasium bildeten.

Turnstunde in der Volksschule Lockweiler an der alten Kirche in den dreißiger Jahren. Leibeserziehung stand ganz oben auf dem Lehrplan, was die Jungen eindrucksvoll beweisen.

Kartenspieler vor dem Gasthaus Johann Maus in Morscholz: Die jungen Männer waren in den dreißiger Jahren auf dem Sprung ins Erwachsenenalter.

Das erste Taxi in Nunkirchen fuhr Mathias Thome im Jahre 1930. Die Automobile wurden zu neuen Statussymbolen – auch für die Jugend, wie auch auf dem Bild gut zu erkennen ist.

Ein Motorrad zu besitzen, war schon etwas Außergewöhnliches. Eduard Kasper vom Dösterhof fährt in den dreißiger Jahren dieses DKW-Motorrad.

Diese Nunkircher Frauen tragen mit ihren Kittelschürzen und Kopftüchern die alltägliche Arbeitskleidung. Das Fahrrad mußte als etwas Besonderes unbedingt mit auf das Bild.

Ein weiteres Fortbewegungsmittel, der Traktor, steht im Mittelpunkt der Familienaufnahme vom Dösterhof. Die Familie Kasper bewirtschaftete den Hof von 1929 bis 1946.

Personenverkehr um 1900: Die Kutsche des Posthalters Peter Lohrig war zwischen Wadern und dem Bahnhof in Dagstuhl sowie zwischen Wadern und Weiskirchen im Einsatz.

Güterverkehr Mitte der fünfziger Jahre: Lkw-Fahrer Berthold Meier unterwegs im Auftrag der Großhandlung Flasche in Wadern. Am Lkw ist das Fahrzeugkennzeichen des eigenständigen Saarlandes zu erkennen.

Der Waderner Bürgermeister Friedrich prägte von 1870 bis 1899 das kommunale Geschehen. Er bemühte sich sehr um den Anschluß der Gemeinde Wadern an die Eisenbahn, welcher erst am Ende seiner Dienstzeit im Jahre 1897 realisiert wurde.

Bürgermeister Max Müller war von 1899 bis 1928 in Wadern tätig. Er widmete sich intensiv der Heimatforschung und verfaßte 1930 die erste Abhandlung zur Waderner Heimatgeschichte.

Selbstbildnis des Waderner Fotografen August Fuisting. Sein Verdienst ist die fotografische Dokumentation des Waderner Lebens Ende des 19. und zu Beginn des 20. Jahrhunderts.

Im sogenannten „Casino" des Gasthauses Schadt am Markt trafen sich die Waderner Großbürger zum Diskutieren und Politisieren. An zweiter Stelle von rechts sitzt der damalige Bürgermeister Müller.

Peter Weishaar aus Überlosheim, „Et Potsche vo Iewerloßem" genannt, geht Ende der fünfziger Jahre seinem Broterwerb nach. Als Topfflicker trug er seine ganze „Werkstatt" immer mit sich und reparierte alles, was in der Küche gebraucht wurde.

Zeugen einer vergangenen Epoche: Die sogenannten „Glorreichen Vier" von Morscholz im Jahre 1952. Von links sind Johann Blasius, Johann Meier, Theo Billen und Peter Schneider zu sehen.

8

Waderner Ortsansichten

Die Mühle Schuhmacher in der Waderner Unterstraße in den vierziger Jahren, im Vordergrund Müller Mathias Schuhmacher.

Blick vom Bellscheid auf den idyllischen Marktflecken Wadern, um 1900. Im Vordergrund die alte Johannisbrücke und die sogenannten Bleichwiesen, auf denen man sogar Bleichwäsche erkennen kann. Gut zu sehen sind in der Bildmitte und unterhalb der Pfarrkirche zwei Höfe der Stockbauern mit Scheunenanbau. Das Haus Dubois in der damaligen Bahnhofstraße, heute Poststraße befindet sich gerade im Bau.

Blick auf Wadern vom alten Krankenhaus aus, das links zu sehen ist. Die katholische und evangelische Kirche ragen aus der Ortslage heraus.

Der Fotograf August Fuisting hat unzählige Postkartenmotive hinterlassen. Hier hat er einige Personen auf der heutigen Egon-Reinert-Straße, der alten Straße nach Wadrill, mit Blick auf Wadern, in Position gebracht.

Die Oberstraße, vom Marktplatz aus gesehen. Die schnurgerade Ausrichtung der Straße geht auf den Grafen Joseph Anton zurück, der seine Amtsleute und angeworbenen Handwerker hier Häuser bauen ließ. Rechts im Bild ist das sogenannte „Grafenschloß", in dem seine Amtsverwaltung untergebracht war.

Die Oberstraße, um 1910, in Richtung der Kirche. Die kleinen Häuser sind gut zu erkennen, und rechts im Bild fällt auf, daß die Elektrizität bereits Einzug gehalten hat.

Blick in die Kräwigstraße, um 1910. Bis heute hat die Straße ihre Grundstruktur bewahrt. Allerdings sind nur einige der kleinen Geschäfte, die es hier bis in die siebziger Jahre gab, übrig geblieben.

Blick in die Kräwigstraße aus Richtung Hasenberg. Die Bäume entlang der Straße wurden gefällt.

Blick in die heutige Poststraße, um 1900. Links ist der spätere „Dagstuhler Hof“ zu erkennen, in dem noch eine Brauerei zu finden ist. Im Jahre 1890 gab es noch vier Brauereien in Wadern. Rechts im Bild ist die alte Kirchenmauer zum Pfarrhaus.

Blick in die Unterstraße aus Richtung Johannisbrücke. Rechts ist die Gerberei Birtel zu sehen, eine von vier Gerbereien um 1900.

HOTEL SCHMITT.

Blick in die Unterstraße vom Markt aus. Der Fotograf Fuisting hat für seine Aufnahme einige junge Statisten aufgestellt. Links ist das damalige Hotel Schmitt und spätere Gasthaus Brücker zu sehen.

Die Ecke Kräwigstraße/Uhlandstraße, um 1910. Anstelle der Autoreparaturwerkstätte Blum steht heute hier die Apotheke Moeckel.

An der Kirche, um 1900. Die linke Straßenseite zeigt sich bis heute kaum verändert, die Häuser auf der rechten Seite gibt es heute nicht mehr.

Die Gastwirtschaft Dubois in der Oberstraße, Aufnahme aus den zwanziger Jahren. Heute befindet sich hier das Starlight-Kino. An der Fassade des Hauses sind die barokken Strukturen erkennbar.

Die Metzgerei und Wirtschaft Johann Feltes am kleinen Markt, um 1900: Stolz zeigt sich der Metzger und Wirt Feltes im Türrahmen, kein Wunder beim Anblick seiner Warenauslage. Das Gebäude ist einem modernen Wohn- und Geschäftshaus gewichen.

Das Haus des bekannten Dr. Kuntschert gegenüber der Kirche, um 1900. Standesgemäß war der Doktor mit einer Kalesche, später mit dem ersten Automobil in Wadern und Umgebung unterwegs. Seinen Berufsstand hat er an der Fassade verewigt.

Das Kaiserliche Postamt. Nachdem die Post zunächst im Hause Klauck neben der Kirche untergebracht war, residierte sie ab 1909 gegenüber dem kleinen Markt. Hier befand sich früher das erste Schulhaus von Wadern.

Die neu erbaute Evangelische Pfarrkirche, um 1900. Am Ende des 19. Jahrhunderts hatte sich durch Zuzug einiger evangelischer Familien eine kleine Kirchengemeinde gebildet. Im Jahre 1897 wurde das erste Gotteshaus erbaut, das im Zweiten Weltkrieg seinen Turm verlor, ansonsten aber bis heute äußerlich unverändert geblieben ist.

Innenansicht der katholischen Pfarrkirche Allerheiligen, um 1920: Das Kircheninnere zeigt sich reich geschmückt, die Neuzeit hat mit den elektrischen Kronleuchtern Einzug gehalten.

Die Schlosserei Heinrich Aatz in der Johannisstraße am Wadrillbach in den zwanziger Jahren. Das Haus steht heute noch etwas zurückgesetzt von der Straße.

Ein kleines Bauernhaus auf dem Christianenberg in der Nähe des heutigen Sportplatzes. Mehrere dieser kleinen Häuser standen hier bis in die fünfziger Jahre.

Ansicht des alten Klostergebäudes auf dem Christianenberg. Im Jahre 1770 war das Kapuzinerkloster auf Betreiben der Gräfin Christiane fertiggestellt worden. Nach der französischen Revolution wurden die Gebäude an Privatleute versteigert. Heute ist das Klosterkreuz auf dem nach der Gräfin benannten Christianenberg die letzte Spur des Klosters.

Brand auf dem Stockbauernhof des „Bechterbauern", dem heutigen Anwesen Simon, auf dem Christianenberg, um 1900. Nach dem Brand ist man mit den Aufräumarbeiten beschäftigt.

In der Unterstraße am Mühlengraben, um 1900. Links ist die Mühle Schuhmacher zu erkennen, im Hintergrund die Schmiede Kläser. Der Fotograf Fuisting hat das Foto wieder einmal gekonnt komponiert: Die Kinder im Vordergrund halten Schiefertafel und Griffelkasten in den Händen,

die Kinder im Hintergrund lesen. Neben dem Leiterwagen sind Eichenstangen zu erkennen, die nach Ablösung der Eichenrinde für das Gerben als Brennholz Verwendung fanden.

Blick von der Poststraße in Richtung Unterstraße, um 1900. Rechts im Bild der bekannte „Dagstuhler Hof", der unter diesem Namen bis in die achtziger Jahre existierte.

Ein Prost auf den Straßenbau in der Unterstraße: Bauarbeiter und Bürger stoßen auf die getane Arbeit an, die tatkräftig von der modernen Dampfwalze unterstützt wurde. Anlaß könnte der Neubau der Bahnhofstraße in Richtung Bahnhof Dagstuhl in den Jahren 1896/97 gewesen sein.

Ein Pferdefuhrwerk gegenüber der Kirche, um 1900. Heute befinden sich hier in einem neu erbauten Haus das Herrenbekleidungshaus Becker und die Zoohandlung Tiator.

Das Backsteingebäude in der Poststraße, welches rechts im Bild zu sehen ist, hat zahlreiche Nutzungen hinter sich. Hier befanden sich unter anderem ein Landmaschinenhandel und eine Limonadenfabrik.

Die Pfarrkirche Allerheiligen und das Haus Post-Klauck in Wadern, um 1900. Die nachbarocke Saalkirche wurde 1817 erbaut, der Glockenturm ist im unteren Teil noch romanischen Ursprungs. Bis nach dem Zweiten Weltkrieg standen auf dem Kirchvorplatz noch stattliche Bäume. Das barocke Bürgerhaus wurde 1789 vom Stockbauer Klaudius Klauck erbaut. Hier befand sich bis 1909 die Waderner Post mit einer Pferdewechselstation. Danach wurden die heutigen Jugendstilelemente an der Fassade angebracht, die an die alte Nutzung des Hauses erinnern.

Blick auf die Johannisstraße, „Gehannsgass" genannt, von der Wadrillbrücke, um 1900. In den kleinen Häusern wohnten im 18. Jahrhundert Gesinde und Bedienstete des Grafen Joseph Anton.

Die alte Johannisbrücke über die Wadrill, um 1900. Nach vielen Umgestaltungen fließt die Wadrill heute an dieser Stelle unter der Straße entlang.

Schlittschuhlaufen im Brühl: Da die Wadrill regelmäßig über die Ufer trat, gab es im Winter immer wieder Gelegenheit zum Schlittschuhlaufen.

Das erste Waderner Krankenhaus um 1900. Ein erstes Hospiz wurde von der „Malergräfin“ Octavie de Lasalle von Louisenthal im Jahre 1856 gegründet. Das Haus wurde immer wieder erweitert, heute befindet sich hier das neu erbaute Altenheim.

Blick in die Unterstraße in den fünfziger Jahren: Die Straße wird ausgebaut. Im Hintergrund ist das Lebensmittelgeschäft von Franz Haas zu sehen, dem Begründer des heutigen Haco-Einkaufszentrums. An den Werbetafeln ist erkennbar, daß noch in Franken-Währung bezahlt wird.

Das erste Freibad von Wadern wurde in den dreißiger in der Nähe des ehemaligen Klosterweihers gebaut. Sogar an Sandkästen für die Kinder wurde gedacht.

Am 16. April 1959 erfolgte der Spatenstich zum Neubau des Hochwald-Gymnasiums am Kaisergarten unter Teilnahme des damaligen Kultusministers und späteren Ministerpräsidenten Dr. Franz-Josef Röder (CDU).

Die Arbeiter haben sich mit Blick ins Wadrilltal vor dem Baumodell des Gymnasiums aufgestellt. Das Gymnasium konnte im Jahre 1961 aus dem alten Amtsgericht in der Oberstraße in den Neubau umziehen.

9

Von Bardenbach bis Wedern

Eine Ansicht von Schloß Münchweiler, Anfang der dreißiger Jahre. Lange Zeit befand sich im Schloß eine Familienpension, für die mit dieser Postkarte Werbung gemacht wurde.

Blick auf Bardenbach in Richtung Primstal, um 1900. Rechts ist die alte Brücke über den Speckenbach zu erkennen. Die Dorflage ist geprägt von kleinen Bauernhäusern. Heute sind nur noch wenige in ihrer alten Struktur erhalten.

Eine der vielen Postkarten des Fotografen Fuisting zeigt den Blick auf Buweiler und Kostenbach im Jahre 1899. Die Pfarrkirche Herz-Jesu bildet den Mittelpunkt der Dorfstruktur. Rund um das Dorf gibt es im Gegensatz zu heute kaum Wald, die Flächen werden landwirtschaftlich genutzt.

Die alte Linde in Rathen war beliebter Treffpunkt für die Dorfbevölkerung. Sie mußte am 14. März 1952 gefällt werden. Aufnahme um 1939.

Ein Blick auf Rathen, vermutlich in den dreißiger Jahren. Im Gebäude links befanden sich die ersten Schulräume des Ortes.

Die Wederner Kapelle, auf dem Bild in ihrem Urzustand zu sehen, ersetzte im Jahre 1950 eine zu klein gewordene Vorgängerkapelle.

Die Gastwirtschaft von Nikolaus Görgen in Wadrill, um 1900. Auf der Rückseite der Fotografie von Fuisting wurde der Postkartentext festgehalten: „Gruß aus Wadrill – Forsthof, Gasthaus Nikolaus Görgen“.

Der Abbruch der Ruine der Burg Büschfeld im Jahre 1900. Die Reste der Burg wurden im Zuge des Baus der Merzig-Büschfeld-Eisenbahn abgetragen.

Blick auf Büschfeld und den Bahnhof, Mitte der dreißiger Jahre.

Der alte Kirchturm von Lockweiler in den dreißiger Jahren. Im Vordergrund ist der verwunschen wirkende Friedhof mit vielen alten Grabsteinen zu erkennen.

Gesamtansicht der alten Pfarrkirche St. Michael Lockweiler um 1900: Der romanische Glockenturm stammt vermutlich von einer Vorgängerkirche aus dem 12. Jahrhundert. Das Kirchengebäude wurde 1844 neu erbaut, ist aber heute verschwunden.

Die Gastwirtschaft von Mathias Lauer im Morscholz, Ende der zwanziger Jahre. Seit jeher ist sie eine der drei Gastwirtschaften im Ort und heißt heute „bei Bäckersch“.

Ansicht von Morscholz mit Umland in den dreißiger Jahren: Von der Verbindungsstraße zu Wadern geht der Blick in die Landschaft, die intensiv landwirtschaftlich genutzt wird.

Das alte Gerbhaus in Nunkirchen im Jahre 1925. Rechts steht der Schuster Karl Weyand, genannt „Reins Karl“, links Maria Thome, genannt „Gerbhauser Marie“. Das Haus steht heute noch, ist aber komplett umgebaut.

Bauer Zöhler, genannt „Schmitz' Hännes“, vor seinem Haus in Lockweiler mit dem Knecht Müller, um 1930. Das Haus steht heute nicht mehr.

Die Brücke zwischen Pfarrhaus und Kirche in Nunkirchen in den fünfziger Jahren. Damit der Pastor nicht über die Straße gehen mußte, wurde dieser kleine Brückenbogen gebaut, den es inzwischen nicht mehr gibt.

Blick auf Nunkirchen, Ende der fünfziger Jahre. Die alte dörfliche Struktur des Ortes wirkt sehr idyllisch.

Blick vom Schloßberg über Schloß Dagstuhl in das Primstal, um 1900. Im Vordergrund ist der Schloßgarten zu erkennen.

Der Ursprung des heutigen Stadtteils Dagstuhl ist die Eisenbahnersiedlung rund um den Bahnhof Wadern. Das Bild zeigt den ersten Bauabschnitt der Siedlung, den Bahnhof und den Lokomotivschuppen um 1900.

Das Hotel Schloßberg wurde im Jahre 1897 im Zuge des Eisenbahnbaus errichtet. Das Bild zeigt das Gebäude in den dreißiger Jahren, rechts ist ein kleiner Kaufladen für die Eisenbahnerfamilien eingerichtet. Das Haus wurde 1987 abgerissen.

Holzverladung am Bahnhof Wadern, Anfang der vierziger Jahre. Ein Eichenstamm aus dem Osburger Hochwald wird verladen, im Hintergrund ist der Lokomotivschuppen zu erkennen. Die Personen von links sind der Fuhrunternehmer Nikolaus Blau aus Osburg, Josef Schirra, Martin Scherf und ganz rechts ein französischer Kriegsgefangener, der beim Fuhrunternehmer arbeiten mußte.

Die Eisenbahnersiedlung und der Bahnhof Wadern, um 1910. Inzwischen sind fast alle Eisenbahnerwohnhäuser gebaut. Das letzte der neun Beamtenhäuser wurde im Jahre 1913 fertiggestellt. Das erste Privatwohnhaus entstand erst in den zwanziger Jahren.

Leben in der Großstadt ...

Tauchen Sie ein in das quirlige Großstadtleben vergangener Tage. Spazieren Sie über breite Boulevards und stürzen Sie sich ins Nachtleben. Erkunden Sie ihre Stadt durch die Fensterscheiben einer Straßenbahn oder des ersten Käfers und bewundern Sie prächtig geschmückte Schaufenster.

Erinnerungen an die Schulzeit …

Erinnern Sie sich noch an die Zeiten von Abakus und Schiefertafel, an Klassenausflüge oder den ersten Taschenrechner? Blicken Sie zurück auf große Klassen und gestrenge Schulmeister, entdecken Sie auf Klassenfotos Freunde und Bekannte von früher!

Gesellige Stunden im Verein …

Fußballclub und Schützenverein, Musikkapelle und Gesellenverein: Schauen Sie zurück auf Volksfeste und Turniere, Chorproben oder Prunksitzungen. Erinnern Sie sich an schöne Stunden und das gesellschaftliche Leben in Ihrer Heimat.